T0384100

Elephants

Behavior and Conservation

This volume compiles more than 20 years of behavioral research on the three living species of elephant in Africa and Asia (African Savanna, African forest, and Asian), and considers the implications of these studies for conserving and managing wild elephant populations. The theoretical background, key terminology, and findings are explained and presented in engaging language accessible to a wide range of nonspecialists, from students to seasoned professionals. By viewing data from numerous studies through a comparative evolutionary perspective, the similarities and distinctions between species and populations come into clear relief, providing insight into the complexities of protecting these charismatic yet highly threatened megaherbivores. Rather than merely providing an exposition of what is known, the book invites readers to reflect on the additional questions and puzzles that are still in need of answers, in the hope of inspiring a new generation of researchers and conservationists.

Shermin R. de Silva is a professor in Ecology Behavior and Evolution at the University of California, San Diego. She is a founder of the US-based conservation nonprofit Trunks & Leaves. She directs the Udawalawe Elephant Research Project in Sri Lanka and is a member of the Asian Elephant Specialist Group of the International Union for Conservation of Nature.

Elephants

Behavior and Conservation

SHERMIN R. DE SILVA
University of California, San Diego

Shaftesbury Road, Cambridge CB2 8EA, United Kingdom

One Liberty Plaza, 20th Floor, New York, NY 10006, USA

477 Williamstown Road, Port Melbourne, VIC 3207, Australia

314–321, 3rd Floor, Plot 3, Splendor Forum, Jasola District Centre, New Delhi – 110025, India

103 Penang Road, #05–06/07, Visioncrest Commercial, Singapore 238467

Cambridge University Press is part of Cambridge University Press & Assessment, a department of the University of Cambridge.

We share the University's mission to contribute to society through the pursuit of education, learning and research at the highest international levels of excellence.

www.cambridge.org
Information on this title: www.cambridge.org/9781107143289

DOI: 10.1017/9781316534380

When citing this work, please include a reference to the DOI 10.1017/9781316534380

First published 2025

A catalogue record for this publication is available from the British Library

Library of Congress Cataloging-in-Publication Data
Names: de Silva, Shermin R., 1981– author.
Title: Elephants : behavior and conservation / Shermin R. de Silva, University of California, San Diego.
Description: Cambridge, United Kingdom ; New York, NY : Cambridge University Press, 2025. | Includes bibliographical references and index.
Identifiers: LCCN 2024022689 | ISBN 9781107143289 (hardback) | ISBN 9781316534380 (ebook)
Subjects: LCSH: Elephants. | Elephants – Behavior, | Wildlife conservation.
Classification: LCC QL737.P98 D478 2025 | DDC 599.67–dc23/eng/20240522
LC record available at https://lccn.loc.gov/2024022689

ISBN 978-1-107-14328-9 Hardback

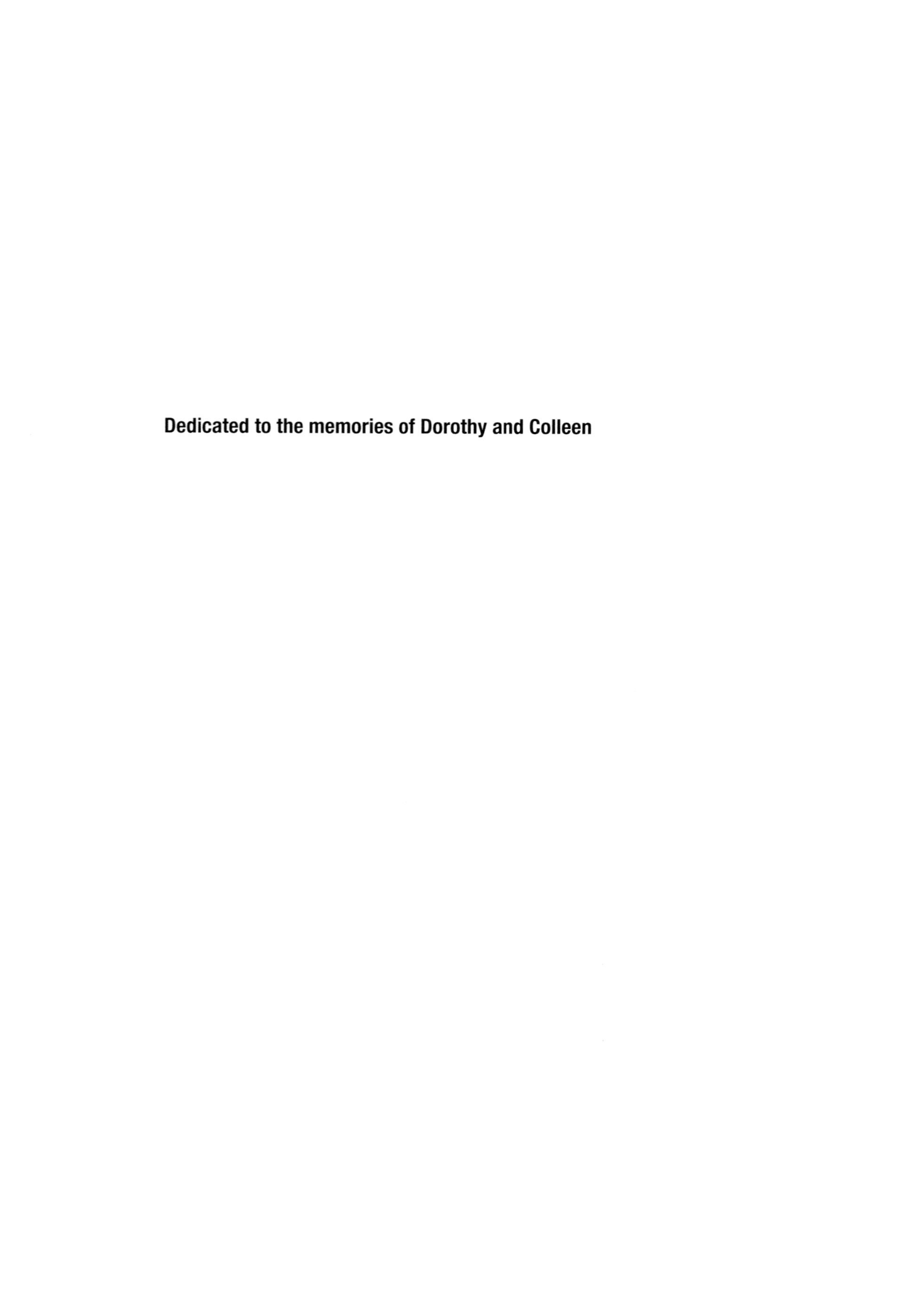

Dedicated to the memories of Dorothy and Colleen

Contents

Foreword

A Comparative Exploration of Many Things "Elephant"

Elephants, living and extinct, huge and dwarf, forest-dwelling or desert, are simply fascinating. This book synthesizes those elements that intrigue as well as concern humans about elephants, but it is more than a study of human–elephant relations (these have their place in Chapter 8). It is an overview of elephants' relationships with their wild environments, social, physical, and edible, using an approach grounded in comparative and evolutionary ecology. It also has great illustrations and photographs, which directly showcase the concepts discussed in each chapter. And these pictures are often worth their weight in scientific and explanatory gold.

I first met Shermin when she came to Kenya in 2007 to experience the Amboseli Trust for Elephants training program run by our Assistant Director, Norah Njiraini. At that time, Shermin was setting up a long-term elephant study in Sri Lanka and also working toward her PhD at the University of Pennsylvania. She brought her considerable knowledge of Asian elephants in Udawalawe to us, and sitting in the elephant research camp in the evenings we all benefited from a lively exchange of comparative information on African and Asian elephants – how to age, sex, and identify individuals, follow in the field, determine families and kinship, monitor families and their aggregations, understand male behavior, and try to tease apart their vocalizations. So many similarities, but many interesting differences too, as we discovered together. Shermin has taken this comparative perspective to the next level in this book.

Throughout the book, she integrates perspectives drawn from her own and collaborators' detailed studies on the Udawalawe individuals, together with a comprehensive reading of a plethora of other studies. For example, she notes that much of our understanding of diets and social systems (Chapters 3 to 5) comes from observational studies of elephants in unusual, possibly atypical, habitats – those with excellent visibility like Samburu, Amboseli, or Udawalawe – rather than from forests or areas of denser vegetation. Yet knowledge about how elephants aggregate, when and how they mate, and how they choose and use plants from a wide range of environments is necessary if we are to be able to predict their population persistence. As elephant observers, we can hopefully look to the next generation of techniques, genetics, e-DNA, remote sensing, and so on to answer some of the questions about elephant futures raised in this book. As an example, how elephants use space (Chapter 6) neatly dovetails historical knowledge and large population surveys with impressive new habitat occupancy modeling. There is so much to learn from an integrated perspective.

As long-term studies of elephants have accumulated across a number of African and Asian countries, one thing we hope we have learned as researchers is humility. Whatever we decide is "an elephant typical pattern" soon becomes obsolete, as the elephants do the opposite of what we predicted – or even had seen for decades. Such reversals, and alternatively, surprising consistencies, are at least in part explained in this book through evolutionary ecology paradigms that allow a better understanding of variation in elephant and other species' sociality, foraging, population distributions, sexual interactions, and sexual segregation. And, of course, these evolutionary questions are the grand themes of the book. The book succeeds in holding our attention for eight excellent chapters and informing us of much that seasoned elephant biologists have overlooked, as I discovered while reviewing chapter drafts. Elephant researchers and general ecologists can only hope that we will have another 50-plus years to continue to observe these wonderful, funny, and sociable animals in their highly diverse wild environments.

Professor Phyllis Lee
Psychology, School of Natural Sciences, University of Stirling
Director of Science, Amboseli Trust for Elephants

Acknowledgments

This book might never have materialized had I not been standing next to Phyllis Lee once at an animal behavior conference. It happened that we were passing by a booth for Cambridge University Press, whose representative asked Phyllis if she would be interested in undertaking a book on elephant behavior. She swiftly (and perhaps wisely) replied with a definitive *no*. "But why not ask these two? They've got lots of energy," Phyllis said, casually gesturing at myself and Vicki Fishlock, who also happened to be standing there (we elephant researchers travel in packs). I also may not have embraced the opportunity had Vicki not been equally enthusiastic about it. We figured that between the two of us, we might just have enough experience with all three elephant species to write something worth reading. As life would have it, Vicki had to bow out of the writing, but much of the thinking behind the topical focus of the book originated in our early discussions. I am indebted to Vicki for giving me the initial confidence to undertake the project and Phyllis for continuing to be a cheerleader, offering her profuse comments on each and every chapter along the way. In addition to bringing her formidable wealth of experience and relevant pieces of literature to my attention, her candid questions, opinions, and occasional skepticism helped to ensure my arguments were well grounded.

It is not enough to merely read about elephants, one really has to witness them in their own environments. It was very exciting to visit Kenya and be able to see firsthand the savanna elephant populations that had been so well studied and written about. I am most indebted to the individuals and organizations who facilitated these visits. Cynthia Moss and the Amboseli Trust for Elephants accommodated me at their camp, where I had the happy fortune of first meeting Phyllis and swapping stories. Norah Njiraini was a wonderful host and it was motivating to see how the Amboseli team conducted their day-to-day monitoring with such dedication, fueled by their obvious love for the animals they knew so well. Years later, I was again fortunate to be able to visit Samburu National Reserve thanks to a postdoctoral fellowship from the National Science Foundation providing me the opportunity to work with George Wittemyer at Colorado State University (CSU). I very much appreciate George making arrangements with the organization Save The Elephants so that I could spend time at their camp. I tagged along with Shifra Goldenberg, a PhD student at the time, alongside other floating interns, and tried not to ask too many annoying questions. The company was appreciated both in the field and back at the lab at CSU. My visit to the Kinabatangan River Sanctuary in Borneo was facilitated by the intrepid Nurzhafarina Othman, whom I'm glad to have as a friend and colleague. Finally, my visit to the remarkable Leuser Ecosystem and bucket-list dream of seeing elusive Sumatran elephants was enabled by the personnel of the Orangutan Information Centre. These were all golden experiences and will remain preserved in the amber glow of memory.

The comparative perspective that lies at the heart of this book is well established in the study of animal behavior, but for me was especially prominent in the work of my graduate advisors, Dorothy Cheney and Robert Seyfarth. The fact that two primatologists would happily recruit a student who wanted to study elephants is one testament to this. I am grateful to them for giving me the chance to get into this research area in the first place, as well as the many conversations with them that shaped my thinking as a student and scientist. I specifically recall an offhand comment Dorothy made while puzzling over why elephants weren't more like rhinoceros, given their dietary overlap and comparable sizes as megaherbivores. *Why be social?* Clearly, she pointed out, elephants seem to have done better in terms of sheer numbers (despite both sets of species, sadly, having been subjected to extreme hunting pressure by humans). I still don't have a definitive answer, but this heavy hint got me reflecting on whether sociality was just a phylogenetic artifact of the elephant lineage or something maintained and encouraged by a relatively new and very powerful selective force – ourselves. These ideas have found their way into the opening chapters of this book and continue to guide my perspective on the evolutionary pressures faced by charismatic but also highly threatened and sometimes "problematic" species.

The prospect of writing about elephant evolution was somewhat daunting, so I very much appreciate the thorough and encouraging feedback provided by Adrian Lister as well as comments by Asier Larramendi. It is an exciting time to be a paleontologist, with new discoveries coming so fast that I can only hope what I have written will not be outdated before long. Likewise, it feels as though research on the living elephants is now exploding. I am excited that there are now so many research groups studying elephants in their many diverse situations, but it does create a challenge for anyone trying to synthesize it all. I am therefore grateful to those who took the time to read and provide their comments on sections of the book: Priya Davidar, Kate Evans, Alex Mossbrucker, Jean-Philippe Puyravaud, and Graeme Shannon. I humbly acknowledge that any inaccuracies and omissions are entirely my own. Likewise, I thank the preceding individuals and several others – Ahimsa Campos-Arceiz, Joshua Plotnik, Joyce Poole, and Chithra Rathnayake – for providing permission to use their images. Joyce was especially generous in mining her enormous archive of photographs in search of good comparison images and kind enough to find time to respond to my sporadic questions over email. I also appreciate the sound recordings provided by Angela Stoeger and the team from the Malaysian Elephant Management and Ecology group: Lim Jia Cherng, Yen Yi Loo, and Ee Phin Wong. I owe a tremendous thanks to Sameera Weerathunga and T. V. Pushpakumara ("Kumara") of the Udawalawe Elephant Research Project field team, who not only supplied several of the photographs in the book, but were instrumental in gathering the data cited in associated studies through their dedicated and tireless efforts for more than a decade.

The writing process itself has at times felt like a companion, something concrete and meaningful to turn my focus to when other parts of life felt less sure and my direction unclear. Still, at times, things can be overwhelming. My heartfelt gratitude to my friends (or colleagues-turned-friends) who have been a sounding board for me over the years as I tried to navigate the professional and personal paths – Nivedita Srinivas, Ewa Szymańska, Maria Rakhovskaya, Grace Wenger, Kinjal Doshi, Carolyn Quam, Lizzie Webber, Matt Banghart, Gulcin Pekkurnaz, Sameera Dharia, and Dan Gebhardt. You've supported my endeavors in more than one way throughout the years. Our networks may be diffuse and global, but what Thoreau said has never rung truer: "Nothing makes the earth seem so spacious as to have friends at a distance; they make up the latitudes and the longitudes."

This book is like a third child, but I have to admit that it in fact took longer to produce than the other two actual children have been around. There's nothing like having kids to really make one take notice of the fact that we are mammals, with all that it entails. It certainly made me appreciate the energetic demands elephant mothers must face in trying to keep their goofy, accident-prone progeny out of trouble, and the importance of grandparents, aunts, and other alloparents (relationships that globalized modern human societies make it difficult to maintain). Their physical presence, and absence, was especially felt during the pandemic. I am so thankful for my family, Ananda and Dane de Silva, Galina Kryazhimskaya, Nisha Suhood, and our nannies Marina Navarette and Liz Thomas, for being there when we needed them. Truly they were the keys to my sanity for many moon cycles. I am especially thankful to my husband Sergey Kryazhimskiy for giving me the space to think, write, and work on this monstrous project (alongside any other idea I was prone to chasing) and believing not only that I could pull it off, but above all that it was an important and worthwhile pursuit. And of course, I cannot forget Scheherazade and Syrene, who give me joy each day and push me to be the best and truest version of myself. Perhaps they'll never read this work, but I hope they do, and are excited to say "My mom wrote that – now I understand what she was doing all that time!"

I gratefully acknowledge the wonderful staff and colleagues at the Institute for Advanced Study in Berlin (Wissenschaftskolleg zu Berlin or WIKO), where I had the pleasure of spending six magical months as a junior fellow in the College of Life Sciences. The structure and contents of this book were largely outlined during this period, taking inspiration from the various lively conversations we had over Dunia's masterful dinners. From the moment I left, I've dreamed of returning to WIKO again with fresh ideas and new projects. In closing, I must thank Floriane Taruc, my gem of an assistant, who formatted figures, tracked down licenses and image-use permissions, and did it with such efficiency, passion, and enthusiasm that I felt there was an angel helping me. Rachel Sprague generously provided her artistic talents for the cover, which I am incredibly thrilled will finally be making it out into the world! And finally, I am deeply appreciative for the initial spark provided by Dominic Lewis, formerly at Cambridge University Press, who held my hand through the early conception of this project, as well as the patience and guidance provided by Aleksandra Serocka of Cambridge University Press, in shepherding me through the writing process and answering the many questions I have peppered her with over the years. I hope the wait was worth it.

1 Introduction

1.1 Aims and Scope

Though now greatly diminished in diversity, living elephants represent the sole survivors of what was once a much more speciose clade, dispersed throughout North and South America, Eurasia, and Africa. One of the primary motivations for this book is to explore possible generalities in the evolution of behavior in social species. The book offers an updated synthesis of research on elephant behavior that has taken off since previous compilations. Often, when searching for the evolutionary underpinnings of human social behavior, language, cognition, and so on, the tendency among anthropologists and biologists has been to look to our primate relatives. Research is therefore often dominated by the quest for *homologous* features, that is, those that might have been derived from our shared common ancestors. This seems often motivated by a desire to identify which features are uniquely human, as opposed to those maintaining continuity with related lineages. To take a morphological example – our bipedalism is unique, involving a realignment of the pelvis and thigh bones as well as our toes, but built upon a skeletal structure that is homologous with other primates. However, limiting the inquiry to any single clade presents the difficulty that there may be evolutionary constraints or predispositions for the entire clade that are difficult to rule out. Occasionally, there have been attempts at identifying attributes that might have arisen independently across lineages owing to similar social or ecological pressures. Such has been the case in studies of communication and cognition. For instance, many different primate and nonprimate species have evolved alarm calls that categorically pick out predators of a particular type. This is often termed convergent evolution, giving rise to *analogous* features. Analogous structures are relatively easy to spot when they involve physical forms (e.g. the wings of bats and birds, which have evolved independently) but more difficult when they concern invisible structures such as dominance hierarchies. But, in principle, it is conceivable that there are environmental pressures, social or ecological, that shape behavior as much as they shape morphology – indeed, this provides much of the motivation behind the field of behavioral ecology. Just as wings are shaped by the laws of aerodynamics, so too must behavior be shaped by the (messier) forces of ecology. The most interesting opportunity presented by collecting the scattered findings of many independent studies into the long format of a book is the chance to compare across different systems. For the same reason, I take extra care to explicitly identify the particular populations and locations being discussed and try not to overextend conclusions. At the same time, I do try to piece different studies together into a cohesive whole and, in closing certain chapters or sections, take the liberty (and pleasure!) of speculating in some imaginative directions.

Elephants are large-brained, long-lived mammals that exhibit a diverse and intriguing array of social and cognitive abilities. There is a tendency to think, among scientists as much as the general public, that an elephant is an elephant is an elephant. By virtue of large ears and long noses they all appear rather similar to one another and distinct from everything else. Indeed, much of the popular perception of elephants has rested for a great many years on research on a single species – African savanna elephants. It is not difficult to see why: They are charismatic, highly visible, and perpetually (unfortunately) the targets of hunting and trafficking. However, this is now changing, as both the research and policy realms catch on to the fact that elephants are far more varied than was initially suspected. Elephants are fascinating subjects of study because they exploit a range of ecosystems and represent social intelligence that is independently evolved from other social mammals such as that of primates, marine mammals, or even other ungulates, offering opportunities to explore what general principles may be applicable across very different systems. The ancestors of elephants also coevolved with our own, locked in the perpetual dance between prey and predator.

A second, urgent, motivation for this book is that elephant populations are under intense pressure globally, all of them being classified as either "Endangered" or "Critically Endangered." Until recently, only two species of elephant were formally recognized. There are now three. The African continent is inhabited by the African savanna elephant (*Loxodonta africana*) and African forest elephant (*Loxodonta cyclotis*). Separated from this genus by an evolutionary time span of approximately six million years (roughly the same amount of time separating lions, tigers, and leopards from their last common ancestor), the Asian elephant (*Elephas maximus*) dwells in a variety of habitats ranging from grasslands to dense forests in Asia. Though classified as a single species, Asian elephants also come in different varieties, the mainland elephant (*E. m. maximus*) – found in India, Sri Lanka, and Southeast Asia – the Bornean elephant (*E. m. borneensis*), and the Sumatran elephant (*E. m. sumatranus*). For brevity, African savanna elephants will be referred to as savanna elephants throughout this book, while African forest elephants will be referred to as forest elephants. Mainland Asian elephants will be referred to as Asian elephants, and the island populations will be referred to as Bornean or Sumatran elephants respectively, recognizing the diversity of habitat types in which Asian elephants are found.

Elephants have never been more threatened than they are now, in the face of our unprecedented population growth and consumption habits. The largest surviving land vertebrate forces us to confront a critical issue – the need to share space with wildlife and wilderness on a finite and crowded planet. The horde of threats elephants face vary across species and populations: Whereas ivory poaching is the most serious threat for *L. africana* and *L. cyclotis*, habitat loss and conflict are of grave concern for *E. maximus*. Though overshadowed by the ivory crisis, another issue African elephants face is increasing and intensifying conflict with burgeoning human populations. Elephant populations in the central African forests have experienced a precipitous decline of 60% in the late 2000s (Maisels et al., 2013). All forest-dwelling elephants on both continents are threatened by the expansion of logging and plantation activities driven by international trade, even as fresh research is uncovering fascinating secrets of their hidden societies. The loss of not just individuals, but populations, impoverishes not only biodiversity, but also our ability to understand the evolution of a unique and already depauperate clade within the animal kingdom (Caro & Sherman, 2011).

Generally, elephants' ability to persist in the face of modern threats depends largely on their capacity to adapt to landscapes increasingly altered by humans. Behavioral flexibility has conferred considerable historic success on elephants, and ensuring their future requires that we understand and conserve that flexibility. Determining where and how elephants might fare outside captivity or small, isolated protected areas is a key conservation priority for elephants and their habitats. In an age of landscape-scale conservation planning, understanding how elephants perceive their world is key for creating appropriate linkages in mosaics of land use and mitigating the potential for conflict. Some areas in the southern African region are already encountering problems following successful conservation efforts, as increasing elephant populations interface more frequently with human populations. Numbers alone cannot ensure future persistence: What matters is not only how many animals are left, but also the relationships and experiences of those survivors.

Elephants have also long held cultural value and fascination for people – the very ones who may at times be involved in conflict with these mammals, as much as the distant consumers of products produced at the expense of native habitats. The future of successful conservation solutions lies in overcoming not only physical but also psychological barriers; the capacity of elephants to evoke human empathy may be the key to protecting people, elephants, and the ecosystems that support both. Thus, elephant conservation depends greatly on human behavior as well. This volume will examine the implications of certain human activities and decisions for elephant behavior, and vice versa, to better understand what has worked so far and what has not. I hope it will facilitate and encourage collaboration between behavioral biologists and conservation practitioners of all stripes, within or outside academia.

I must first place boundaries identifying what will not be found in this volume, which is necessarily limited in scope. I do not address elephant evolution or physiological adaptations in depth, as these topics are discussed elsewhere (Shoshani & Tassy, 1996; Sukumar, 2003), except as a launching point for understanding elephant behavior as it manifests today. I provide only light coverage of a body of earlier research pertaining to the interaction of elephants and their habitats in savanna environments, which has been reviewed in Sukumar (2003), focusing instead on more recent work that provides a comparative perspective. I use primary peer-reviewed literature and limit the inclusion of material found in other books (e.g. *The Amboseli Elephants* by Moss et al. (2011a)) only to observations that are not reported in primary literature and which, again, are necessary for making comparisons among populations or species. I avoid discussing unpublished data with the exception of published theses and some of my own lab group's work in progress, as I cannot otherwise attest to their quality. It is also not my aim to discuss the welfare and ethics of elephants in captivity; those interested in these topics are directed to Wemmer and Christen (2008). Lastly, some elephant species and populations have been studied far more extensively than others. To cover each in proportion to the number of studies would result in a fairly lopsided discussion with heavy bias toward particular geographies. In trying to balance the content across taxa and locations, I therefore necessarily trade off being comprehensive.

1.2 Behavior in Conservation: Frameworks for Integration

Animals interact with the conspecifics, other species, and the environment. This truism constitutes the sum total of behavior, and understanding this multifaceted complex is the aim of those who study animal behavior. Such studies, however, are unfortunately viewed by some as a luxury of little consequence to the practical challenges of conservation – especially for species or populations that do not lend themselves easily to observation. That wildlife management has historically relegated behavior to a black box, focusing instead on more tangible numerical quantities such as population abundance, growth rates, habitat extent, and so on, is therefore perhaps not so surprising. On the academic side, the role that behavioral research can and does contribute significantly to solving actual conservation challenges has been viewed with a mix of tempered skepticism (Angeloni et al., 2008; Caro, 2007; Caro & Sherman, 2011, 2013) or hopeful optimism (Blumstein & Fernández-Juricic, 2010; Caro & Sherman, 2013; Greggor et al., 2016); substantial overlap among the two sets notwithstanding (full disclosure: I include myself in the ranks of skeptical hopefuls). The divide may have been reinforced over time by differing priorities, not only among the practitioners but also scientific publishers (Angeloni et al., 2008).

And yet there has been a gradual increase in the recognition that behavioral variables are pivotal to the success or failure of conservation efforts over preceding decades, even if the rate of increase in scientific publications that jointly mention behavior as well as conservation seems an imperfect reflection of this (Angeloni et al., 2008; Nelson, 2014). It is my hope that by consolidating the recent behavioral research for elephants specifically, this volume makes this research more accessible to those who would wish to use it while highlighting areas in which more work is needed. Here, I briefly provide some background to the study of behavior for those who are unfamiliar with this. I then outline frameworks that have been developed in recent years that link behavioral research with applied conservation and management, to put this book in context and clarify terminology.

1.2.1 Levels of Causality and the Comparative Method

Readers unfamiliar with the study of animal behavior may find it helpful to bear in mind one of the first sets of organizing principles outlined in the field by Niko

Tinbergen, one of its founders. They are often referred to as *levels of causation*, paralleling the schema of causalities laid out by Aristotle (Hladký & Havlíček, 2013). Tinbergen (1963) explicitly identified four possible categories of explanation for any and all behavior:

- *Mechanistic*: The physical processes that go on within an organism that directly control its behavior (sensory stimuli, hormonal pathways, neuronal processes, etc.).
- *Ontogenetic*: Processes that occur over the development (ontogeny) and lifetime of an individual.
- *Functional*: The consequence a behavior has on an individual's (or population's) propensity to leave offspring, that is, its effect on fitness or its adaptive value.
- *Phylogenetic*: The context of an organism's evolutionary history, which can both constrain and enable evolution in a certain direction.

A well-studied example that is often used is that of birdsong. The question "What makes a bird sing?" can be answered in four different ways. A mechanistic explanation may discuss how changes in daylight or day length stimulate certain hormones or neurotransmitters in the brain, which then fires off signals to the syrinx, resulting in song production. An ontogenetic explanation may discuss the process by which a bird learns to sing through observation of other older individuals, imitating and developing its own species-specific repertoire. A functional explanation may outline the ways in which songs are necessary in order to attract mates and display one's own health and vigor. A phylogenetic explanation situates a songbird's ability to sing in the fact that it is a member of a closely related group of species, all of which have the ability to learn vocal production. Likewise, the last level of explanation offers some insight as to why other bird species, which are *not* songbirds, do not also sing. The glory of song, even if adaptive, is not really an option for an ostrich (at least, in the near future).

From this, it should be evident that the four levels of explanation are complementary rather than competing, although in specific instances it might be difficult to avoid muddling them together. In particular, the first two are typically subsumed under the heading of *proximate* causes (with the term "proximate" sometimes becoming synonymous with "mechanistic"), whereas the second two are often described as *ultimate* (with the term "ultimate" sometimes being used interchangeably with "functional"; see Bateson & Laland, 2013; Mayr, 1961). The distinctions are not merely semantic, however, because confusing one type of explanation for another can hinder understanding of the interplay between levels and lead to unnecessary disputes (Laland et al., 2011; MacDougall-Shackleton, 2011). Behavioral ecology concerns itself largely with questions pertaining to the ultimate causes. So too, in this book, I will be focused more heavily on the functional and phylogenetic explanations than on the mechanistic or ontogenetic (though with some exceptions, such as in thinking about learning, cognition, and communication). I mention these qualifications here as I will not continuously flag the terms explicitly.

Just as with physical structures, the current utility of a behavior need not necessarily reflect the historic pressures that drove its evolution (Bateson & Laland, 2013). To paraphrase the population geneticist Theodosius Dobzhansky (1973), nothing in behavior makes sense except in the light of comparison.[1] The *comparative method* is therefore an invaluable means of trying to understand the function or adaptive utility of a behavior through comparisons of species or populations. Different populations or closely related sister taxa that share a phylogenetic history but occupy different environments may exhibit corresponding behavioral differences. Conversely, distantly related taxa may show convergent patterns in response to similar selective pressures. Comparisons offer a way to extract general principles that would be impossible to infer through the study of single populations or species. A frequent error is to forget phylogenetic explanations or constraints entirely and view all behavior as shaped by its immediate utility (Bateson & Laland, 2013). On the other hand, some behaviors may not make sense without understanding the broader historic context. Unfortunately, these confusions are likely to increase as species and populations become extinct, eroding the capacity for comparative study and thus threatening the some of the very means by which the discipline of animal behavior progresses (Caro & Sherman, 2011, 2012).

1.2.2 Conservation and Behavior

Those interested in a thorough treatment of conservation behavior as an emerging discipline may refer to Berger-Tal and Salz (2016). Blumstein and Fernandez-Juricic (2010) provide an accessible starting point for integrating the study of behavior with conservation, defining the subject of "conservation behavior" as "the application of knowledge of animal behavior to solve wildlife conservation problems." They lay out a detailed rationale, characterization, and tools for this enterprise, taking the very applied perspective that because behavior is "at the root" of many conservation problems, behavioral knowledge at the level of single species should be centrally integrated into conservation plans, while recognizing that large-scale conservation measures of course involve many species. What are the types of conservation problems that have behavior at their root?

Captive breeding to rescue small populations is a paradigm example of an applied problem in which behavioral understanding seems invaluable. One has to ensure that the ordinary behaviors that enable an individual to successfully rear young are successfully duplicated by humans, while being careful not to introduce further disruptions to the animals while they are in human care, especially if the aim is to eventually release these individuals back into the wild. The mechanistic and ontogenetic levels may be of greatest concern. For instance, the abilities to recognize other members of their own species and, in the case of gregarious species, to actually prefer their company to that of humans, are critical. Once released, individuals must know how to forage, recognize and avoid predators (including humans), and eventually find a mate (and compete for one, if required). If they are successful, they may need to rear offspring. Any individual of a species can only acquire those skills and abilities that

[1] "Nothing in Biology Makes Sense Except in the Light of Evolution" is the title of a 1973 essay by Dobzhansky.

are possible within the constraints of its evolutionary history, a consideration that may not be at the forefront of breeding efforts but which may latently affect the results; thus, breeding efforts would benefit from a comparative perspective. For instance, might the behavior of a more common species be instructive in designing the breeding program? If surrogates from a different species are used, might their behavior be appropriate models for the species being reared? One might ask whether sister species require similar or different social environments early in life, given their natural histories? When a young animal is brought into the highly artificial environment that captive breeding represents, fulfilling each of these requirements may be a tall order, yet nevertheless essential.

More broadly, reproduction is a domain of behavior that is accompanied by a host of complex behaviors in many species, ranging from competition during courtship to the act of mating itself. For some species, these might involve relatively hardwired displays. But for elephants and many mammals in the wild, learned tactics allow individuals to successfully compete for mates or establish dominance, such as the experience young males might obtain by sparring. Likewise, for females it is important to learn to assess high-quality males and behave appropriately during reproductive periods so as to avoid injury to themselves. In social species, especially those which live and breed communally, the "Allee effect" (Allee et al., 1949) represents a well-known phenomenon whereby populations exhibit nonlinear declines as a function of decreasing group size, which are often underpinned by the accompanying social interactions. For example, the need for assistance when rearing offspring or simply the opportunity to learn how to be a good parent through the example set by others.

For elephants, it is seldom the case that captive breeding programs are designed with the intention of eventual release; more often, the aim is to maintain a captive population. However, programs for rehabilitating individuals brought in from the wild face similar challenges. For those programs that aim to replenish the population of elephants in captivity, it is worth asking whether there is a behavioral rather than purely physiological basis for why elephants reproduce so poorly in many captive settings yet appear to do well in others. Some may also raise ethical and moral issues regarding elephants in captivity, on the grounds of whether their behavior is suitably "normal" in the sense that they do not experience undue suffering. Here the question of what elephants in captivity truly contribute to conservation is one worth deeply considering. If their role is to serve as ambassadors for their species, then it is imperative to understand how best to meet their needs relative to what they would experience in the wild.

Another set of conservation challenges centers on land-use planning. At a basic level, one would want to know what the habitat requirements of an individual are before designing reserves and corridors. In doing so, it would be well to recognize that wildlife cannot be expected to simply stay in the areas allocated to them – they will need move to meet their feeding and reproduction requirements. Often landscapes are modified with economic or development goals in mind, with reserves and corridors occupying compromised spaces. "Conflicts," or more generally, negative interactions, may therefore arise at the boundaries – such as over agriculture or livestock or through collisions between wildlife and human transportation devices. As climate changes and landscapes continue to be modified through human activity, we will be faced with the enormous challenge of making our spaces conducive to such movement while minimizing disturbance to both humans and wildlife. Movement is therefore among the most problematic and yet essential behaviors with which conservationists and wildlife managers must contend. Perhaps for this reason, it has the distinction of being a behavioral domain with entire subdisciplines devoted to it (movement ecology and spatial ecology).

Preserving long-distance migration and preventing human–wildlife conflict are related concerns that preoccupy many practitioners and policymakers. Initially assumed to be an innate ability, there is increasing evidence that at least for some species, migration also depends heavily on learning and cultural transmission of knowledge (Festa-Bianchet, 2018; Jesmer et al., 2018). Although individuals of a species may have exhibited such behavior over millennia constituting hundreds of generations, these are clearly fragile processes that are incredibly vulnerable to disturbance, especially if components of the behavior have to be learned anew with each generation. This involves recognizing specific timing cues such as changes in temperature or day length, or in food quality or availability. It then requires coordinating with conspecifics during the actual migration event, learning the routes and resources along the way along with vigilance against the potential dangers one may encounter. Lack of appropriate experience, the absence of knowledgeable leaders to guide movements, even over shorter distances, and the disappearance of critical resources along movement paths can all potentially disrupt such behavior. The ability to successfully participate in migration events will again have consequences for an individual's fitness and phylogenetic effects insofar as the behavior persists in future generations. On the flip side, when animals are forcibly moved as part of a management initiative, not only spatial but social factors (such as acceptance by local conspecifics) may influence their ability to successfully establish themselves at new locations. Problematic behaviors may also be spread from one population to another via social transmission.

As managers and policymakers, it is important to anticipate any unintended adverse consequences that a management action might have. Reproduction and movement are just two domains of behavior in which the relevance to conservation and management are so obvious that they have benefited from early recognition; the list is much longer (those interested may begin with Greggor et al., 2016). Yet other aspects of behavior, notably areas such as personality, cognition, and culture, have been more neglected until recently. If, for example, if one continuously removes (by lethal or nonlethal means) problem animals without letting them reproduce, one may be removing precisely those individuals that are bold and curious. If such attributes have a genetic component, we may effectively be removing these attributes from the gene pool, leaving behind a population that is on the whole less exploratory and less able to cope if conditions change, such as the preferred food items go into decline. In the long term, this threatens the evolutionary potential of the population in the face of environmental changes.

1.2.3 Behavioral Variation and Flexibility

Behavior is a phenotype, broadly defined as the outward result of interactions between genotype and environment. Because the phenotype is what natural selection can directly act on, and because selection requires standing variation, it follows that in order for organisms to evolutionarily adapt their behavior, they must have preexisting behavioral variation (see also Bateson & Laland, 2013, and references therein). As alluded to in preceding sections, it is crucial to recognize that behavior is itself something that might need protecting (Caro & Sherman, 2012). If we view organisms as complex biological machines, it is evident that disruptions at any level of behavior can have consequences not only for individual members of a species but also programs aiming to protect or restore populations. Long-distance migration is again a perfect example of this, in which the loss of skilled and knowledgeable individuals can extinguish the behavior itself within a population. The upshot of knowing this is that one can appreciate the importance of social contact in the reintroduction of a species and can facilitate it if necessary. Berger-Tal et al. (2011) provide a framework for thinking about behavioral variation in relation to conservation aims. The central issue is the need to recognize that, along with genetic variation, individuals within or among populations of the same species exhibit behavioral variation.

For instance, individuals may exhibit regional differences in food preferences, and correspondingly manifest behavioral adaptations suitable to tackling those food preferences. Given time, those preferences may lead to morphological changes such as changes in the physical structures involved in feeding (Odling-Smee et al., 2003). Of course, it is also possible that behavioral shifts follow physical changes – the more traditional view. Projecting toward the future, the ability of populations to continue adapting in the face of changing environments will require that there is not only sufficient genetic diversity (where the term "diversity" refers to not just any neutral variation, but variation that is *structured by natural selection*) but also behavioral diversity – the willingness to explore and exploit new food sources.

These inclinations may be tied to certain types of personalities, which again represent standing variation. Some individuals may be more inclined to be exploratory than others and thus more readily try novel food sources, potentially enjoying less competition from others. Novel food sources might include recently introduced non-native species, or simply changes in the demographics of existing species such that once-common food sources are scarce and once-rare food sources are more common. The flip side of this is that novel food sources might be toxic. Human refuse (i.e. garbage) is a particularly problematic novel food source that may be readily exploited by some individuals, bringing them into contact with both harmful substances such as heavy metals and plastics as well as pathogens. It is clear from this example that what is adaptive under some circumstances may not be in others.

Behavioral variation is different from flexibility (also termed "plasticity"), where a single individual might employ a host of different strategies (which may or may not be learned) or modify strategies through learning, a process that occurs at the ontogenetic level, as already mentioned. Although behavioral flexibility may sometimes be confusingly referred to as "adaptability," it does not necessarily refer to *evolutionary* adaptability. In common parlance, the term "adaptability" might refer to an individual's ability to modify his or her own behavior in an advantageous manner. Nevertheless, behavioral flexibility refers to an individual-level characteristic, whereas true adaptability refers to a population-level characteristic. To go back to the example of feeding, it might mean being able to exploit different food sources at different times of year. This could itself be a genetic trait. Over the long term, if individuals that are more behaviorally flexible are also more successful at leaving offspring, and this trait is itself heritable, one could say that behavioral flexibility is adaptive.

In summary, individuals may demonstrate flexibility or plasticity in their own behavior; there may be variation among individuals in the different types of strategies they employ and the degree of flexibility they demonstrate; and finally there may be variation among different populations of the same species as a result of selection or culture. These ideas are exceedingly relevant for protecting and managing elephants. The cognitive and behavioral complexity of elephants – male and female – presents both challenges and opportunities for conservation. The challenges are that populations can be sensitive to disturbances and management strategies (e.g. culling) in unforeseen ways; individuals may find ways around methods designed to reduce conflict by learning from one another. The opportunities include novel solutions to conservation challenges based on our expanding understanding of elephant sociality, communication, and motivation. Likewise, as habitats diminish it becomes crucially important to recognize that the depletion of distinct subpopulations represents not just loss of genetic diversity but also behavioral diversity, which might hold the key to maintaining resilience in this long-lived species, which is destined to witness a generation of rapid change. The recognition that conserving species requires more than simply managing numbers is finally gaining traction, particularly in arenas such as reintroduction programs, conserving migratory species, and addressing human–wildlife conflict. One indicator of this shift is the movement toward considering cultural units as being significant for conservation, in addition to the standard focus on species and genotypes (Brakes et al., 2019).

1.3 How This Book Is Organized

This volume takes behavior as the central organizing principle, with different domains of behavior contextualized with the relevant theoretical background. Rather than take up specific conservation or management issues and examine which behaviors are relevant to them, which would result in a very disjointed discussion of behavior or artificially separate the academic study of "behavior" from applied "conservation" via distinct sections or chapters, I discuss the conservation implications of particular behaviors as they naturally arise. The result should be that the two are naturally interwoven in readers' minds, just as they are in reality. The themes and stances sketched out in the preceding sections will therefore appear throughout, but not be invoked explicitly. Those coming from a more applied perspective are advised to start with a concrete problem first and then refer to the relevant behaviors (Blumstein & Fernández-Juricic, 2010; Caro, 2007), whereas students looking for research ideas in particular domains might first notice the gaps that are identified as needing exploration and then, given what is learned, consider how they might inform practical conservation or management activities. I also found that there is no single format that works equally well for all chapters. Therefore, instead of organizing each individual chapter the same way, they are somewhat idiosyncratically carved up into sections that make sense for the topic at hand.

Chapter 2 sets the stage by discussing the likely origins of the living elephants and their relationships to one another, together with the ecological conditions that may have prevailed upon their lineages. Predation and resource availability, the forces of selection that operate on all life forms, are no less important for the largest land mammals ever to have graced the earth than they are for the smallest. And yet the living elephants have somehow managed to avoid the wave of extinctions that overcame other large-bodied vertebrates throughout the globe. I examine why this may be, and what it tells us about the prospects for elephants' continued persistence in the face of rapid change.

Chapter 3 builds on this background to describe what is known so far about the structure of female societies. Because elephants exhibit sexual segregation upon reaching adolescence, the social contexts of males and females have to be treated distinctly. Females, having to invest heavily in the production and care of offspring, are sensitive to three pressures: resource availability, predation, and competition from other individuals. In this chapter I examine how why these factors may interact to shape female social relationships in the three species. Hand in hand with this, I consider how social and ecological disturbances have been shown to impact elephant societies and explore the responses of various elephant populations.

Chapter 4 explores male social life, which is to some extent inseparable from discussion of reproduction and mating. Freed from the duties of parental care, male elephants are primarily concerned with gaining access to sufficient nutrition for themselves and improving their prospects of reproducing successfully. I discuss how male social life begins to diverge from that of females upon reaching sexual maturity and how relationships among males manifest. I then examine reproductive strategies in both sexes, highlighting some of the similarities shared by all elephant species by virtue of their basic physiology, irrespective of their ecologies. This in turn shapes reproductive tactics uniquely employed by male elephants that distinguish them from other species. I conclude by examining some consequential differences between the Asian and African species both in terms of evolution and conservation, arising from one conspicuous dissimilarity: the absence of tusks in many Asian elephants.

Chapter 5 considers the role that elephants play in ecosystems. Elephants are variously referred to as flagship species, keystone species, ecosystem engineers, and even "mega-gardeners" (Blake & Hedges, 2004; Campos-Arceiz & Blake, 2011). These monikers all allude to elephants' ability to modify habitats and participate in a variety of ecological interactions, for the most part thanks to their sheer size. I examine their interactions, both with the species they exploit and those that exploit them. Aside from this, I consider their dependence and impact on natural resources, as well as the effect of artificial resources. I then introduce the concept of the "landscape of fear," which provides an intriguing lens through which to view behavior insofar as it is a response to perceived threats in the environment. The consequent use or avoidance of certain areas may thus depend not only on what is present in those areas, but also on what is absent.

Chapter 6 explores elephant movements and space use in greater detail. This chapter is grouped by habitat and landscape type, subdivided further by geographic regions. I use this structure to try and present a somewhat cohesive picture of the factors driving space use, where different species might share commonalities but also underscore conservation challenges relevant to their particular context. The examples often demonstrate why it is important to consider departures from assumptions made by classical theories, such as optimal foraging theory and the ideal free distribution, when contending with materially embodied species facing real problems physically embedded in space and time, as opposed to theoretical constructs or simulations.

Chapter 7 considers communication and cognition as flipsides of the same coin, recognizing the close relationship between dominant modes of communication and dominant modes of perception. I begin with different sensory modalities, which include the chemosensory apparatus, tactile perceptions, use of vibrations in air and ground, and visual domain. I discuss sensory signals and cues in terms of how they may be actively or passively employed. Among specialists of these topics, the term "signal" generally refers to something that the animal actively produces in order to broadcast itself to other conspecifics (such as a vocalization or controlled chemical secretion), whereas the term "cue" can refer to something produced either inadvertently or signals for which the subject was not the intended target (e.g. vocalizations of other species, seismic vibrations due to footfalls), as well as environmental stimuli (e.g. the scent of ripening fruit). I then move deeper into what we know of the elephant mind, considering various kinds of awareness, knowledge, and emotion. Lastly, I circle back around to elephant societies and the prospect of culture, which represent a sum total greater than the constituent individuals.

In the final chapter, I consider the long history and inherent challenges of human–elephant interactions on shared landscapes. In particular, I reflect on the question of whether we have an adequate understanding of the types of landscapes on which elephants have long lived and the processes that maintained them. Humans may have had a crucial role to play. Despite our role as predators in many ecosystems, we have also served as ecosystems' stewards and guardians. These relationships were in many cases disrupted throughout the world during a multicentury process triggered by the radical transformation of landscapes and societies that took place around the globe during colonialism. The reverberations of this era persist and continue to be felt. But as our own population keeps rising to dizzying heights, accompanied by the unpredictability of a changing climate, the ability of the past to guide the future seems limited. Yet, in looking forward, I think it is imperative to learn from our past in order to ensure a just and sustainable future for both our species and the many others that live alongside us.

2 Elephant Evolution

2.1 Ancestral Attributes

2.1.1 Dawn Elephants

The tendency for elephants to group together appears to be quite an ancient trait, as revealed by one of the most fascinating pieces of fossilized behavior uncovered to date – a trackway. We begin our story in the Miocene, a geological period stretching between 5 and 23 million years ago, and it is within this sizeable window of time that we have the earliest glimpse of sociality among now-extinct proboscideans that may have been closely related to the ancestors of modern elephants. The lineage that would eventually lead to the mastodons had already diverged off into their own evolutionary path by this point (Rohland et al., 2007). The site, known as Mleisa, is located in what is today Abu Dhabi, in the United Arab Emirates. At the time, it was likely covered by grassy open areas. From the elevated vantage point of a kite, one sees the remains of an ancient riverbed or mudflat, traversed by a collection of beautifully preserved disc-shaped footprints of varying sizes running nearly parallel, with another single set of large footprints diagonally intersecting them (Figure 2.1). There are 14 individuals altogether. By measuring the size and depth of these imprints, the researchers estimated the approximate body mass of the animals that left them (Bibi et al., 2012). The range of sizes includes individuals within the current distribution of sizes for adults of both Asian and African elephants (between 2,000 and 6,000 kg). The largest of the animals appear to be closer to 6,000 kg, comparable to or slightly greater than the size of the larger male African savanna elephants. The tiniest imprints may have belonged to calves, weighing less than 1,000 kg. The authors therefore conjecture that the group of tracks might represent a herd of mixed age and sex, while the single large set of prints is that of a lone male. Because the deposit occurs within the approximate time period when the common ancestor of the mammoths and extant elephants would have been present (Rohland et al., 2007, 2010; Shoshani & Tassy, 1996, 2005), it is natural to wonder if the makers of these tracks, the putative common ancestor, also lived in family groups as modern elephants do (Bibi et al., 2012).

Bibi and colleagues put forward that the most likely candidate for having made the tracks is *Stegotetrabelodon*, one of the earliest true elephants, known from around 7–5 million years ago and probably close to the ancestry of the living species, but which survived until the time when African and Asian elephant lineages and the mammoth

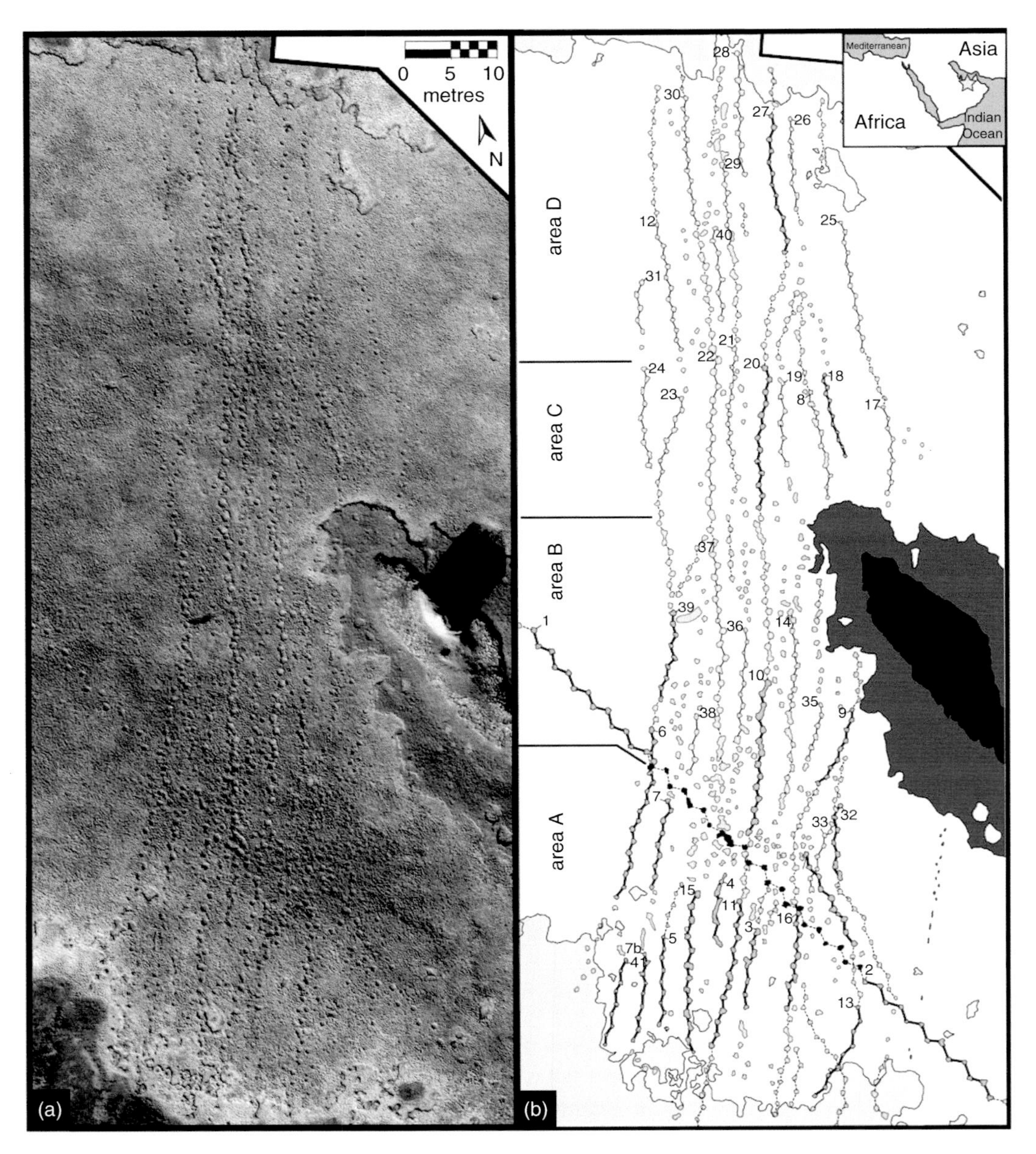

Figure 2.1 The Mleisa trackway and candidate species. (a) An aerial photomosaic of the site. (b) Individual tracks measured in the study (reproduced from Bibi et al., 2012, © 2012 The Royal Society, with permission).

were already diverging. This suggests that social behavior seen in the living species was already present in their common ancestor. However, it cannot be ruled out that the footprints belonged to more proboscidean species such as *Deinotherium*, *Mastodon*, or *Amebelodon*, which diverged much earlier, which would indicate a much earlier origin for these behaviors. Although the tracks are very suggestive that these ancient proboscideans might have had a similar social structure to modern elephants, the image is of course based on what is already known of the living species and remains a tantalizing but incomplete story. More recent studies have revealed differences in the finer structure of social relationships among females of the different extant species and have also shown that there is more to the social lives of males than previously supposed. As all living elephants derive from some common ancestor, an understanding of evolutionary relationships can potentially help us to understand which attributes are ancestral traits and which evolved more recently. Easier said than done.

2.1.2 The African Clades

Understanding the origins of the *Loxodonta* genus, which gave rise to the African elephants, has been a problem for zoologists and geneticists as well as paleontologists. The group is named after the lozenge-shaped cusps (lamellae) that are the raised surfaces found on the molars, and may have emerged at least 7 million years ago at the end of the Miocene based on the earliest teeth (Tassy, 1995). By this point the group had already diverged from the lineage that would lead to the *Elephas* and *Mammuthus* genera, the Asian elephants and mammoths respectively (Brandt et al., 2012; Rohland et al., 2007). Although *Loxodonta* therefore must have lived in Africa throughout the early Pliocene (5 million to 1.6 million years ago), the species is scarce in the African fossil record until the Pleistocene. Instead the continent was dominated by a species by the name of *Palaeoloxodon recki* (which has previously also gone by the name of *Elephas recki*), whose relationship to the Asian and African clades has been debated (Meyer et al., 2017). With a typical shoulder height around 4 m (14 ft) and weighing in at about 12 tonnes, it was one of the largest proboscideans on Earth (Larramendi, 2016). *Palaeoloxodon* presents a bit of a puzzle because its skeletal attributes seem akin to that of Asian elephants and mammoths (Figure 2.2). By the structure of its skull and hyoid bones, which anchor the musculature of the tongue, it was grouped together with the *Elephas* and *Mammuthus* genera (Shoshani et al., 2007). However, the reconstruction of evolutionary relationships using DNA has thrown up a complication. A species widespread in Eurasia known as *Palaeoloxodon antiquus* (the “straight-tusked elephant”) turns out to cluster more closely with the African forest elephant *Loxodonta cyclotis* on the basis of mitochondrial and nuclear DNA sequences (abbreviated mtDNA and nDNA respectively; Meyer et al., 2017).

Mitochondrial DNA can be used in tracing the relatedness of living individuals as much as reconstructing the evolutionary relatedness of species of over various timescales. Because mitochondria are found in the body of the egg cell but not the nucleus, both male and female offspring can only inherit it maternally, if there is sufficient diversity in the mtDNA lineages (referred to as *haplotypes*) it gives a certain matrilineal family history. While it is a very useful technique, it is of course only half the story. Nuclear DNA, which is the genetic material of the nucleus and thus inherited from both the maternal and paternal sides, can sometimes create a conflicting picture. Taken together, the two sources of DNA can provide more complete insights into how the two sexes dispersed and contributed to gene flow in the past. Looking more closely at genetic structure once again, there was mounting evidence that not one but

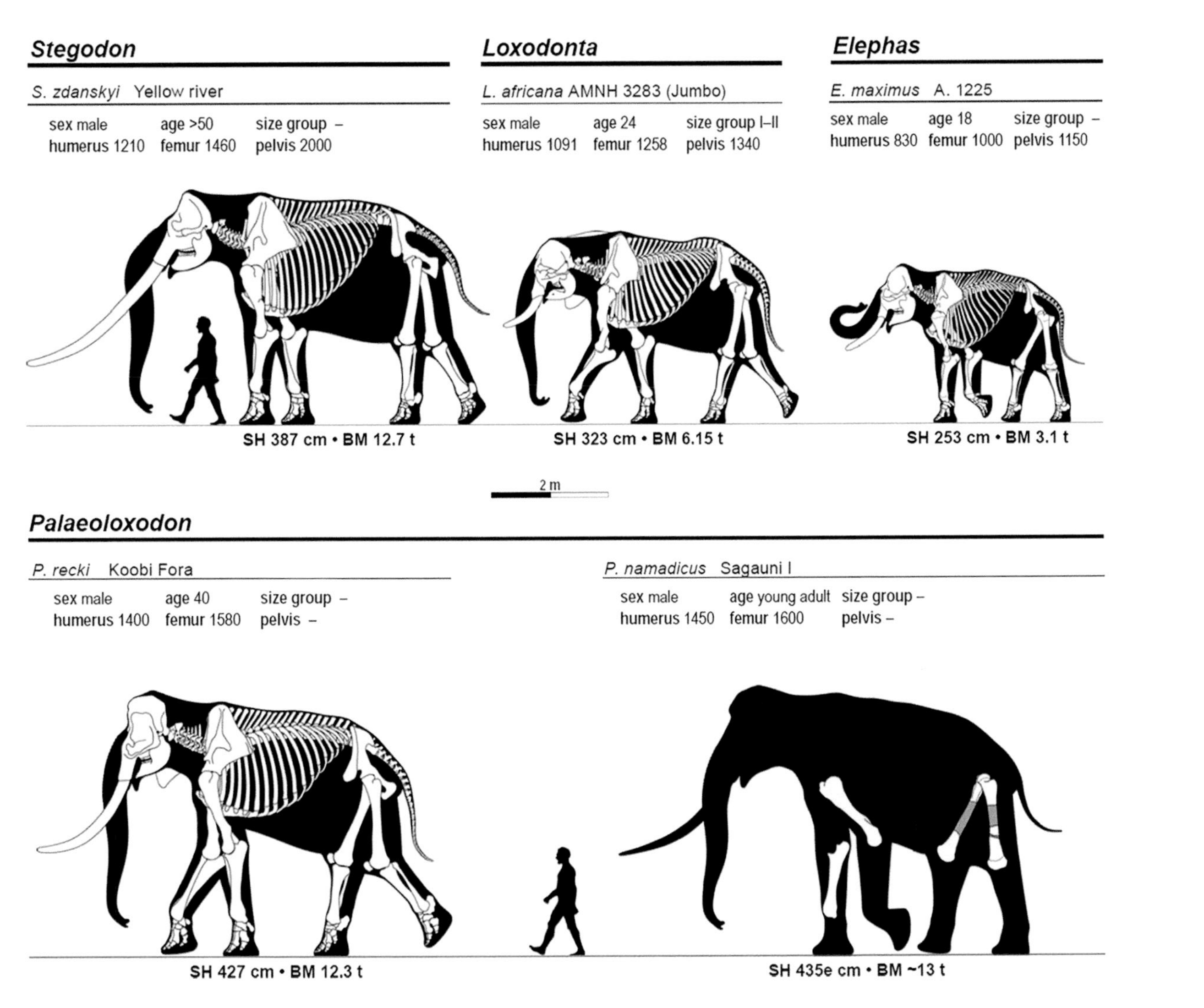

Figure 2.2 A comparison of the dimensions of several proboscideans, extinct and extant (reproduced from Larramendi, 2016, under cc-by-4.0).

two sister species of elephant remain in Africa, although this view has been contested. Here, I briefly sketch the twists and turns of this tale, without dwelling too deeply on the methodological details, as a striking example of how an organism's behavior can leave a complex and indelible mark on its evolutionary history, even as evolutionary history constrains behavior itself (for a sweeping discussion of this topic, see Lister, 2014). The case also highlights the value of comparison (between taxa and sources of data) for understanding an organism's evolutionary path.

In 2001 Roca and colleagues proposed distinguishing the African savanna and forest elephants (*Loxodonta africana* and *Loxodonta cyclotis* respectively) as distinct species on the basis of nDNA, whereas they had previously been assigned subspecies status (*L. africana africana* and *L. africana cyclotis*) based on visible differences in their ecology and morphology (Sukumar, 2003, and references therein). Examining short repetitive DNA motifs known as microsatellites, Comstock et al. (2002: 2496) stated that the genetic distances between forest and savanna populations "are almost as great as the difference between Asian and African genera." However, an analysis of mtDNA by Eggert et al. (2002), which additionally included West African samples that had been previously omitted, suggested that if anything there were at least three "deeply divergent" populations corresponding to different geographic regions rather than ecotypes: the forest elephants of Central Africa, the forest and savanna elephants of West Africa, and the savanna elephants on the rest of the continent. Based on these findings and reconstructions of the climate and vegetation of the time, Eggert and her colleagues suggest that *L. africana* in fact evolved out of a species inhabiting the Central African forest, which ventured out some time before 3.5 million years ago to colonize areas consisting of forest in the northernmost part of the range and progressing into grass and scrub mosaics further south. They further suggest that the prevalence of forest habitat, which does not favor the preservation of remains, might account for the poor fossil record of *Loxodonta*. However, grasslands were becoming the increasingly dominant vegetation at around this period (Cerling et al., 2011; Levin et al., 2011). The dispersal of *Loxodonta* might initially have also been held in check by their competitor, *Palaeoloxodon*, which occupied more exposed habitats. *P. recki* persisted until at least 500,000 years ago on what was then a very well-watered Arabian peninsula (Stimpson et al., 2016), when *Loxodonta* in its currently recognizable form made a comeback and displaced it from the African continent. The sister taxon *P. antiquus* continued to persist in Eurasia.

Debruyne (2005) conducted another study using mtDNA that added new specimens from the Democratic Republic of Congo, which seemed in disagreement with the two former studies. Once again two divisions were emphasized, between clades labeled F and S; however, these did not appear to precisely correspond to forest/savanna morphological or ecological types either. The F clade predominated among forest elephants, but its mitochondrial lineages were also found in a substantial proportion of savanna elephant samples. The S clade, however, was represented only among savanna elephants. Debruyne concluded that the two taxa were incompletely differentiated, given what he interpreted as frequent interbreeding in areas wherever the ranges of savanna and forest-dwelling populations intersected. He instead attributed the apparent groupings from the prior studies to have resulted from incomplete sampling of the various elephant populations (Comstock et al., 2002; Roca, 2001) or particular choices made in statistically modeling the molecular phylogenies (Eggert et al., 2002).

Studies up until this point had looked only at one source of genetic material or another, but not simultaneously at *both sources from the same individuals*. Naturally this was the next step, and in the same year as Debruyne published his results, Roca and coauthors published the intriguing finding that mtDNA and nDNA obtained from the same individuals could nevertheless show different evolutionary histories (Roca et al., 2005). They looked at three sources of DNA that have different paths of inheritance: the mtDNA, which is only passed down via the female line as we have already noted, the nuclear Y chromosome, which is only passed down via the male line as well as restricted to males, and the nuclear X chromosome, which can be passed down by either sex and acquired by either sex. They also explicitly considered which sites represented true forest habitats, which represented savanna habitats, and which represented intermediate transition zones. The X and Y chromosomes showed near complete separation of the forest and savanna populations. However, mtDNA typical of forest elephants did show up in savanna elephant populations inhabiting transition zones as well as two populations much further south, in Botswana and Zimbabwe. In the latter, they attributed locally wetter conditions evidenced by the presence of one of the largest lakes in Africa, which might have allowed forest elephants to colonize the area in the past. They also conjecture that the same conditions might have allowed savanna elephants to make headway in areas previously dominated by *Palaeoloxodon* (cf. *Elephas*), creating a temporary opportunity for hybridization that left a small imprint of forest elephants' genetic legacy despite their current absence.

Though these scenarios remain speculative, the major pattern is striking. If hybridizations were frequent in both directions, one would expect nuclear and mitochondrial genetic structures to be similar. Instead, it appears that female forest elephants have contributed to savanna gene pools, but not the reverse. Likewise, the male contributions do not appear to cross between the two groups. Because forest elephants of both sexes are smaller than savanna elephants, the authors suggest that male savanna elephants likely outcompeted male forest elephants and any hybrid males with respect to mates. Therefore, female forest elephants may have bred with male savanna elephants, thereby passing on their mitochondrial genomes to savanna populations. There are two problems with this explanation, however. First, one would then expect a fair amount of Y-chromosome material from savanna populations to also pass into forest populations, which is not the case. The second issue, which Roca et al. acknowledge, is that this would require female offspring to disperse into savannas periodically and transfer into savanna social groups, which seems highly unlikely given that females tend to remain with maternal kin (see Chapter 3). Alternatively, the persistence of forest-type mtDNA in savanna populations may be due to past reproductive events at a time when both ecotypes frequently occupied the same habitats, after which at least some hybrid families remained behind on savanna landscapes to become absorbed by the latter populations.

Table 2.1 Genetic divergence and heterozygosity estimates for the elephantids

First taxon	Second taxon	Genetic divergence (heterozygosity if within taxa) normalized by savanna–Asian genetic divergence	±1 Standard deviation
Across taxa			
Savanna	Forest	74%	6%
Savanna	Mammoth	92%	5%
Savanna	Asian	100%	N/A
Forest	Mammoth	96%	7%
Forest	Asian	103%	5%
Mammoth	Asian	65%	5%
Within taxa (heterozygosity)			
Savanna	Savanna	8%	2%
Forest	Forest	30%	4%
Mammoth	Mammoth	9%	2%
Asian	Asian	15%	3%

Note: Genetic divergences based on 549 sites that are polymorphic among elephantids, normalized by savanna–Asian elephant genetic divergence (for which a standard deviation therefore cannot be calculated). Savanna and forest elephants are >4 standard deviations less diverged than savanna and Asian elephants, while mammoths and Asian elephants are >6 standard deviations less diverged than savanna and Asian elephants. Recreated from Rohland et al. (2010).

The genetic code changes over time, and certain segments of DNA, appear to accumulate changes at a somewhat regular pace, which has led to the notion of a molecular "clock." By comparing the differences in genetic sequences between related species, one can infer how long it took these differences to accumulate. In this manner, Roca (2001) had initially "clocked" the divergence point between the African savanna and forest elephant to at least 2.5 million years on the basis of mtDNA; however, more recent studies have pushed the divergence of nuclear genomes back even further to between 4 and 5.5 million years ago (Li & Zhang, 2010; Rohland et al., 2007). This is likely owing to the different pathways of inheritance for the two sources of DNA.

Rohland, Roca, and other colleagues in 2010 once again argued in favor of the distinctness of forest elephants, this time from a comparative basis. Comparing the nuclear genome sequences of forest and savanna elephants against those of mastodons, woolly mammoths, and Asian elephants, they reported that the African forest and savanna elephants were "as or more divergent" as mammoths and Asian elephants (Table 2.1). The genetic divergence (±1 standard deviation) of savanna and forest elephants is estimated to be 74 ± 6%, whereas the divergence between Asian elephants and mammoths is around 65 ± 5%. As a group, forest elephants also harbor greater genetic diversity (measured in terms of heterozygosity) within their samples (30%) than either Asian elephants (15%) or savanna elephants (8%).

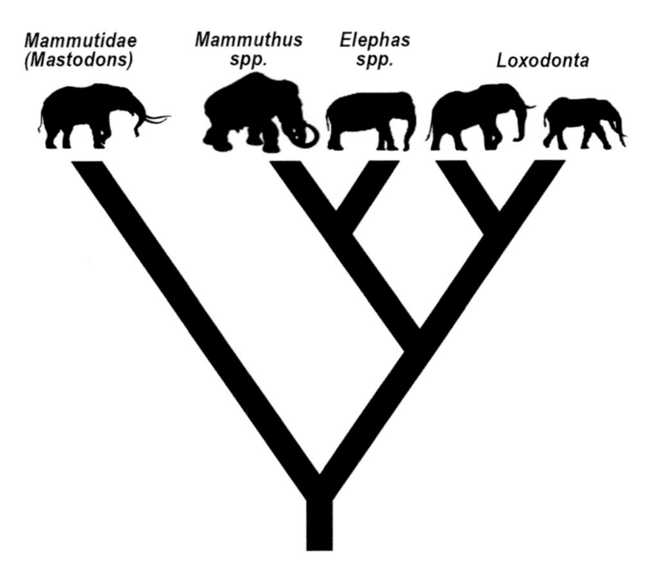

Figure 2.3 The evolutionary relationships among the elephantidae with mammoths as an out-group (adapted from Rohland et al., 2010, under CC license).

Since we distinguish Asian elephants and mammoths as distinct species (and indeed they are currently considered to belong to distinct genera),[1] by the same standards one must also distinguish the two African elephants (Figure 2.3). Fortunately for me, this book took longer to write than I anticipated and, as of the time of this publication, the IUCN (International Union for the Conservation of Nature, which maintains the legally relevant catalogue of species and their conservation status) has at long last recognized the distinction between savanna and forest elephants. We will consider what this means for understanding their behavior and conservation requirements.

2.1.3 The Asian Clades

During the Pleistocene (2.6 million to 12,000 years ago), northerly relatives of *Elephas* evolved to tolerate colder temperatures in the form of several mammoth species. At least two of these, *Mammuthus meridionalis* and its likely descendant

[1] The debate concerning "species" status in this entire discussion rests upon the *biological species concept*, in which species are biological units that are reproductively isolated (Dobzhansky, 1937; Mayr, 1942). Hybrids present a problem, yet they abound in nature. The Linnean system of stratifying phylogenies into discrete tiers (species, genera, family, etc.) is also now understood by most evolutionary biologists to be more imprecise than originally intended (not least because there are a plethora of different species concepts) and some systematists would prefer to dispense with the Linnean hierarchy altogether in favor of simply using the singular term "clade." Nevertheless, these designations remain in popular use as well as referenced in taxonomic naming systems. Suffice it to say, "species" are not set in stone, even if one must identify them for legal purposes if nothing else; but what we really care about ultimately is whether these distinctions matter biologically.

Table 2.2 The distribution of adult sizes for some living and extinct proboscideans

Estimated percentage of the total population size		Lower ~5%	~90% (mean)	Upper ~ 5%
Loxodonta africana ♂	shoulder height	304 <	304–336 (320)	337–352
	body mass	5.2 <	5.2–6.9 (6)	6.9–7.7
Loxodonta africana ♀	shoulder height	247 <	247–273 (260)	274–286
	body mass	2.6 <	2.6–3.5 (3)	3.5–3.8
Loxodonta cyclotis ♂	shoulder height	209 <	209–231 (220)	232–242
	body mass	1.7 <	1.7–2.3 (2)	2.3–2.6
Mammuthus primigenius ♂	shoulder height	266 <	266–294 (280)	295–308
(North Siberian form)	body mass	3.9 <	3.9–5.2 (4.5)	5.2–5.7
Mammuthus primigenius ♂	shoulder height	299 <	299–331 (315)	332–346
(European form)	body mass	5.2 <	5.2–6.9 (6)	7.0–7.5
Mammuthus columbi ♂	shoulder height	356 <	356–394 (375)	395–413
	body mass	8.3 <	8.2–10.9 (9.5)	11–12.1
Mammuthus meridionalis ♂	shoulder height	380 <	380–420 (400)	421–440
Mammuthus trogontherii ♂	body mass	9.6 <	9.6–12.7 (11)	12.8–14
Elephas maximus ♂	shoulder height	261 <	261–289 (275)	290–303
	body mass	3.5 <	3.5–4.6 (4)	4.6–5.1
Elephas maximus ♀	shoulder height	228 <	228–252 (240)	253–264
	body mass	2.3 <	2.3–3.1 (2.7)	3.1–3.4
Mammut borsoni ♂	shoulder height	389 <	389–431 (410)	432–451
	body mass	13.7 <	13.7–18.4 (16)	18.5–20.4
Mammut americanum ♂	shoulder height	275 <	275–305 (290)	306–319
	body mass	6.8 <	6.8–9.2 (8)	9.3–10.2
Deinotherium proavum ♂ and ♀	shoulder height	347 <	347–383 (365)	384–402
	body mass	9.1 <	9.1–12.1 (10.5)	12.2–13.4
Palaeoloxodon antiquus ♂	shoulder height	380 <	380–420 (400)	421–440
	body mass	10.8 <	10.8–15 (13)	15.1–16.6

Note: Shoulder height is given in cm, body mass in tonnes, and ranges show minimum–maximum. Modified from Larramendi (2016).

Mammuthus trogontherii, are contenders for the largest proboscideans ever to have evolved (Table 2.2). There is evidence that the more recently extinct Columbian mammoths in North America (*Mammuthus columbi*) also likely traveled in small family units, as determined by analysis of the stable isotope signatures of sets of remains containing adults and calves. Hoppe (2004) analyzed different types of sites in North America with the hypothesis that remains representing family groups should have lower isotopic variability than those that did not, owing to similarities in diet and movements. At one location in Waco, Texas, a group of 14 unfortunate mammoths appears to have been rapidly buried together, perhaps during a flooding event. They consisted of adult females and calves, with a single adult male. Another site in Texas, which was likely a cave occupied by the scimitar-toothed cat, contained several hundred mammoth teeth, mostly of juveniles, likely to have been prey accumulated over time. She then compared these to three other sites in Texas, New Mexico, and Colorado, which yielded collections of 5, 6, and 13 individuals respectively, at least some of which bore signs of human butchery. Because animals moving together as a unit should experience the same environments and thus consume the same food and water, she reasoned that any collections of remains representing families should match one another more closely in their stable isotope composition than those representing unrelated individuals frequenting different localities. The carbon isotope values supported this view with regard to the putative family group represented at the Waco site, whereas the others are likely to have been unrelated animals that died independently (Hoppe, 2004). A comparison of the observed range of oxygen and strontium isotope values additionally suggested that these individuals were unlikely to have habitually moved over distances greater than 600 km.

The *Elephas* genus, which gave rise to the Asian elephants, is also thought to have originated in or near Africa, with the earliest known members, *E. ekorensis*, appearing between 4 and 5 million years ago, and, subsequently, *E. planifrons*, occurring in Asia from ~3 mya. A later species dating between 500,000 and 200,000 years old found in Jordan and Israel shares some similarities with *E. hysudricus*, originally found in northern India, which was an apparent ancestor to the living species (Lister et al., 2013). This positions the modern-day species, *Elephas maximus*, at less than 500,000 years old. In the intervening time, Asia had its own compliment of giants to rival those on other continents. Another member of the closely related *Palaeoloxodon* genus, *P. namadicus*, might be a contender for the largest land mammal to have ever evolved on earth (Figure 2.2). Limb bones excavated from Sagauni, India, indicate that some individuals could have achieved a shoulder height exceeding 5.2 m and body mass exceeding 22 tonnes when fully grown, which, if true, would have made these both taller than some species of the popularly recognized sauropod dinosaurs such as *Camarasaurus lentus* and *Apatosaurus louisae* and more than twice as heavy as *Diplodocus carnegii* (Larramendi, 2016)! However, this remarkable characterization is contested by others.

The present-day Asian populations (all of which are still considered a single species) are hypothesized to have also survived a process of range constriction thought to be driven by climatic events, namely glaciation (Vidya et al., 2009). A range-wide assessment of mtDNA from more than 500 individuals showed two major lineages among the living populations. The older group, termed the α-clade, appears to have originated in the northernmost part of its range and spread southward until it occupied much of the available territory. A newer β-clade seems to have arisen in what is now southern India or Sri Lanka (the two being connected at some point) and subsequently dispersed back into the subcontinent. Following an intervening glacial period that split the population into eastern and south Asian subsets, the β-clade seems to have further diversified independently in Sri Lanka and the Sunda region, which includes peninsular Malaysia and Sumatra. Today, the greatest mitochondrial diversity is found in Sri Lanka and Myanmar where both α and β groups occur in close proximity, while one or other of the clades dominate the other populations (Figure 2.4).

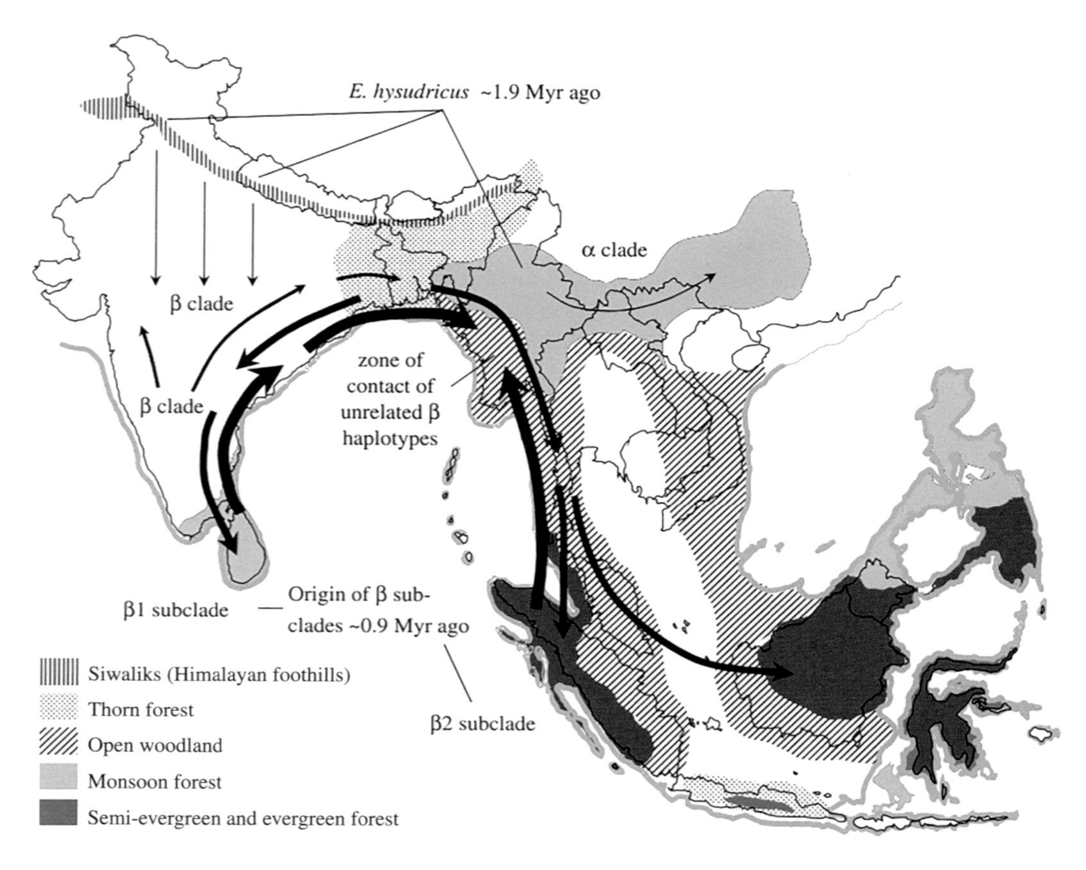

Figure 2.4 A schematic representation of the evolutionary routes and environments that gave rise to Asian elephants, *Elephas maximus*. Thicker arrows indicate more recent dispersal events (reproduced with permission from Vidya et al., 2009, © 2008 The Royal Society).

2.2 Evolutionary Drivers of Elephant Sociality

2.2.1 The Predation Hypothesis

Why might these ancient proboscideans be gregarious at all, given that living in a group creates inevitable competition? A ready hypothesis is that groups confer protection against predators. The idea of the "selfish herd" proposed by William D. Hamilton (1971) is a purely numerical argument, namely that the more potential prey there are around you, the less likely you are to be eaten yourself. This is also known as a "dilution effect," and explains why not just ungulates but animals of many different species quite often congregate in large numbers, especially when found in open spaces. Being in a large aggregation can also mean extra pairs of eyes keeping vigilant, as well as chaos and confusion for predators when they do make a move. Merely being spotted can often be a deterrent for predators, which would rather not waste energy on an ill-timed ambush. Among bovids (cattle and their relatives) it has been shown that gregariousness is positively associated with species favoring more open habitats (Caro et al., 2004), although the question of why remains unanswered. But family groups are distinct from the anonymous congregations represented by schools of fish, flocks of birds, or migrating ungulates (Moffett, 2020). If one is among family members, particularly offspring, the dilution effect seems an inadequate explanation, since the loss of relatives is costly in genetic terms (though, of course, not as costly as being captured oneself). Indeed, family groups among mammals often, though not always, consist of related females (Clutton-Brock & Lukas, 2012), thus elephants fit a familiar, recurring pattern. Families may assist one another in active defense, and those that do so successfully are likely to leave more survivors. Kin selection for cooperative defense may therefore be an alternate pathway to gregariousness.

Did the elephants of prehistory have anything to fear from predators, given their formidable size? Large body size automatically confers protection to adults against predation; however, juveniles remain vulnerable until they grow sufficiently. As already mentioned, elephants evolved alongside equally formidable predators. Given that sociality is likely to be an ancestral trait that has been preserved across many lineages rather than something that has been more recently derived, what matters are the size and ecological requirements of the ancestor. The leading candidate for this ancestor is *Palaeomastodon*, which inhabited wetlands in North Africa 35–36 million years ago in the Eocene-Oligocene (Figure 2.5). It had a recognizably proboscidean appearance and stood about 2.2 m (7 ft) tall. It was a tapir-like creature about whose social life one can only guess. Modern-day tapirs, which seem similar not only in appearance but also lifestyle, are mostly solitary. On the other hand, hippos, which again are a variation on the same ecological theme, are very gregarious. Though these distantly related species may not be good guides for how *Palaeomastodon* might have behaved, they show that there is enough variation in the possibilities to make the reality rather difficult to guess. Nor do we gain much clarity by looking to the closest living relatives of elephants, the *Sirenians* (dugongs and their ilk) and *Hyracoidea* (hyraxes), which themselves show a variety of social structures including both solitary and gregarious tendencies, having diverged from the evolutionary branch containing *Palaeomastodon* at a still earlier date and that are now situated firmly outside the Proboscidean clade.

Meanwhile, in these same African landscapes, another lineage was evolving. It led to the great apes, with the first split leading to *Pongo*, the next to *Gorilla*, then finally *Pan* and *Homo*. This last took place somewhere in the vicinity of the time of divergence between *Elephas* and *Loxodonta* lineages, approximately 5–7 million years ago. Whether elephant kills were opportunistic or dietary staples for early hominins depends on when and where the evidence happens to be located. As already mentioned, there is plentiful evidence that our own species hunted elephants, but this is relatively recent. However, at Olorgesailie, an 990,000-year-old site in Kenya, there is also evidence of an elephant having been butchered by *Homo erectus* (Potts, 1989). It's impossible to know if the animal had been scavenged or actively hunted, since such finds are exceedingly rare. There is evidence from multiple sites that *Homo heidelbergensis*, which is thought to have lived between 1.3 million and 600,000 years ago (Qiu, 2016)

Figure 2.5 *Palaeomastodon beadnelli*, a putative ancestor of the proboscidean clade including the mastodons and elephantids. It possessed tusks on both the upper and lower jaws as well as a nasal cavity suggestive of a trunk (image: Nobu Tamura, cc-by-4.0).

and is likely an ancestor of modern humans, hunted *Palaeoloxodon* alongside other big game such as horses, hippos, deer, and rhino, according to the Smithsonian National Museum of Natural History. Although the species is named after the town of Heidelberg, Germany, where it was first discovered, its oldest members have been uncovered in Ethiopia. *H. heidelbergensis* was not only a skilled hunter, but appears to have begun to tame fire, according to the Smithsonian National Museum of Natural History. These innovations and the abundance of large-bodied prey allowed this species to spread far and wide beyond its putative origin in Africa where it is believed to have given rise to the other hominin lineages known as the Neanderthals of Eurasia and their close relatives, the Denisovans (Brown et al., 2021; Reich et al., 2010). Elsewhere in Asia, where fossil discoveries continue to challenge our understanding of human evolution and migration, hominins were busy proliferating into a baffling variety of species. Specimens dating as far back as 900,000–700,000 years have been recovered in China and Indonesia respectively, with remarkably modern-looking teeth being recovered from a site in China's Hunan province that nevertheless appear to be 80,000–120,000 years old (Qiu, 2016, and references therein).

One bold hypothesis goes so far as to contend that proboscideans had just the right balance of protein and fats to support the development of ever-more powerful brains among the hominins, culminating in the eventual replacement of *Homo erectus* with our own species (Ben-Dor et al., 2011), and thus our early ancestors might have actually favored elephants as prey. Moving forward in time to Paleolithic North America, mammoth and mastodon hunting appears to have been relatively frequent (Hoppe, 2004; Surovell & Waguespack, 2008), whereas in Siberia it is described as sporadic (Nikolskiy & Pitulko, 2013). However, the earliest anatomically modern human settlers of Eurasia, known informally as the "Cro-Magnons," specialized in big-game hunting, possibly with the aid of dogs (Shipman, 2015). A fascinating site in the Czech Republic named Předmostí, with abundant remains of woolly mammoths, reveals extensive use of all parts of the animal for toolmaking, decoration, and even dwelling construction. The marrow and brain matter might have also been consumed at some locales for their fat content (Yravedra et al., 2012). Echoing the statements of Ben-Dor and colleagues, Shipman (2015) argues that the ability to cooperatively hunt big game such as mammoths might have been crucial to our successful invasion of new territories, and indeed our displacement of other human lineages.

All of this would imply that humans in some sense owe our very existence, survival, and proliferation to the ancient proboscideans. In turn, might it not be possible that particular aspects of sociality (and associated cognition) in elephantids coevolved at least partially in response to (and perhaps even to outsmart) the ever-present threat of our ancestors? Early proboscideans would have grouped together when traveling over vast open grasslands to overcome the threat of nonhuman predators, but the emergence of humans as ever-more lethal hunters may have helped maintain and enhance these tendencies. It is a subject of ongoing debate whether humans were the direct cause of proboscidean extinctions throughout the world (as opposed to climate change), along with that of other megafauna, in the late Quaternary (Barnosky, 2004; Nikolskiy & Pitulko, 2013; Sandom et al., 2014; Surovell et al., 2005). The timing of extinctions of proboscideans in particular, coinciding with human range expansion, has been offered as evidence for the so-called overkill hypothesis (Sandom et al., 2014; Surovell et al., 2005). One naturally wonders why, if humans were such successful elephant hunters, did the African and Asian lineages persist where humans are likely to have been present the longest? Quite possibly, this very fact may have been the key to their survival.

History shows that in areas where humans encountered any species that were naïve to the threat we posed, these were quickly dispatched. But where humans and proboscideans coevolved side by side, elephants may have been pressured to acquire the necessary characteristics to evade humans – either through cooperative defense or crypticity. It's possible that in heavily forested environments elephants may have been spared the degree of hunting experienced by those in more open environments, with humans subsisting on smaller game and plant matter. If this is true, the effects of predation and other environmental attributes on elephant behavior would be confounded and difficult to distinguish. What is clear though is that the paths of our two lineages often crossed throughout our evolutionary histories, long before the ancestors of either of our species became what we are today. Human and elephant histories trace the interwoven stories of predators and prey. Humans continue to be the most dangerous predators of elephants, wherever they may be, a subject to which I shall return. We also continue to push earth's remaining megafauna toward the same fate as those of the past (Ripple et al., 2015).

2.2.2 The Offspring Care Hypothesis

Other possible explanations why females of some species, mammals especially, might choose to be gregarious are worth considering. I mention these only briefly, as they will be revisited in later chapters. One is that baby elephants can be quite a handful.

Not only are they quite vulnerable to getting lost or eaten, they are rather clumsy and accident-prone. Personal observations suggest very young calves, confronted with a befuddling forest of legs, might sometimes even mistake who their mothers are, following the wrong adult (or even an overzealous subadult). This touchy situation persists for at least two years, during which calves are dependent on milk. For a female, this can mean that the energetic costs of keeping a vigilant eye (or ear, trunk, or tail as the case may be) on her offspring must be traded off against her foraging requirements, both for herself and for lactation. Females with dependent offspring stand to lose their body condition more quickly than males or nonlactating females. Classic life history theory contends that any investment a female makes in offspring, especially one that requires parental care, necessarily exerts a cost on her own survival and the possibility of producing additional future offspring. Enter the alloparent – a nonparent that undertakes some of the caregiving responsibilities, thereby at least partially relieving the actual parent of such. The helpers are typically (though not necessarily) nonbreeders that are related to the breeders, and thus gain inclusive fitness benefits, which is another argument for gregariousness based on kin selection (Emlen, 1982, 1984).

Many bird and mammal species exhibit alloparental care, especially when individuals are hard-pressed to go off and become breeders themselves, as might be the case when habitats are saturated and territories are scarce, when dispersing is risky, and/or individuals tend to be long-lived owing to high adult survival (i.e. when species with extended life histories encounter ecological constraints; see Emlen, 1982, 1984; Hatchwell & Komdeur, 2000; Vehrencamp, 1983). The term "cooperative breeding" is used to describe such systems, and this can involve a dazzling array of species with distinct social organization (Hatchwell & Komdeur, 2000). But elephants are not textbook examples of cooperative breeders in the sense that this term is usually applied to species with specific home sites (nests, dens, burrows accompanied by territories) around which family life is built, and indeed this is seen as the limiting resource that governs whether new breeders can become established. Elephants are not territorial in any classic sense, need not be anchored to a particular site in order to breed, and indeed most females will themselves eventually breed. However, the number of alloparents present were a significantly positive influence on calf survival in at least one population of African savanna elephants, but only after accounting for major ecological factors such as droughts and demographic effects such as maternal age (Moss & Lee, 2011a). Although their longevity and extended time to reproductive maturity sets the stage for cooperation in offspring care, also evidenced in woolly mammoths (Lister & Bahn, 2007), such cooperation may not be a universal attribute among proboscideans and therefore its relative importance as a driver of sociality remains unclear.

2.2.3 The Harassment Hypothesis

Another driver of female gregariousness to consider is the degree of harassment they experience from bulls. Eisenberg et al. (1971) observed of Asian elephants that younger females seemed to experience a lot of harassment from males, especially those that were also young themselves. They suggested that females might be able to have respite from such harassment thanks to the dominance of older females over young bulls. My colleagues and I have not observed similar harassment in the population we have studied in southern Sri Lanka, except when young females are entering oestrus for the first time and appear especially attractive to all males in their vicinity. Lee et al. (2011) describe that although within-family aggression is rare among the savanna elephants at Amboseli, males that are dispersing independently and starting to grow bigger than adult females may harass them; the females in turn respond by forming coalitions to drive them off, often with the aid of the matriarch.

Studies of other species show varying responses to harassment based on the underlying social system. Among schooling guppies, in which females are capable of developing associations based on individual recognition, the presence of a male seems to actually undermine the development of female relationships (Darden & Watts, 2012; Darden et al., 2009). Male Grévy's zebra defend territories with important resources, through which females must pass. Females have relatively unstable relationships with one another but can avoid harassment by sticking around a single male (Sundaresan et al., 2007). Female South American sea lions, however, reduce their own odds of being harassed by congregating in larger groups, thereby taking advantage of a "dilution effect" similar to mechanisms for avoiding predation (Cappozzo et al., 2008). Among chimpanzees, in which males remain in their natal groups but females disperse, females in oestrus risk injury or death from aggressive efforts by dominant males to restrict their movements (van Schaik et al., 2004, and references therein). Bonobos are closely related to chimpanzees, but unrelated adult males and females nevertheless form bonds (Furuichi & Ihobe, 1994). Harassment by males in the mating context is rare among bonobos (Hohmann & Fruth, 2003), but young immigrant females must form alliances with older females in order to avoid aggression from residents of either sex (Idani, 1991). Thus, not only sexual harassment, but prospective conflicts with conspecifics might simultaneously motivate bonds with others.

Elephants have a reproductive physiology that distinguishes them from the majority of species, and this is that females come into oestrus for a short window once in several years. Therefore, as they are candidates for sexual harassment only for a relatively brief period, this alone is insufficient to explain why they might be gregarious the rest of the time. Conflict over resources with other conspecifics is ever present, however, and may similarly facilitate the retention of social bonds among familiar individuals, as discussed later in this chapter and in coming chapters.

2.2.4 The Resource Dispersion Hypothesis

The classical view of social evolution in ungulates gives primacy to the role of resource distributions in governing female movements, with males then responding to the distribution of females (Emlen & Oring, 1977). Although this view has since been revised to account for social feedbacks such as infanticide and harassment by males themselves, in addition to the ecological factors (Sundaresan et al., 2007; van Schaik et al., 2004), it remains a fundamental idea. Because of sheer size, elephants are slaves to their stomachs. They must cover sufficient ground to meet their water and nutritional requirements. The resource dispersion hypothesis predicts that animal group sizes are predicted by the total heterogeneity and richness of resources (Johnson et al., 2002). If resources in the ancient environment were unpredictable and far flung, it would behoove a wandering elephant to follow the guidance of more experienced individuals. A long-lived animal would also have to learn which other individuals are friendly and which are not, thus avoiding energetically wasteful confrontations. From an ecological perspective then, grouping together can potentially confer fitness benefits both for oneself and one's kin, not only through cooperative defense but also through information and knowledge transfer. A major difficulty in understanding how important the configuration of resources might be for shaping social behavior in any species is that it is so fundamental – given that resources are never uniformly distributed in space or time – it's difficult to see how it could *not* be relevant. Yet other more immediate and obvious concerns, such those outlined in the preceding sections, can mask its influence (Johnson et al., 2002).

Another challenge is that the diets and ecological conditions experienced by ancestral elephants are difficult to pin down (with rare exceptions, such as those frozen in permafrost), even if broad patterns in vegetation are known. The Baynunah Formation, which contains the Mleisa trackway, is thought to have occurred amid open grasslands perhaps not unlike that of savannas today (Bibi et al., 2012). But what is a savanna? This is a very loose term used to describe diverse habitats, being applied to grasslands, woodlands, and the mosaics in between. The central feature, though, is that grasses dominate the understory (Ratnam et al., 2011). Asian elephants likely experienced the entire gradient of habitats over their evolutionary history, from open grassland to dry deciduous woodland, to dense closed-canopy forest (see Vidya et al., 2009, and references therein). Although the dentition of elephants can be suggestive of particular diets, with those of Asian elephants being better adapted to a grazing lifestyle than that of African elephants, behavioral studies show that neither is particularly constrained by diet and will graze, browse, or eat fruit depending on whatever is most convenient (reviewed extensively in Sukumar, 1989, 2003). Indeed, Lister (2013) makes the fascinating and persuasive case that Proboscideans (along with many other herbivores) actually switched from a primarily browsing diet to a predominantly grazing diet several million years before their dentition adapted! The implications of this are profound to consider, and we shall return to them in Chapters 5 and 8.

None of these hypotheses are mutually exclusive – it is entirely possible that there is more than one explanation for why elephants evolved their particular form of sociality and why certain populations continue to maintain it. But this means that the relative importance of each may be difficult to disentangle, especially today when many populations are small, fragmented, and highly disturbed by human activities at a scale unprecedented in history. If the question of how elephants lived is one side of the coin, the other is how they died.

The closer we get to the modern era, the better our ability to discern what climate and vegetation proboscideans are likely to have experienced. Stuart (2005) proposed a reconciliation of the climate change and hunting hypotheses for proboscidean extinction by comparing the patterns of extinction of woolly mammoths (*Mammuthus primigenius*) and straight-tusked elephants (*Palaeoloxodon antiquus*) in Europe. The former were reliant on open steppe-tundra habitats containing grassland vegetation whereas the latter favored woodland. Accordingly, woolly mammoth ranges were at their greatest extent during the last glacial maximum when steppe-tundra habitats were widespread, whereas straight-tusked elephants enjoyed their heyday during the preceding, warmer, interglacial period. Both species saw range expansions and contractions during glacial–interglacial cycles but were able to withstand them. He notes that although both species underwent severe range constrictions prior to going extinct, island-dwelling representatives of both taxa persisted for quite some time after mainland extinctions as relict populations. These should have been even more vulnerable to climatic changes given their restricted movement. Stuart therefore finds it unlikely that climate and corresponding vegetational change alone could have driven their extinction (see also Lister & Bahn, 2007).

The major difference between the end of the Pleistocene and the prior challenges these clades faced was the emergence of human hunters as already mentioned (Barnosky, 2004; Nikolskiy & Pitulko, 2013; Sandom et al., 2014; Stuart, 2005; Surovell et al., 2005). Pockets of suitable habitat free of humans, such as that represented on the islands, would have served as temporary refuges. Human hunters, posits Stuart, may have interfered with the ability of proboscideans to follow resources as they otherwise would, contributing to their ultimate collapse. Therefore, it is not necessarily one or the other, but the interaction of both predation and resource stress that contributed to the extinctions of these species.

Modern elephants find themselves in a still more unpredictable world, with access to resources now being rapidly modified through the actions of humans and their movements (along with that of many other species) much more heavily constrained (Tucker et al., 2018). Whereas once they were able to traverse the length and breadth of continents, today they struggle to march the few kilometers that may separate one forest patch from another. At the same time, recent rethinking of the role of humans in shaping ecosystems that properly acknowledge the ancient ties between indigenous communities and the land are blurring our understanding of what "wild" landscapes truly represent. These historic practices underwent a massive upheaval over the past few centuries as colonialism swept the globe and introduced new social and economic forces that led to rapid land-use changes that transformed many biomes (Ellis et al., 2010). In the present, "savannas" and "savanna-like" systems may represent ecosystems containing native grass and tree species, as well as habitats altered by human logging, burning, and cultivation activities that have created grasslands where forests once stood – perhaps through the additional introduction of nonnative species, as will be examined more deeply in Chapter 8. For all these reasons, characterizing what habitat type an elephant has experienced over its lifetime, let alone its evolutionary history, can be quite a challenge. What we do know of the past, however, might be instructive in helping us learn what to expect in the future.

2.3 A Possible History

When all is considered, one might imagine the following scenarios. It is likely that gregariousness in itself had arisen within the proboscideans before their emergence from the African continent, originating at some unknown time when the ancestral species occupied a relatively open environment, body size was smaller, and the threat of predation accordingly greater. As elephants dispersed and diversified, some lineages grew large enough to the point that adults may have no longer needed to be concerned about nonhuman predators, although calves would have remained vulnerable and thus provided some motivation for collective defense. Additionally, the vicissitudes of environmental conditions, especially in areas where climatic conditions or rainfall regimes were more unpredictable, may have conferred additional advantages to group living by facilitating resource acquisition, either through the knowledge of experienced individuals or simply numerical advantage over other competing elephants (of the same or different species). Thus, the ancestral trait persisted and evolved in complexity.

If phylogenetic history exerts a stronger influence on elephant behavior than recent ecological conditions, then one expects forest and savanna elephants to be more similar to one another than to Asian elephants. However, if the opposite is true, then elephants in similar habitats should show similar behavior irrespective of lineage. Elephants may also show behavioral flexibility depending on the situation. Given both the diversity of habitat types occupied by contemporary Asian elephants and their degree of genetic divergence from either forest elephants or savanna elephants, one might expect their behavior to show a great potential for local variation in response to the site-specific ecological context. An additional corollary to this is that there are likely necessary tradeoffs in physical stature depending on the favored behavior and ecology: Large body size and group defense is favored in open environments (*L. africana*); however, smaller body size, crypticity, and smaller group size is likely to be favored in more sheltered habitats (*L. cyclotis* and *E. maximus*; Stankowich & Caro, 2009). The social environment itself imposes its own selective pressure: social selection, which is especially of relevance for females (Tobias et al., 2012). Foraging strategies, communication, and cognition in turn must all evolve in concert with the social and ecological context or socioecology.

On the Asian and African continents, hominins evolved and radiated in parallel (to one another as well as to elephants), scavenging carcasses at first but eventually rising to occupy the position of top predator. But unlike the proboscideans, early hominins did not make it to the Americas. Just as living predators follow their migrating prey, so too our more recent ancestors may have trodden well-worn elephant trails out of Africa. Large body sizes, unfortunately, come with a demographic cost – the inability to reproduce very quickly (discussed in Chapter 4 and revisited in Chapter 8). Having traded reproductive capacity for longevity and large body sizes, proboscidean populations would have been hard-pressed to cope with the dual pressures of human hunting and climate change (Cantalapiedra et al., 2021). By the end of the Pleistocene, only those populations hidden within the forests of Asia or Africa would persist, together with elephants adapted to the expanding drier savanna ecosystems.[2] Where there was more cover, elephants may have relied on crypticity and the avoidance of people, whereas elephants in exposed environments drew upon social defenses and their wits to survive the rise of humans. Because proboscideans are not today a speciose clade, and all are constrained by their collective phylogenetic history as well as present-day habitat loss, we cannot fully test the scenarios outlined here regarding how elephant behavior evolved in the first place. As mentioned before, many of the variables act in concert and therefore confound one another. However, elephant responses to present-day conditions in varying ecological contexts might give us an inkling as to why they evolved certain behaviors, as well as the ability to understand and predict how they may be affected by the current changing world. The closest we may be able to get to understanding what shaped elephant societies may be through understanding how modern elephants respond to the pressures they face across different locations and ecological regimes.

[2] The forest-favoring American Mastodons and savanna-dwelling European straight-tusked elephants also became extinct, as previously discussed. Mastodons did not coevolve with humans, thus I would contend living in forests did not save them. Straight-tusked elephants were faced with the dual pressures of hunting *and* changing ecology.

3 Female Social Life

As in many mammals, the social lives of male and female elephants largely diverge around the time that males reach adolescence. This sexual segregation is fundamentally driven by the differing nutritional requirements of males and females owing to their reproductive needs. A breeding female is usually feeding for two, whereas a male must invest only in his own physical growth. From this follows everything else about how an individual interacts with members of its own sex as well as those of the opposite. First, we will focus on the social lives of females.

It is by now likely to be familiar to many readers that "herds" of elephants are composed primarily of females and calves, while bulls tend to be more solitary or loosely associated with one another. Our understanding of elephant social life is derived primarily from seminal studies of African savanna elephants (Buss & Smith, 1966; Douglas-Hamilton, 1972; Moss & Poole, 1983) and the studies that built upon the paradigms that these began. Because some elephant populations have been intensively studied in subsequent years (e.g. at Amboseli, see Moss et al., 2011), there is a great wealth of understanding about these particular populations that tends to shape how all elephants are viewed. But elephants can show a great deal of behavioral variation and flexibility, even between closely neighboring populations. Our conception of elephants is of necessity a composite made up of individual studies, each conducted within its own specific context. The findings of one locale should not automatically be assumed to apply at another, even for members of the same species, let alone across species. Seemingly general statements may require revision as more studies from diverse geographies continue to be added. Different studies tend to use varying terminology, so I clarify my usage of some terms in Table 3.1.

3.1 African Savanna Elephants

3.1.1 Social Structure

The first tier of female society for African elephants consists of the individual female and her dependent offspring, which for all intents and purposes are treated as a single inseparable unit (Buss & Smith, 1966; Douglas-Hamilton, 1972; Eisenberg et al., 1971; Moss & Poole, 1983; Wittemyer et al., 2005a). Female savanna elephants are very gregarious, forming very cohesive multigenerational family groups, typically

Table 3.1 Definitions of various grouping categories

Term	Definition	References
Group or aggregation	A set of individuals observed in the field moving, resting, or interacting nonaggressively within an approximately 500 m radius of one another.	(de Silva et al., 2011; Wittemyer et al., 2005a)
Cluster or community	A set of individuals that repeatedly associate together such that they form a distinct unit that is revealed by an objective analytical clustering method.	(de Silva et al., 2011; de Silva & Wittemyer, 2012; Wittemyer et al., 2005b)
Social unit	A general term describing sets of socially affiliated individuals. Social units can form at different levels of organization, corresponding to community structure at that level. They may therefore apply to a single family but also be extended to encompass entire populations that represent similar cultural attributes.	(Brakes et al., 2019; de Silva et al., 2011; de Silva & Wittemyer, 2012)
Matriline	The descendants of a particular female. Social units may consist of one or more matrilines.	(Archie et al., 2006b; Fernando & Lande, 2000; Vidya & Sukumar, 2005; Wittemyer et al., 2009)
Family group	A social unit consisting of related females that also associate together at high enough levels to form a cluster. Typically these are second-tier units in savanna elephants, but they may not be as distinctive in the other species.	(Douglas-Hamilton, 1972; Moss & Poole, 1983; Wittemyer et al., 2005a)
Bond group or clan	Stable third- and fourth-tier social units respectively in savanna elephants, unconfirmed in forest elephants, and graded social units of changing composition in Asian elephants.	(Douglas-Hamilton, 1972; Moss & Poole, 1983; Nandini et al., 2017; Wittemyer et al., 2005b)

consisting of related females and their offspring (Archie et al., 2006b; Douglas-Hamilton, 1972), though this is not always the case, as we shall see. The descendants of any particular female make up a matriline, with a "family" consisting of one or more matrilines. These matrilineal kin units (also called core groups) can be very cohesive, with members spending at least 50% of their time together and typically over 70% (Archie et al., 2006a, 2006b; Moss & Lee, 2011b). They comprise the second tier of female society, which shares similarities with other mammalian societies, including that of primates such as chacma baboons and carnivores such as lions or hyenas. This tier is what people are most likely to encounter if they come upon wild elephants in a landscape. Typically, they may contain between one and seven females of reproductive age and their calves, with most having about three to four adult females (de Silva & Wittemyer, 2012). Larger groups occur, but less frequently, and may at times consist of more than one family, as discussed later in this chapter (Figure 3.1).

Families are very cooperative units that are the basic building blocks of savanna elephant society, assisting in group defense as well as oversight of offspring. However, there is also competition for resources within and between groups, thus female savanna elephants exhibit clear, linear (transitive) dominance hierarchies. Unlike certain primates such as chacma baboons, where dominance hierarchies are nepotistic and females of a particular matriline within a troop occupy linearly adjacent ranks

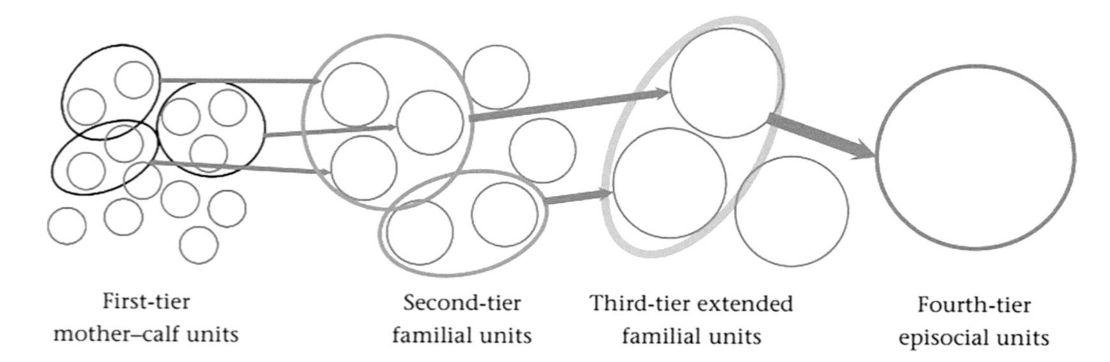

Figure 3.1 A schematic representation of the four hierarchical levels or "tiers" of female African elephant society. Association rates decrease at each subsequent tier, with smaller groupings completely nested within larger ones. Fission–fusion processes typically occur between second and higher-order tiers (reproduced with permission from Wittemyer & Getz, 2007, Copyright © 2007 The Association for the Study of Animal Behaviour. Published by Elsevier Ltd. All rights reserved).

(Smuts et al., 1987), age is the determinant of rank within an elephant social unit, even when it consists of multiple matrilines (Archie et al., 2006a). Chacma baboons not only receive maternal support in establishing ranks themselves (Cheney, 1977), but are also invested in maintaining rank relationships within the entire troop to preserve social stability (Bergman, 2003). By contrast, among savanna elephants rank is determined primarily on the basis of an individual's body size, which also correlates with age (Archie et al., 2006a). As a result, the oldest female in a family unit, which is likely to be its biological matriarch, is also its most dominant individual. This dominance is evident to human observers simply by observing clear signals of aggression, threat, or intimidation (Archie et al., 2006a; Wittemyer & Getz, 2007). Familiarity offers more opportunity for asserting dominance, which is perhaps why, at Amboseli, individuals with the highest rates of association also had the highest rates of agonism (Archie et al., 2006a). Although dominance interactions between different social units are observed far less frequently, matriarchs of different social units do appear to form hierarchies, though the interactions among them are rarer, making linearity harder to quantify (Wittemyer & Getz, 2007). There is also a weak effect of social unit on the outcomes of dominance interactions of nonmatriarchs, possibly again driven by matriarch age (de Silva et al., 2017).

Although there may not be overt reproductive suppression or skew among elephants (Moss & Lee, 2011a; Vehrencamp, 1983), being on the receiving end of such aggression as a subordinate comes with costs. Early observations at Tsavo National Park showed that dominants tend to have calves before peak rainfall, and thus had better rates of offspring survival than subordinate females, which tended to give birth after this peak (Dublin, 1983). This lends support to the idea that dominance ranks impose social selection pressure (Tobias et al., 2012; West-Eberhard, 1979, 1983). Long-term data from Amboseli also show that first-time mothers under 20 years lost on average 31.5% of their offspring to natural causes, whereas those over 20 lost only

12% (Moss & Lee, 2011a). Given that families contain individuals of varying ages, all of which travel together much of the time and access the same areas, this difference cannot be explained in terms of gross space use. Rather, dominants may indirectly influence the reproductive capacity of subordinates through their priority of access to resources. However, Moss and colleagues explicitly reject the idea that competition within families could be the reason. Instead, they suggest it is possible that this is due to younger females being unable to support offspring owing to physical immaturity or simply inexperience. The younger group contains first-time mothers that are simply "inept" as Moss and colleagues put it, and survival improved by the time a female had three or more calves, owing to increased experience at suckling offspring. This is difficult to disentangle from the effect of rank however, since experience as well as size increase with age, which are in turn tied to dominance.

3.1.2 Foraging Competition and Predation

Which begs the question, why do elephants tolerate being socially subordinate at all? Classic socioecological models predict that dominance hierarchies should form as a means of minimizing conflict when resources are contestable (Koenig, 2002; Koenig et al., 2013; Sutherland, 1996; van Schaik & van Hooff, 1983); however, as generalist herbivores it might not at first appear that elephants should have much to contest. From the perspective of an individual deciding whether to remain with a group in which it is subordinate, or to disperse, the benefits of doing so must outweigh the costs. Competition between social groups may constrain the ability of individuals to disperse. Savanna elephants must navigate vast landscapes with scattered and sometimes scarce, unpredictable resources, subject to severe droughts now and then. They must learn to react to predators that could actually pose a threat and areas that are likely to be dangerous as well as those that are safe. They must also contend with the presence of other elephants that want access to the same resources (especially localized resources, such as shade, water, or mud) and safe havens. These challenges might be overcome through group living, with the easiest means of achieving this being to maintain maternal ties.

In particular, it has been argued elephants may benefit from the experience derived from age (McComb et al., 2001, 2011); therefore, being socially subordinate might be the price to pay. During droughts, calves of younger mothers have a higher risk of mortality than those of more experienced individuals (Foley et al., 2008). It's not clear whether this is due to better knowledge of the environment and conspecifics, as opposed to simply the individual's own physical condition or parental skills, and this alone does not establish that simply having older individuals around enhances calf survival. But it has been shown that families with older matriarchs are better at recognizing and responding to threats presented by different numbers of predators, such as lions (McComb et al., 2011) and also respond more appropriately to familiar versus unfamiliar vocalizations from other social groups (McComb et al., 2001). This could translate into fitness benefits, as individuals need not waste energy worrying about nonexistent threats, while reacting appropriately to real ones. The experience an elephant acquires with age is therefore certainly useful; however, the evidence that elephant families or groups derive fitness benefits from the knowledge and experience of matriarchal leaders (i.e. that one individual benefits from the experience of another) remains indirect, as discussed further later in this chapter.

Families can, of course, grow or shrink through the birth and death of individuals, and more rarely, through fission or fusion processes whereby certain individuals permanently alter their social affiliates (Moss & Lee, 2011b). These constitute long-term changes. On shorter timescales of days, weeks, or seasons, family groups unite and split with one another, a characteristic of the social dynamic termed "fission–fusion" (Kummer, 1968; Moss & Lee, 2011b). Through this fission–fusion process, female savanna elephants may also have up to two additional tiers extending beyond that of the immediate family (Wittemyer et al., 2005a). Some families have close relationships with other families, which may be related but are not necessarily. Douglas-Hamilton (1972) termed these higher-order units "kin groups," also referred to as "bond groups" by Moss and Poole (1983). Families that overlap across much of their range are termed "clans" and may also be more likely to encounter one another (Moss & Poole, 1983; Wittemyer et al., 2005a). When many families aggregate in wet seasons into much larger groups, they can number hundreds of individuals (Wittemyer et al., 2005b), which then move together, with particular families joining and leaving these megaherds as they please. These large groups can be extremely excitable, given the high levels of energy individuals have when there is plenty of food and water (Figure 3.2).

What motivates these large aggregations seems puzzling, as they do not seem to be driven by immediate foraging necessity. Indeed, big groups mean more competition between individuals within them. One possibility is that these large groupings themselves are not necessarily adaptive, instead being a form of runaway sociality (Wittemyer et al., 2005a). This could merely be a byproduct of the fact that savanna elephants are inclined to be gregarious to begin with and are so in ever-increasing numbers when they happen to find themselves more densely packed into safe zones that can only support such a number during wet seasons. But alternately, these large groups may provide opportunities for exchanging social information, learning about one another, and establishing relationships among families. When encounters between family groups occur, it is also possible to order relationships between the matriarchs of the families into a linear dominance hierarchy (Wittemyer & Getz, 2007). An intriguing possibility is that these large aggregations reinforce bonds among individuals, even those that do not frequently interact, enhancing their resilience in the face of disturbances.

Two elephant populations in Kenya that have been the subject of long-term research offer complementary perspectives on elephant sociality. The first, at Amboseli National Park, consists of a large population that had experienced some ivory poaching in the 1960s and 1970s, but for the most part survived relatively intact. When elephant groups split, the lines along which they do so can be predicted in terms of the genetic structure of the population. More closely related individuals will stay together. It has been estimated that fewer than 10% of families within Amboseli

Figure 3.2 Elephant aggregations. (a) African savanna elephants at Samburu and Buffalo Springs National Reserves, Kenya. Savanna elephants tend to form large groups consisting of multiple families in wet seasons, typically in protected areas, but move through the landscape and are not necessarily fixed around any single resource (photo by the author). Family groups join and leave the aggregations as they wish. (b) African forest elephants at Dzanga Bai,

contain a female immigrant (Archie et al., 2006b). Approximately 500 km north of Amboseli, around Samburu and Buffalo Springs National Reserves, another elephant population has had to contend with much higher poaching levels. By some estimates, the population may have been knocked down by as much as 85% (Wittemyer et al., 2009). As poachers target animals with the largest tusks, the first wave takes the biggest bulls, followed by the largest females, which are also likely to be the older individuals. This had resulted in large-scale social disruption during the same era of ivory poaching as that experienced by Amboseli, with families losing some or many of their elder members, and with them the group's ecological and social knowledge.

As poaching levels dropped off owing to international bans on ivory sales, elephant populations recovered, albeit with heavy damage to the intricate social fabric. Subsequently, it was found that some "family" groups, or second-tier units, had reconstituted themselves through the bonding of unrelated females (Wittemyer et al., 2009). Behaviorally, these females were indistinguishable from true relatives, revealing an astonishing capacity and propensity to establish social bonds among African elephants. Yet at another location, Mikumi National Park in Tanzania, where the elephant population was reduced by as much as 75% during that period, the consequences appear to have been even more severe. A study published the same year found that just under half the social groups qualified as being genetically disrupted, showing social relationships that were much weaker relative to what would be expected from a simulated unpoached population (Gobush et al., 2009). There was also a lot of variation within the population – some of the females lacking close relatives grouped with others in a similar fix, but some remained alone and were unable to form any stable bonds with other adult females. This indicates that elephants are behaviorally flexible and can reconstitute close relationships to some extent, but this might depend on the severity of the disturbance.

Several elephant populations unfortunately endured a second poaching epidemic in the 2000s, once again leaving behind orphans from decimated families. In Samburu, all adults in some families had been killed. Although many young calves die as a result, some do manage to survive. Researchers this time witnessed the process as it occurred, seeing that daughters rebuilt their social networks by reconnecting with individuals associated with their mothers and her extended network of contacts (Goldenberg et al., 2016). Genetic analysis of mtDNA among elephants at Queen

---

Figure 3.2 caption (cont.)

Central African Republic. Forest elephants congregate in small groups within forest clearings that represent scarce point resources. Aggregations may represent splinter groups that associate temporarily within the clearing, which occur without any distinct seasonality (photo by Victoria Fishlock). (c) Asian elephants at Minneriya Reservoir, Sri Lanka, where elephants annually gather in very large groups. Asian elephants often congregate in dry seasons around water, but typically remain within smaller distinct subgroups. In many parts of South Asia, such sites consist of relatively new man-made reservoirs (on the scale of decades to centuries old). Thus, large aggregations composed of multiple social units may represent a relatively recent behavioral innovation in Asian elephants in response to anthropogenic changes to the landscape and range constriction (photo by the Udawalawe Elephant Research Project).

Elizabeth National Park in Uganda similarly showed evidence of individuals from differing matrilines having been adopted into social groups. It is also suspected that in most cases at Amboseli involving the close integration of unrelated females, a similar process may have occurred (Archie et al., 2011). The integration of orphans into non-natal families therefore appears to be not uncommon following disturbances. Nevertheless, this process does not appear to be easy, especially for unrelated individuals. Orphans experienced more aggression on average than nonorphans, being relegated to the periphery of social groups (Goldenberg & Wittemyer, 2018). Perhaps this is why the elephant population in Mikumi exhibited weaker social ties following the first wave of poaching. Nevertheless, in an otherwise grim situation, savanna elephants demonstrate their potential for resilience through social connectivity. This suggests that there are immediate benefits to gregariousness beyond inclusive fitness, which make the inherent challenges worthwhile (Wittemyer et al., 2009).

What might such benefits be? In Chapter 2 we sketched several possible hypotheses. The circumstances of orphans underscore a practical problem in distinguishing which of these might be the primary drivers of observed social behavior. Under undisturbed conditions, elephant social structure would conform nicely with the animals' genetic structure, and hence it would be natural to assume that inclusive fitness is a strong driver of this structure. Inclusive fitness then becomes somewhat of a null hypothesis, difficult to reject. But the very fact that this pattern is so pervasively commonplace makes it difficult to set aside, and it may mask other considerations that are equally if not more important. Only a severe disruption of this genetic structure, one that is so detrimental to elephant populations that it is not possible to condone manipulating intentionally, reveals that other factors may be equally or more important. In the case of African elephants, it would appear that inclusive fitness is not a *necessary* condition for the close match between genetic and social structure, although it may be a *sufficient* one.

3.1.3 Range Familiarity

If associating with conspecifics is ecologically useful, one might expect that the propensity to do so would be enhanced when animals are placed in an unfamiliar environment. In 2005 the translocation of 150 elephants from Shimba Hills National Reserve to Tsavo East National Park in Kenya, offered an opportunity to evaluate this possibility. The translocation was an attempt to reduce human–elephant conflict, an instance in which a management intervention provided a means to study a biological question that is effectively also a conservation question: What would these elephants do? Pinter-Wollman and colleagues hypothesized that translocated elephants might benefit from the social learning opportunities afforded by associating with resident elephants, but that these benefits should diminish over time (Pinter-Wollman et al., 2009). If so, on the one hand, they expected that the number of conspecifics a translocated animal associates with should decrease over time, from the day of arrival. On the other hand, if social benefits tended to accrue over time, they might expect the number of associates to increase. They also examined whether association rates themselves

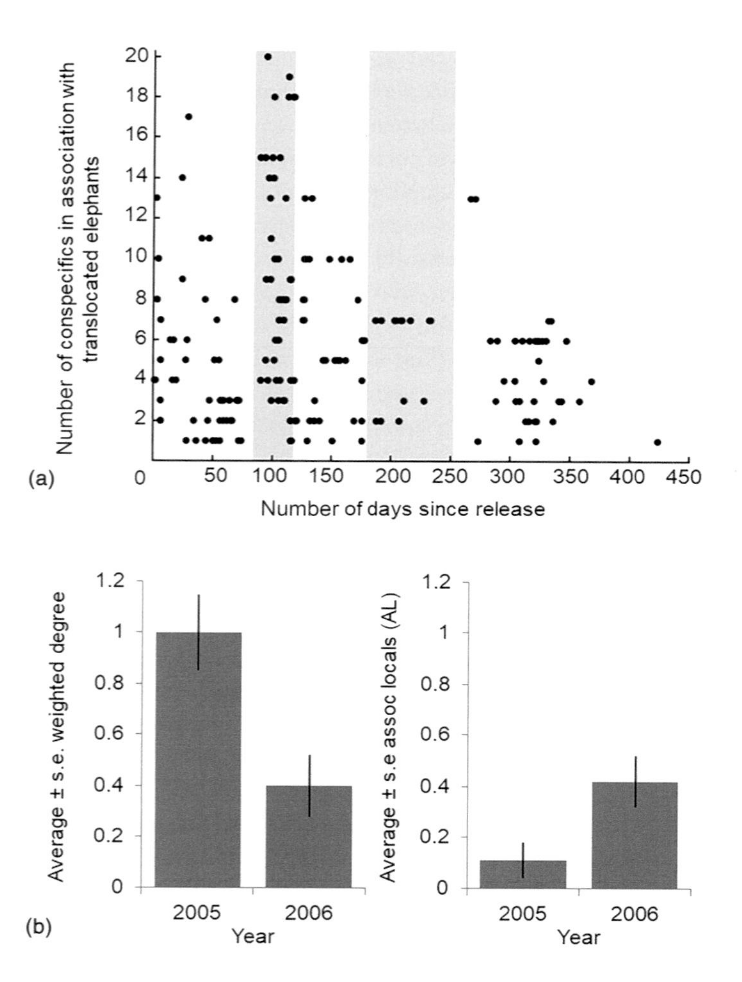

Figure 3.3 (a) Translocated elephants associate with fewer conspecifics post-release as time goes on (gray bars represent dry seasons). (b) This is reflected in a reduction in the weighted degree, quantifying associations among themselves from one year to the next. However, the proportion of associations consisting of locals (AL) goes up, though never exceeding association levels among translocated individuals among one another (figures redrawn from Pinter-Wollman et al., 2009, reproduced with permission from The Royal Society, © 2009 all rights reserved).

increased or decreased. For two years, they observed the interactions among the translocated animals, which were identified individually either through paint markings or natural features, and the resident elephants, which were not individually recognized.

It must be noted that two years is a relatively short period over which to assess the impact of such trauma, and their findings were somewhat mixed, requiring a nuanced interpretation. First, the overall number of other elephants that a translocated individual associated with did show a decrease over time (Figure 3.3a), in support of the first hypothesis. Elephants that associated with more conspecifics also seemed to be in better body condition. But initially at least, translocated elephants tended to associate most highly with other translocated elephants rather than with residents. On one hand, this has to be true if some of the associations were with family members, but such associations also extended between unrelated families that had also been translocated. This is contrary to what one would expect elephants to do if they were seeking to gain ecological benefits from those more familiar with the surroundings. Over the two-year period, resettled elephants did associate more with residents, but not more than they did with other resettled individuals (Figure 3.3b). This phenomenon, which the researchers compared to the formation of a social enclave, is curious. Given that all translocated animals derived from the same area, it is not unreasonable to think that they might have already known one another at some level before being relocated. Social familiarity thus seems to be a more important determinant of association tendencies than whether or not individuals can gain ecological knowledge.

A reason for this might be that although elephants are not territorial in the classical sense, meaning that they do not actively defend areas to which particular individuals or families have exclusive access (Koenig, 2002; Sterck & Watts, 1997), and although they do congregate occasionally in suprafamilial units as already mentioned, different social groups may not be entirely tolerant of one another. Elephant movements are not only governed by the presence or absence of resources, elephants are also responding to the presence of other elephants and of humans (Boettiger et al., 2011). Indeed, when elephants are *not* in large aggregations, which is most of the time, families may undertake elaborate steps to avoid running into one another. Dominance hierarchies persist despite the relatively low rate at which contests over resources are actually observed because elephants are primarily engaged in scramble competition (Wittemyer et al., 2007; Wittemyer & Getz, 2007). This means that different individuals or families may have access to different point resources at any given time, but it is not these specific locations one must focus on – it is the entire landscape.

Certain types of resources are more valuable and spatially constrained at particular times of year; namely, water and safe space. Dominant individuals or families are able to occupy prime real estate in the elephant world – locations nearer to water sources and preferably within protected areas such as reserves or national parks. Subordinates have to risk moving further away from both, especially in dry seasons when resources are more scarce (Figure 3.4; see also Wittemyer et al., 2007). We will take another look at these interactions in detail in Chapter 5. Because dominance increases with age, and the relative ranks of families appear to be determined by the ages of their matriarchs, it follows that families with older matriarchs also stand to benefit from access to the best parts of the range in terms of safety and water (and forage) availability. Dominants expend less energy and may be less vulnerable to poaching as a result, which might enhance both adult and offspring survival. When elephant groups convene in large numbers during wet seasons, these gatherings likely present opportunities for interactions between families that may not otherwise happen, owing to the implicit resource competition. Likewise, when elephants are placed in an unfamiliar environment, they may be reluctant to face residents precisely because social relations have not been established, and thus they may not be welcome.

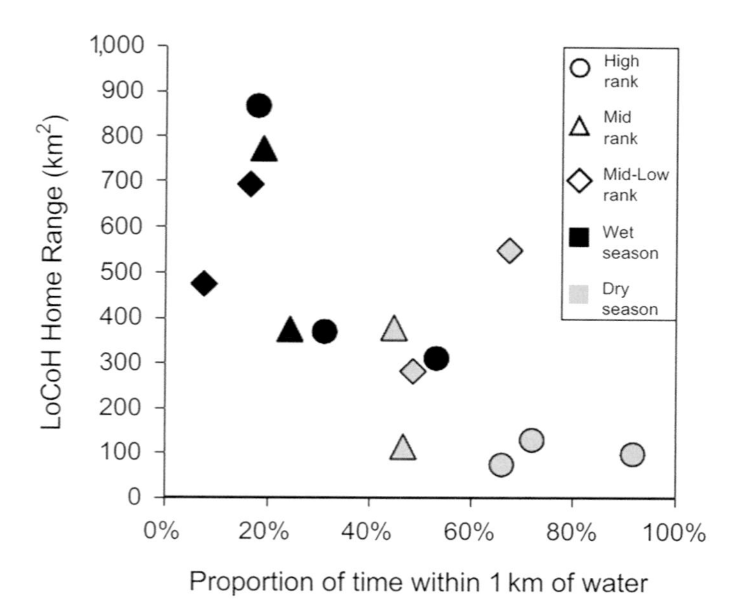

Figure 3.4 Dominance and space use: Use of areas near water by elephant groups. In wet seasons, social groups of all ranks spend less time near permanent water (horizontal axis) and range more widely (vertical axis). But in dry seasons, higher-ranking groups maintain smaller home ranges that are closer to water compared with mid- or lower-ranking groups (reproduced from Wittemyer et al., 2007, © 2007 Springer-Verlag).

3.1.4 Constraints on Gregariousness

Not all African savanna elephants actually live on savanna-type landscapes. The easternmost population of elephants in the southern part of the African continent ranges into arid zones, particularly in Namibia, where they encounter the Namib Desert. Although the range itself appears to be contiguous with more pleasant climes in Angola and Botswana, the so-called desert elephants of Namibia must make a living in an environment where the average monthly rainfall does not exceed 30 mm and daily temperatures in hot dry seasons exceed 40°C (Leggett, 2008). Precious little is known about the social lives of these populations, but they nevertheless might provide critical insights into the factors shaping social behavior, as they inhabit the most resource-stressed end of the spectrum of habitats occupied out of all the living elephants. Having undergone severe postcolonial era population declines from which they have never recovered, these elephants now show much weaker social relationships than savanna-dwelling elephants, with some degree of association among non-relatives but without higher-order structure (Leggett et al., 2011).

Studies of elephants in the of the Kunene region in northwestern Namibia offer a glimpse into these hardy populations. Here there are three types of seasons – wet, cold-dry, and hot-dry. As with other savanna elephant populations, adults spend much of the daytime either feeding or moving, which can add up to nearly 80% of the activity budget. The remainder is split between resting and cooling off with water. Adult females seem to spend only at most 2% of their time on social interactions, at least during the day (Leggett, 2008). The amount of energy devoted to different activities varies slightly depending on the season, with adult females spending slightly more time on social activity during the hot-dry seasons than either the wet or cold-dry seasons. Males, however, seem to be more socially active in cold-dry seasons, mainly owing to reproductive efforts (Leggett, 2008). The population has historically been divided between the eastern and western catchment areas of the Hoanib River, separated by areas of human settlement.

Reported aggregation sizes are quite small, as might be expected given how little time is devoted to social interaction, containing on average between 1 and 12 individuals in the Eastern and 1 and 10 individuals in the Western parts. The larger aggregations were recorded in the 1980s, whereas they had diminished by the 2000s. In both subsets, larger aggregations usually occurred in wet seasons (Leggett et al., 2003; Viljoen, 1988), showing a trend similar to that of savanna elephants in less arid regions. But interestingly, the Eastern population in more recent years also aggregated in slightly larger numbers during the cold-dry periods whereas the Western tended to do so during the hot-dry periods (Leggett et al., 2003). This difference seems to be driven primarily by the propensity of Eastern adult females to have been on their own with their calves more frequently during the hot-dry season than at any other time (25% vs. 0), whereas the Western females tended to be on their own during 22–38% of observations irrespective of season. The Eastern elephants were also described as "generally shy," being inclined to flee from observers at greater distances than the Western elephants, which were "far more approachable" (Leggett et al., 2003: 310). The Eastern elephants also, at the time, had access to vegetation and water sources over a far more dispersed area than the Western elephants, which were more restricted to the riverbed.

3.2 African Forest Elephants

3.2.1 Social Structure

Forest elephants inhabit a relatively aseasonal habitat, with highly dispersed, patchily distributed fruit, forage, water, and mineral resources. Historically, forest elephants would have faced very little predation as no large predators occupy African forest habitats. Thus, one expects there should be relatively little motivation for forest elephants to be gregarious. Much less is known about the social world of forest elephants, given the difficulty of observing them freely in dense forests. Observations by people have mostly been limited to large, open clearings known as *bais* (also referred to as salines), at which forest elephants congregate to gain access to soil, salts, and mineralized water (Turkalo, 2001; Turkalo et al., 2013). From studies at such sites, it seems that, although females and their dependent calves move together, social groups are much less cohesive than observed among savanna elephants (Fishlock & Lee, 2013; Turkalo, 2001). Group sizes (inclusive of both calves and adults) observed when elephants enter *bais* typically range from two to four individuals, which seems to be true at multiple sites (Schuttler et al., 2014b; Turkalo et al., 2013; see also Figure 3.5). At Lopé National Park in Gabon, groups just consisted of one or two adult females if calves were excluded (Schuttler et al., 2014b). Solitary individuals comprise between a quarter and a third of observations across different sites (Fishlock & Lee, 2013; Schuttler et al., 2014b).

Another study conducted at Odzala-Kokoua National Park in the Republic of Congo, showed individuals do tend to mingle with others upon entering a clearing, and seem especially attracted to larger groups that are already present (Fishlock & Lee, 2013). Those that used the opportunity to socialize – mostly females and calves, constituting just over half of all parties observed – stayed significantly longer at the clearing than those that did not (Figure 3.6). Younger females, below the estimated age of 35 years, were more likely to join parties with older females than those older than 35, which could either be due to avoidance of more dominant females or simply because there were fewer individuals in consecutively older age classes from which to choose. Thus, forest elephants are also somewhat inclined to exhibit fission–fusion social behavior. Fishlock and colleagues propose that *bais* may represent more than just physical resources in the form of water, soil, or minerals. They may also serve as arenas for social interactions, with individuals taking the opportunity to actively engage with others that they may or may not already know. However, these associations usually do not persist, with group compositions being the same upon exiting the clearing as they were when they entered (Fishlock & Lee, 2013). Although large *bais* such as Dzanga in the Central African Republic may be visited by elephant throughout the year and host between 40 and 100 elephants at any given time, some individuals might appear relatively frequently, whereas years may pass between the visits of others (Turkalo et al., 2013).

Another possibility that has not yet been thoroughly investigated is that *bais* serve as large centers of information exchange – even among individuals that may not be there simultaneously – for instance through indirect cues such as odor trails and dung piles. White rhinoceros use middens in this manner, in which individuals' defecation behavior signals sex, rank, and reproductive state. Individuals assess one another by investigating dung deposited in communal latrines (Marneweck et al., 2018). Although elephants do not seek out particular locations to defecate as white rhinoceros do, *bais* are nevertheless likely to be rich in a high concentration of chemical cues. Such locations may make mate search and competitor assessment more efficient than random assessments over large tracts of forest, where such signals are probably far more dispersed and diluted. This remains to be demonstrated.

3.2.2 Genetic Structure

Observations at clearings provide a valuable but very limited view of sociality because individuals need only visit clearings in response to their own physical needs. There is no reason to expect that groups of social affiliates should do so together, especially if the forest environment itself physically inhibits movement of larger groups.

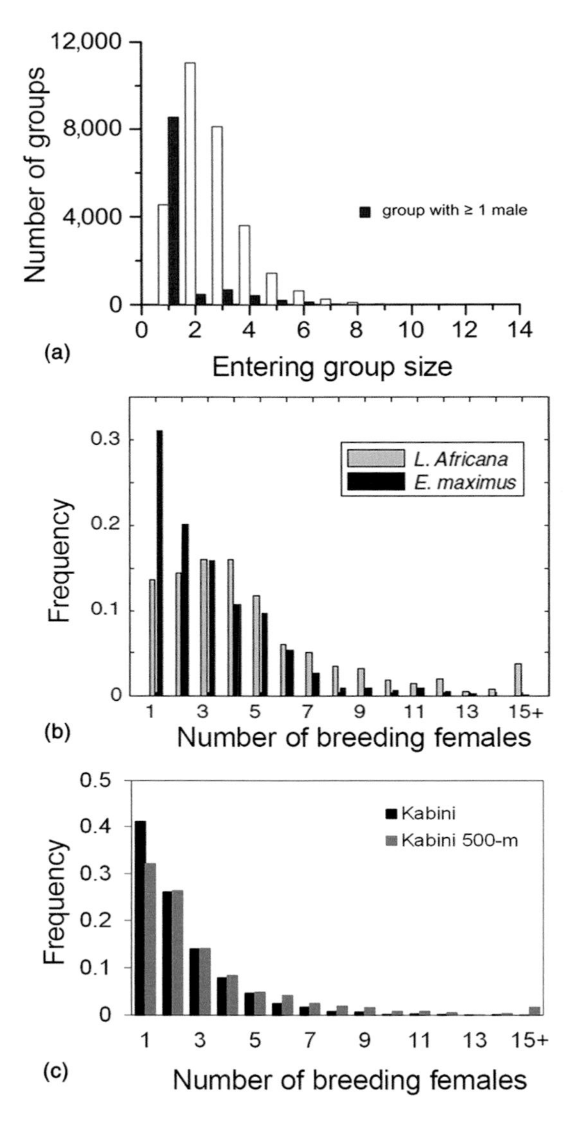

Figure 3.5 Elephant group sizes. (a) Group sizes in African forest elephants at the Dzanga Bai upon entry of the clearing. This histogram includes males and calves (from Turkalo et al., 2013, cc-by 4.0). (b) Group sizes of African savanna elephants at Samburu-Buffalo Springs National Reserves (Kenya) and Asian elephants at Udawalawe National Park (Sri Lanka) only in terms of the number of breeding females, exclusive of all other age–sex classes (from de Silva & Wittemyer, 2012, reproduced with permission from Springer Business Media LLC, copyright © 2012), are larger than that of forest elephants, although group sizes in Asian elephants tend to be smaller. (c) Group sizes at the Nagarahole and Bandipur National Parks (India), near the Kabini Reservoir, resemble those in Sri Lanka. Black bars represent groups defined by observers based on coordinated behavior whereas gray bars represent groupings of individuals within 500 m of one another (reproduced from Nandini et al., 2018, copyright © 2018 the International Society for Behavioral Ecology).

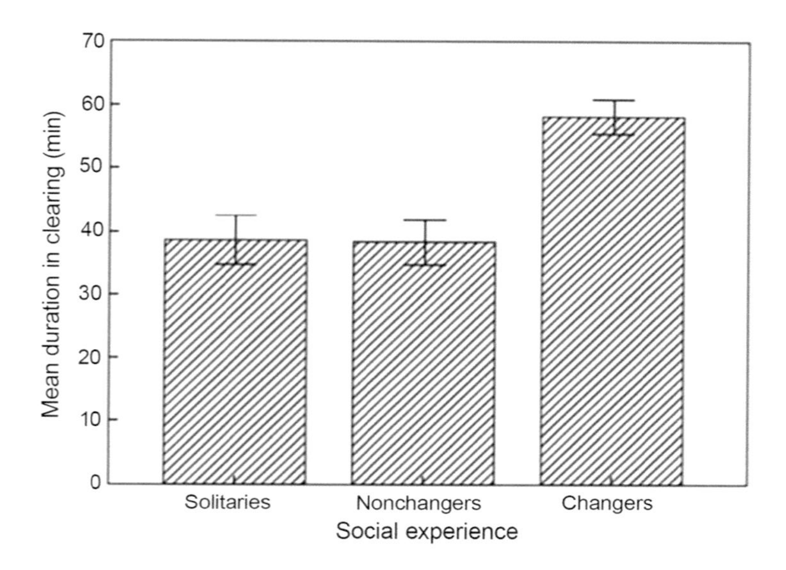

Figure 3.6 Propensity of African forest elephants to mingle socially at clearings. "Solitaries" (N = 155) and "Nonchangers" (N = 110) were individuals and parties respectively that did not associate with others during their time at the clearing, whereas "Changers" (N = 285) mingled with other elephants. Changers stayed significantly longer than the other two (reproduced from Fishlock et al., 2013, © 2007 the Association for the Study of Animal Behaviour).

Moreover, because forest elephants tend to be more frugivorous than the other elephant species (Campos-Arceiz & Blake, 2011), and fruiting trees are highly dispersed rare resources, it is expected that forest elephants may not be as inclined to associate in large groups anyway. Unfortunately, it is very difficult to determine how forest elephants behave when they are *not* in clearings, since they are extremely wary of people and cannot be observed as safely. Researchers therefore have to rely on indirect evidence of their behavior, namely their dung. Because genetic samples can be extracted from dung, it offers the possibility of a glimpse into the hidden social world of forest elephants.

Studies in Lopé National Park, Gabon, reveal how forest elephant genetics are spatially structured at a fine scale (as opposed to regional or continental scales), something invisible to direct observation (Schuttler et al., 2014a, 2014b). Schuttler and colleagues expected that if elephants were drawn together by resources, the genetic composition of dung collected together or in close proximity would be random. However, if elephants did associate in matrilineal family groups as savanna elephants do, then they expected samples to differ more greatly as the geographic distances between them increased. This could also reflect the dispersal distance of forest elephants: The shorter the distance they dispersed from their natal area, the greater the expected differences among samples with increasing geographic distance. They used the size of the dung bolus to assess individuals' age classes, mtDNA to determine relatedness, and assumed that samples that occurred within 250 m of one another and had the same freshness belonged to individuals within the same group. While the last

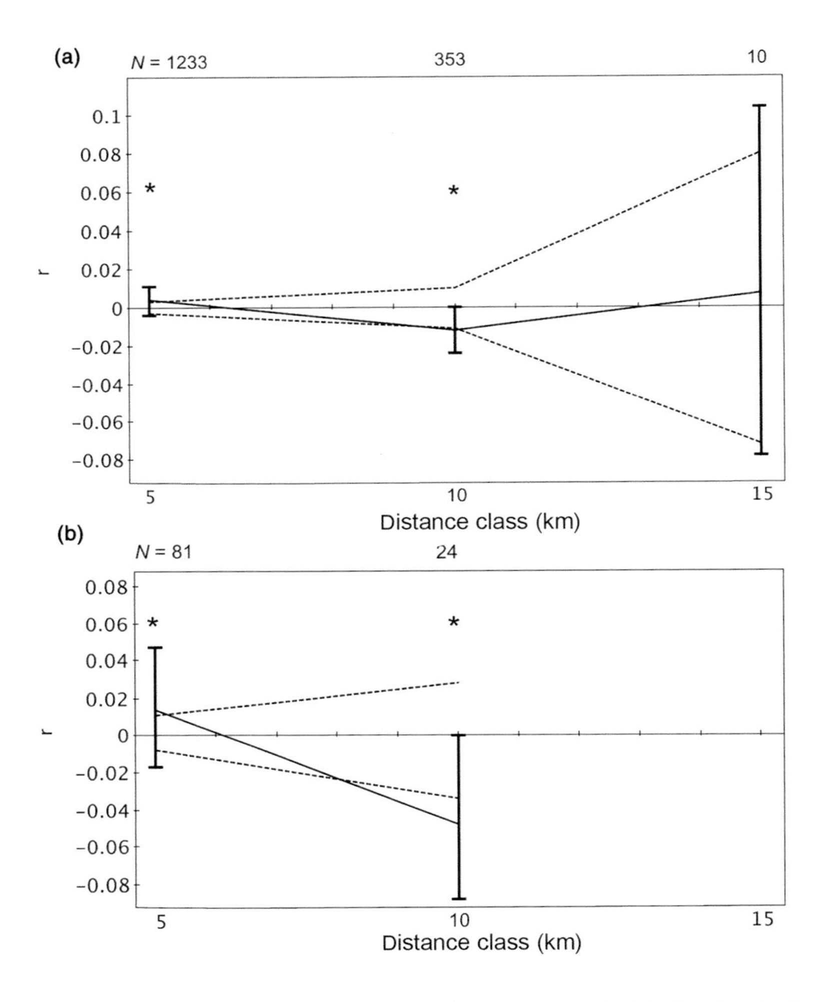

Figure 3.7 Spatial autocorrelation between the occurrence of adult female pairs by year (a) and by day (b). Both show significantly positive genetic structure at 5 km but significantly negative structure at 10 km distances (starred), with ambiguity at longer distances. The error bars and dashed lines represent 95% confidence intervals obtained through two different randomization tests (reproduced from Schuttler et al., 2014, cc-by-4.0).

assumption seems bold, researchers noted that dung piles of varying freshness were rarely found within a small area.

They found that more closely related adult females had a higher propensity to be detected at shorter distances, typically averaging between 1.69 and 2.98 km apart. If samples happened to be collected on the same day, this grew even shorter, averaging less than a kilometer. Females detected within 5 km of one another were more closely related than expected by chance (Figure 3.7). Visualizations of the networks constructed by grouping individuals together based on the distance and dung freshness also illustrates that in general individuals within an interconnected unit tended to have the same mtDNA haplotype, of which there were 10 in total. However, forest elephants do not appear to have exclusive ranges among unrelated individuals, as dung from unrelated individuals could also occur in close proximity, yielding cases in which some interconnected components consisted of different mtDNA haplotypes. This occurred only in 5 out of 28 networks. Conversely, the same haplotype could also be found in several different networks, with the most common one (labeled Lope7) dispersed among 9 of 28 discrete networks. The largest connected component consisted of 22 individuals and was predominantly composed of Lope7, but also contained two other haplotypes. The second-largest connected component contained 12 individuals, all of which were of the Lope7 haplotype.

The fact that the same haplotype could be found among different networks is not surprising, since this may either reflect prior fission events or simply the fact that individuals within the same social group need not defecate within close spatiotemporal proximity. Also, the ability of mtDNA haplotypes to discriminate among families depends on the diversity of haplotypes occurring in a population relative to the number of social units. For instance, the Amboseli population contains only three haplotypes (Phyllis Lee, personal communication). Without direct observation it is difficult to say whether the occurrence of individuals with *different* haplotypes within close spatiotemporal proximity is due to actual associations, the chance passing of two different social groups within a given area, or even chance encounters between individuals of different social groups, as defecation can sometimes be a response to overtures of dominance. A great challenge with this paradigm is that in order to detect a signal of association, individuals had to not only remain within relatively close proximity to one another, but also defecate close together in time. Therefore, any signal of structure the study uncovered very likely underrepresents actual association rates and also cannot firmly resolve whether forest elephants have associates beyond their immediate family members. Researchers did actually observe the elephants themselves in a concurrent study, which found at least three networks in which individuals of different haplotypes associated together and did so on more than one occasion (Schuttler et al., 2014b). There was low indication of preferred (nonrandom) associations, but this could also have been because the study was unfortunately limited by the very low resighting rate of same individuals. While there is still much to be learned about the social lives of forest elephants, these studies hint there may be more to these elephants' social affiliations than meets the eye, even if these affiliations are not as extensive as that of their savanna counterparts.

3.3 Asian Elephants

3.3.1 Social Structure

Asian elephants range across diverse habitats ranging from grasslands to rainforests and everything in between. Historically, the largest terrestrial predators Asian elephants would have faced are Asiatic lions and tigers. Yet today lions have been eliminated throughout much of the range and, while tigers do persist, it is not clear how frequently they actually prey on elephants (or, more likely, calves). As ever, humans remain the outstanding threat, both as hunters and competitors, for land and resources. The physical and gross dietary similarities as well as the shared evolutionary history among the Asian and African species are therefore superimposed on nontrivial ecological differences within and between the two continents. These manifest in some visible ways – both African forest elephants and Asian elephants are, on average, of smaller body size than African savanna elephants, which is suggestive of adaptation to denser habitats, but not proof of it. Other observations, however, are even more subtle. While Asian elephants have generally been assumed to behave similarly to savanna elephants, early observers noted that Asian elephants also seemed to exhibit more "fluid" social relationships than their savanna counterparts (Eisenberg et al., 1990). Indeed, while groups of females and calves do most likely represent related matrilines (Fernando & Lande, 2000; Vidya & Sukumar, 2005), telemetry data showed that close kin associate together only about 60% of the time, a far lower rate than described for African elephants (Fernando & Lande, 2000). Superficially, this might appear more similar to the behavior of forest elephants.

My research group followed individually identified adult females at Udawalawe National Park (UWNP) in southern Sri Lanka over a period of two years, showing that social relationships were indeed very dynamic on a day-to-day basis (de Silva et al., 2011a). Lone females with or without calves were observed far more frequently than among savanna elephants (de Silva & Wittemyer, 2012). While individuals associated nonrandomly with some subset of the population, the companions with whom they most frequently associated changed from season to season – an individual's most frequent associates seldom remained the same over all five seasons (de Silva et al., 2011a). Rates of association among pairs of individuals more often peaked in dry seasons than wet seasons. Females also showed a lot of variation in their social tendencies. Some were very gregarious but had low fidelity to particular companions, while others were less gregarious but showed higher fidelity. In fact, the greater the number of associates an individual had on average, the less time they tended to be associated with them (Figure 3.8a). An analogous pattern held in southern India at Nagarahole and Bandipur National Parks (NBNP; Figure 3.8b), where it was shown that the average association index among clanmates (i.e. individuals comprising a social unit) declined with clan size. Both patterns suggest a tradeoff between the number of associates and the amount of time one can spend with an associate while foraging. Interestingly, NBNP also encapsulates a similar ecological context as UWNP, being adjacent to the Kabini Reservoir. However, the landscapes within them had endured less historical disturbance than that of UWNP.

3.3.2 Foraging Competition

Why would such a tradeoff occur? One possibility is that group sizes are somehow limited by how much forage is available at any particular location or patch (Chapman et al., 1995). Individuals with fewer associates can remain among these nearly constant companions as they forage, because they are able to do so in relatively close

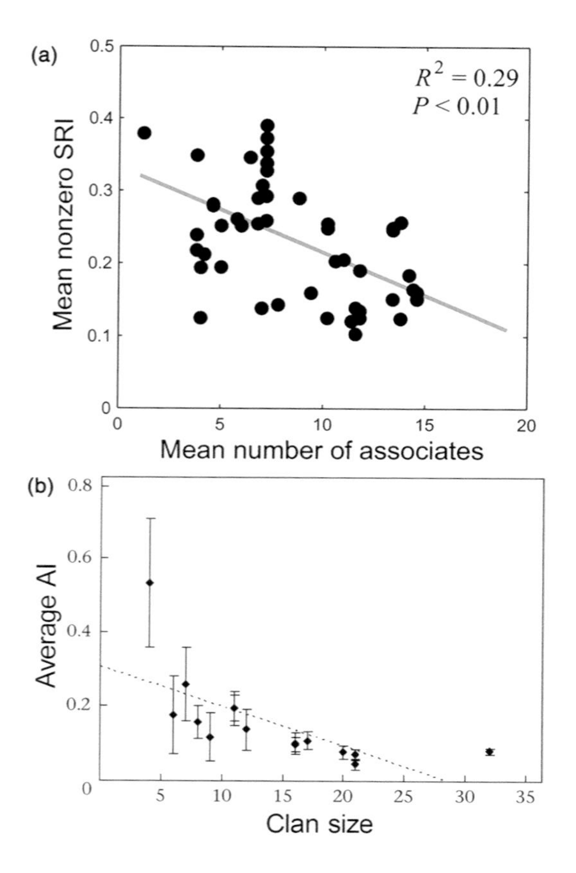

Figure 3.8 A negative relationship between number of social associates and strength of association suggests a tradeoff between the two. (a) At Udawalawe National Park, the mean number of associates an individual had over five seasons is significantly negatively correlated with the average nonzero association index value of that individual (reproduced from de Silva et al., 2011, cc-by-4.0). (b) At Nagarahole and Bandipur National Parks the average association index among clanmates is negatively correlated with the size of the clan. Both are complementary views of the same fission–fusion phenomenon, the first being from the perspective of the individual and the second being from that of the social unit (reproduced from Nandini et al., 2017, © 2018 the Association for the Study of Animal Behavior).

proximity. But as groups become larger, individuals necessarily become more dispersed while foraging and thus must shuffle among companions more frequently. This seems to be supported by studies both in India and Sri Lanka, where average aggregation sizes remain fairly constant, despite a high degree of variation in the size and structure of social networks or clusters (de Silva et al., 2011a; de Silva & Wittemyer, 2012; Nandini et al., 2017, 2018). However, one would then predict the same to be true of African savanna elephants, which are arguably even more resource-limited. This does not seem to be the case, since we have already seen that savanna elephants tend to be found in larger groups (Table 3.2). Nandini and colleagues propose an interesting idea: that fission–fusion processes in Asian elephants maintain relatively

Table 3.2 Social tier characterizations across studies

Population	Average number of females in a family group (range)	Average number of individuals in a family group (range)	Average number of females/family groups in a bond group	Average number of individuals in a bond group (range)	Number of females in a clan/fourth-tier unit/most inclusive unit	Number of individuals in a clan/fourth-tier unit/most inclusive unit
Kabini	–	–	–	–	Mean: 13.31, SD: 7.78, Median: 11.5, IQR: 7.5–18.5, Max: 3	Mean: 29.19, SD: 19.76, Median: 21, IQR: 17–39.5, Max: 83v
Kabini 500m ≥ 20 sightings	–	–	–	–	Mean: 15.57, SD:9.74, Median: 18,IQR: 9–19.5, Max: 31	–
Udawalawe	–	–	–	–	Mean: 11.67, SD:7.47, Median: 12,IQR: 5–17, Max: 23	–
Ruhuna	–	–	–	–	7.75 (4–11)	14.75 (7–24)
Amboseli	2.35 (1–9)	7.22 (2–23)	2–5 family groups		5–9 family groups	Range: 50–250
Samburu	2.2 (1–5)	7.64 (1–15)	2.0 family groups (4.4 females on ave., based on col. 2)	16 (6–40)	Mean: 13.75, SD:7.46, Median: 11,IQR: 8.8–16, Max: 28	Median: 33.5, IQR: 28.8–80.3; Median: 32, IQR: 23.5–38 during a subsequent period.

Data from Nandini et al., 2017 and references therein. Modified from Nandini et al., 2017, Table 1, with permission from Oxford University Press.

constant aggregation sizes in response to resource constraints, rather than expanding and contracting aggregation sizes dramatically at different times in response to resource availability, which is the typical assumption about the purpose of fission–fusion processes, and to which African elephants appear to conform.

This dynamism at the simplest possible level of organization, the relationship between pairs (also termed dyads), prompted looking at elephant social contacts at different levels of organization. The next level up was that of the ego network, which, for any given individual, is made up of all other individuals to which he or she is connected. At UWNP, although ego networks changed from season to season, over the course of two years it became apparent that an individual typically spends time with other individuals that it already knows, even if they are not in each other's company in every season. At the population level, networks could be statistically clustered into distinct communities, but, within these, relatedness remains an open question. Some units, presumably families, chose to remain relatively isolated from other elephants whereas others had more extended structures resembling bond groups; however, these emerged from contacts between only some individuals, not from entire families moving together (de Silva et al., 2011a). Quantifying network structure from the top down as opposed to bottom up, Nandini et al. (2018) likewise found that distinctly clustered communities in NBNP, referred to as clans, showed varying degrees of substructure (Nandini et al., 2017). In both contexts, some of the seasonal differences in the social structure at the population level were attributable to the presence of rarely seen individuals in dry seasons, suggesting that the study locations were attracting additional (or different sets of) occupants in dry seasons.

Yet there were differences between the Indian and Sri Lankan populations studied. In southern India, associations among clanmates appeared to be stronger and better defined in wet seasons, whereas in Sri Lanka associations between dyads were stronger in dry seasons. Moreover, connections between communities were nonexistent in the former, whereas they were fairly extensive in the latter. At NBNP, Nandini and colleagues also found that group sizes were significantly larger in dry seasons (an average ± SD of 2.41 ± 1.837) compared to wet seasons (averaging 2.21 ± 1.786). Sukumar (2003) reported similar observations from southern India. However, no such effect was seen at UWNP, where average dry and wet season groups were 3.07 ± 2.34 and 3.03 ± 2.33 respectively. However, just about 170 km away at Minneriya National Park in central Sri Lanka, elephants annually congregate by the hundreds at the Minneriya Reservoir, again only in dry seasons (Figure 3.2). These large gatherings most closely resemble those of savanna elephants, in that they likely consist of many family groups merging together; however, they clearly center on scarce dry season resources – fresh grass and water. Study of such gatherings may add additional levels of social organization in at least some populations of Asian elephants, which are yet to be quantified.

It would be fascinating to know where seasonal differences manifest, especially where large aggregations occur, and the attributes of such landscapes, which have yet to be systematically documented. One of the difficulties with quantifying seasonal differences in the behavior of social groups is that one must have access to *the same individuals at all times of year*. Where there are pronounced and obvious seasonal effects, such as at Minneriya, study of the phenomenon is impeded by the very fact that elephants can be scattered among small groups not easily observed during much of the year when they are not out in the open, designated protected areas. Protected areas are, of course, themselves nonrandomly placed, often in hilly areas unsuitable for agriculture (Joppa & Pfaff, 2009), and seek to accomplish multiple goals of conserving not only wildlife but also other resources such as water or particular ecosystem services (Corson et al., 2014). Nevertheless, when large gatherings of elephants do occur, their location and timing suggest that such gatherings are resource-driven, tending to occur in dry seasons, typically near large bodies of water (Nandini et al., 2017; Sukumar, 2003). It follows that "herds" of Asian elephants do not represent fixed entities. They are more comparable to foraging parties of lions or hyenas than to herds of buffalo or horses, being splinter groups of a larger pool of social associates that most likely consist of related matrilines. While individuals can be distinctly identified, "herds" cannot. Asian elephants seem to be largely responding to the concentration of food (fresh grass growth as water levels recede) and water occurring in certain locations during dry seasons, rather than the presence of conspecifics, which is quite unlike the behavior of either African species.

One of the greatest challenges in making quantitative statements across species and study sites is in replicating field methodology – whereas what constitutes an elephant group is visually more obvious in a savanna setting with clearer visibility and spatial segregation among elephants, defining groups can be tricky for Asian elephants both owing to visual constraints imposed by the habitat and the drifting behavior of elephants themselves. Direct quantitative comparisons of social behavior in savanna versus Asian elephants in a limited number of studies that tried to apply standardized methodology shows that association rates among the latter indeed tend to be lower, less interconnected than former, and lacking quantitatively distinct "tiers" (de Silva & Wittemyer, 2012). Intriguingly, the actual communities formed by elephants in southern India, southern Sri Lanka, and Kenya, defined by an algorithm that hierarchically clusters the observed social networks, were not of significantly different in size (Nandini et al., 2018). The key differences lay in the patterns of connectivity among individuals, rather than the numbers of individuals within communities. Asian elephant populations do vary, with the population from NBNP exhibiting connectivity and associated network measures that were overall more similar to that of Asian elephants at UWNP, but also showing some similarities to savanna elephants at Samburu and Buffalo Springs National Reserves, depending on how groups were defined (Nandini et al., 2018; see Figure 3.9).

Although it is tempting to conclude that these patterns are universal, here I must reiterate that the studies from South Asia largely describe populations occupying the drier and more seasonal end of the range of habitats used by Asian elephants. If anything, the environments encountered by the South Asian populations are far more similar to African savannas and woodlands than to the central African forests. The latter are more (grossly) similar to the ecosystems of Southeast Asia. Unfortunately, elephants are severely threatened in Southeast Asia as well as being exceptionally

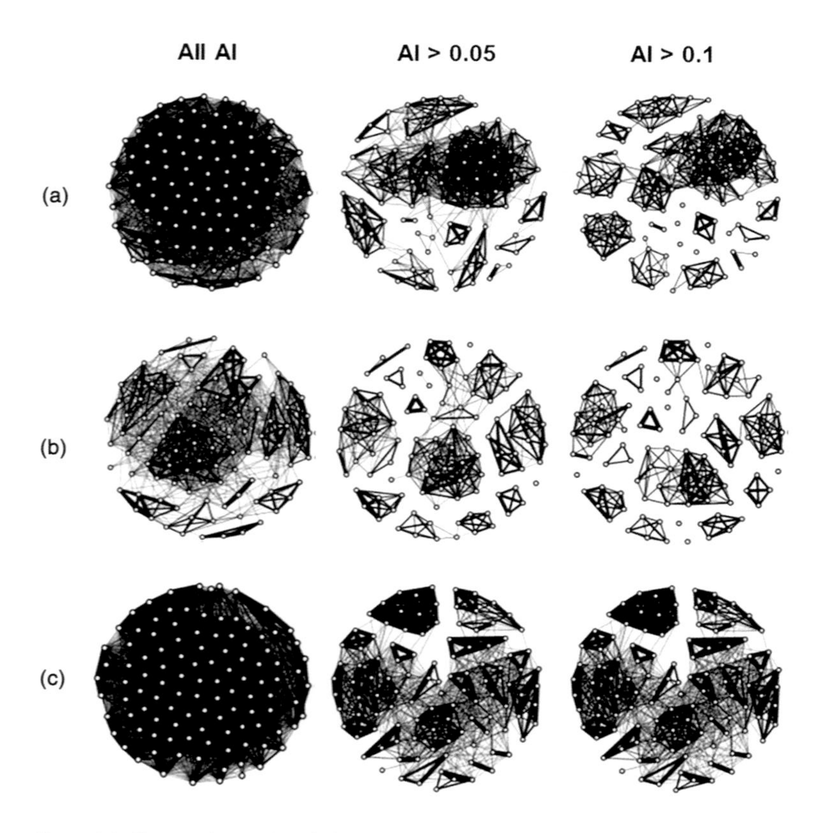

Figure 3.9 Comparison of social networks among Asian elephants (a, b) and savanna elephants (c), where each node is an adult female and edges (ties) represent the strength of association indices between pairs of individuals. Sequentially removing edges below a threshold value reveals network structure at different levels (second and third columns). (a) The Nagarahole–Bandipur Asian elephant population; (b) the Udawalawe Asian elephant population; (c) the Samburu–Buffalo Springs savanna elephant population (modified from Nandini et al., 2018, copyright © 2018 the International Society for Behavioral Ecology).

difficult to observe. However, a small number of studies from Malaysia hint that the individual-level fission–fusion process also occur in the peninsular and Bornean populations, which have evolved separately from one another for several thousand years (Sharma et al., 2018).

One study used camera traps strategically placed at mineral licks frequented by elephants in the Belum Temenggor Forest Complex within the northern state of Perak (Ning, 2016). While a central feature of this complex is a water body, similar to the study sites described in South Asia, the "lake" is in fact a large and sprawling feature, directly bordered by lush forests as opposed to grassy woods or scrubland. As elephants seldom need to venture into the open, they are difficult to observe except through such devices. Unlike the *bais* of central Africa however, soil and mineral consumption sites in Asia tend to be small and hidden, allowing only a few animals to use them at a time, and preventing researchers from constructing viewing platforms that oversee a large area. Cameras are accordingly fixed at very close range. Observers therefore cannot see any great distance, nor what is going on in areas outside the field of view. Therefore, one must make careful and systematic assumptions regarding what constitutes a "group" – for instance, by assigning social "associations" according to elephants' appearance on camera within some time limit in addition to direct proximity. Despite these constraints, it is indeed possible to observe social behavior and make some inferences about the underlying affiliations. The study identified 76 elephants, of which 26 were adult females and 15 were adult males, with the rest belonging to immature individuals of both sexes in varying age classes. Groups contained between 2 and 10 individuals, with a median of 6. From repeated association it was determined that there were at least seven social units and four smaller indeterminate units consisting of two to three individuals. A network analysis indicated the possibility of six social units. Association indices were typically far less than one, as social group members did not always appear together. This suggests that even with such relatively small numbers, there is nevertheless likely to be some fission–fusion occurring.

On the island of Borneo, there exists an elephant population that is not yet officially recognized as a distinct subspecies, but nevertheless carries the designation of *E. m. borneensis*, with the current consensus among specialists being that it very likely should be differentiated. Though fossil evidence is so far still lacking, genetic analysis suggests that the population has ancient origins distinct from the other Asian elephants (Sharma et al., 2018). Moreover, although they are frequently casually referred to as "pygmy" elephants, it has not been formally established whether they are in fact much smaller than other subspecies on average. Using genetic sampling, the entire elephant population on the island was placed at between 1,200 and 3,670 individuals (Alfred et al., 2010), with Sabah containing ~2,000 of these (Goossens et al., 2016); however, a later estimate claiming to use more robust methods contended the population size to be far smaller, at just 387 individuals (Cheah & Yoganand, 2022). These numbers are rather alarming, not least because of their discrepancies, which raises serious concern for the long-term survival of this distinct subspecies.

To date there is unfortunately little published with respect to social behavior among Bornean elephants. Just one study, which focused on a small population using riparian swamps and forested areas around the lower Kinabatangan River, provides some preliminary glimpses of social behavior (Othman, 2017). Forty adult females were individually identified and monitored through direct observation, on foot, at the relatively close range of 10–30 m. These observations were performed each day for 2 weeks per month over a period of 32 months. Twelve individuals, including males, were also tracked with collars for variable lengths of time. Unlike studies in India and Sri Lanka, in which social associations were defined in terms of hundreds of meters, a "group" was defined as individuals within 5 m of the largest individual in the aggregation, possibly owing to constraints in visibility. Results therefore cannot directly be compared. Nevertheless, the study did report significant variability in group sizes, which could range between five and nine individuals, based on time of day (larger groups later in the day), habitat type (larger groups in riparian habitat relative to seasonal swamps

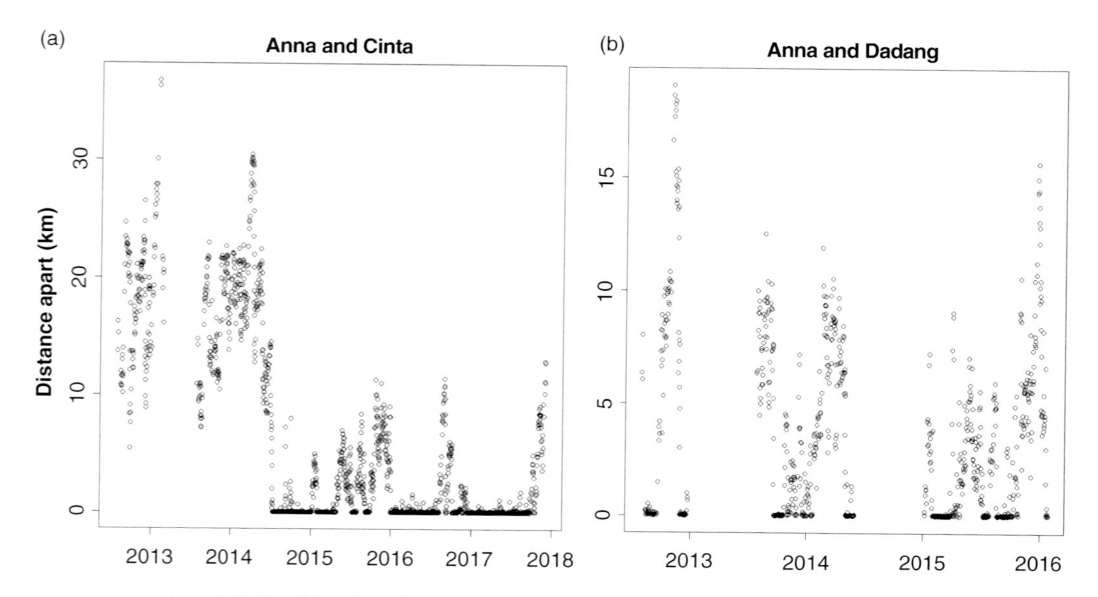

Figure 3.10 Tracking data shows traces of fission–fusion behavior in Sumatran elephants. Anna and Cinta are adult females and Dadang is a young male; all were collared at separate locations. (a): Anna and Cinta are initially apart for a period of two years, ranging mostly between 10 and 30 km from one another. Although they're occasionally less than 10 km apart, they do not meet during this period. However, after their encounter in mid-2014, the pair stay together for at least three years before separating around 2018. During this "together" period, they are rarely more than 10 km apart. (b): Anna has several sporadic encounters with Dadang (with blanks where there is missing data), each of which lasts from days to several weeks or even months. They are also rarely more than 10 km apart; thus it would appear that they are social affiliates despite being of opposite sexes (data courtesy of the Frankfurt Zoological Society, visualization by Matthew Sabin).

or plantations), and season (larger groups during flooding season). This variability indicates there is some fission–fusion dynamic occurring that would be fascinating to explore in greater detail. An issue, however, is that the studied population is small and relatively isolated from others on the island despite relatively high genetic diversity (Goossens et al., 2016); therefore, it is not clear how representative it is.

Peer-reviewed studies of social behavior in the Sumatran subspecies are entirely lacking but data from a few collared individuals outside the Bukit Tigapuluh protected area in Central Sumatra suggests that they too exhibit aperiodic and highly irregular fission–fusion behavior. At times this can involve socialization among individuals that originate from social groups that were initially many kilometers apart (Figure 3.10). However, given the longevity of elephants, it is possible they had known each other previously – these data alone do not suffice to establish that. Genetic analysis reveals extremely small population sizes within the vicinity of Bukit Tigapuluh, at the time consisting of two subsets with just 99 and 44 individuals respectively, which have been subjected to illegal killings (Moßbrucker et al., 2015). With such small numbers, it is quite possible that all individuals within the subpopulations are somehow related. The landscape this population uses, largely located outside the extremely hilly protected area, consists of forest patches interspersed with areas of deforestation and oil palm plantations, as well as numerous roads and rivers. One can therefore consider this population heavily disturbed so it is difficult to know whether elephants would display such behavior under more natural circumstances. Moreover, the observations originate from conservation activities conducted by Frankfurt Zoological Society, which was tracking elephants as a means of understanding ranging behavior and preventing conflicts with people; therefore, the glimpse it provides into social dynamics was entirely unintentional. It demonstrates the potential of tracking data to reveal aspects of social behavior in these hidden elephant populations.

A grave concern is that many of the remaining populations in Southeast Asia may be so small and fragmented that they may be barely viable (more on this in Chapters 4 and 5), let alone able to express the social complexity observed in larger wide-ranging populations. To lose these populations would therefore not only mean the loss of unique genetic history, but possibly the loss of locally adapted suites of behavior, and along with them the opportunity to better understand the social evolution in elephants. Though challenging, studies of these hidden populations are urgently needed for both their biological and conservation implications.

3.4 Fission–Fusion and Dominance

Asian elephant populations appear to have a more graded social structure than savanna elephants, where some families or social units form relationships analogous to bond groups, but others do not (de Silva et al., 2011a; de Silva & Wittemyer, 2012; Nandini et al., 2017, 2018). This results from the fact that in Asian elephants the fission–fusion process occurs at the level of individuals, with females making their own movement decisions in choosing whether or not to keep company with others in their own social unit. This is probably also true of African forest elephants, the distinction being a matter of degree, not of kind. By contrast, for African savanna elephants, the fission–fusion process takes place mostly at higher levels, where entire families merge or split with other families. These species can therefore be seen as occupying different parts of the spectrum of nestedness representing varying degrees of multilevel structure (Figure 3.11).

A consequence of these dynamics is that, for Asian elephants, dominance hierarchies have become difficult to enforce. The comparative approach is once again illuminating. Relative to savanna elephants at Samburu, Asian elephants at Udawalawe exhibited far less behavior indicative of dominance – aggressive or otherwise (de Silva et al., 2017). Individuals could not be arranged into orderly linear hierarchies, unlike savanna elephants. This is because linear hierarchies are necessarily the result of transitive three-way interactions: If A dominates B and B dominates C, then A must also dominate C. If C dominates A, the result is circular. While complete three-way interactions (triads) were rarely observed in either species, we could infer from the number and pattern of incomplete interactions that the Asian population had a greater potential for circularities (Figure 3.12). Moreover, while older individuals tended to

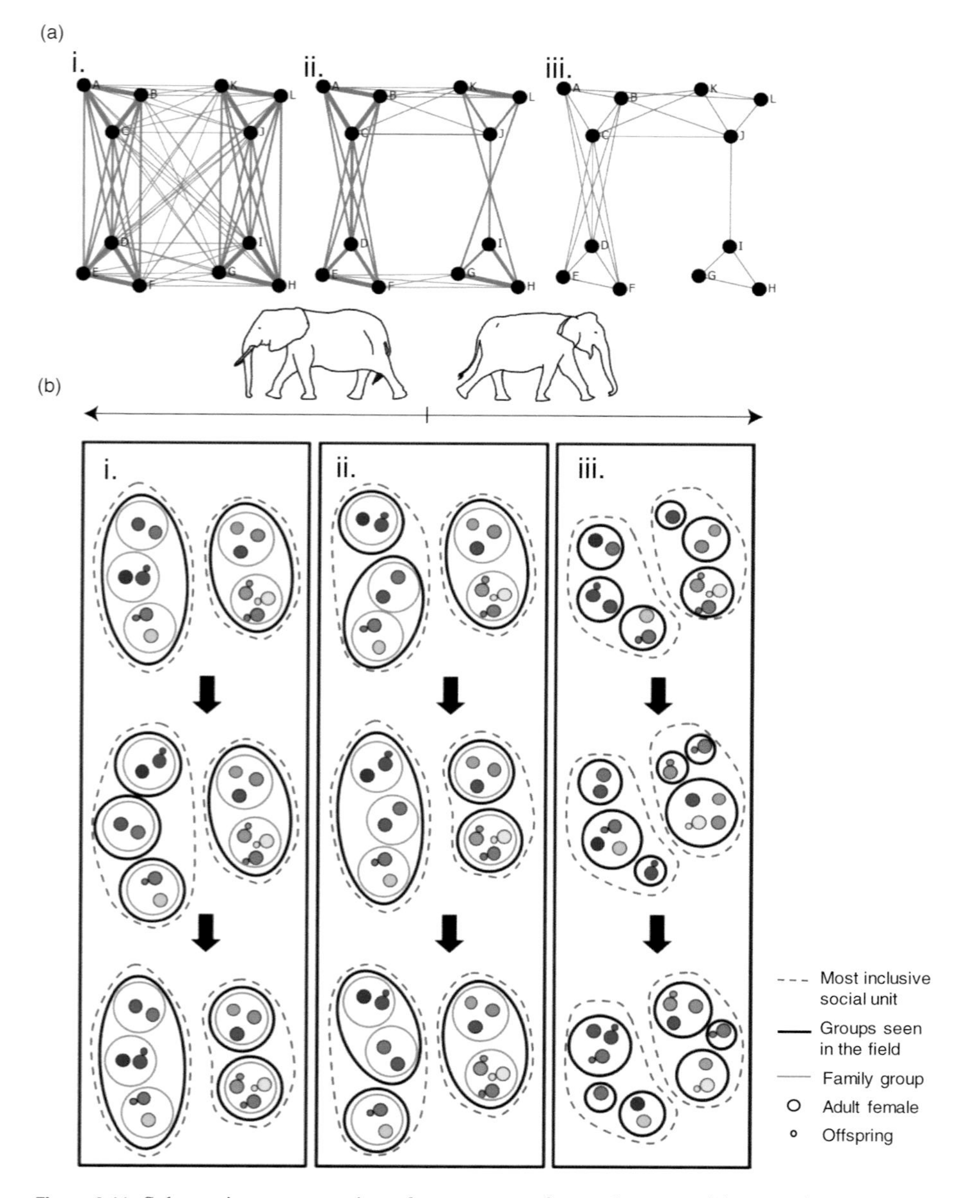

Figure 3.11 Schematic representation of a spectrum of nested communities, ranging from those that are perfectly hierarchically nested, where fission–fusion occurs among higher-order social units (i), to those that are not, where fission–fusion occurs among individuals or mother–calf pairs (iii) ((a) reproduced with permission from de Silva & Wittemyer, 2012. Copyright © 2012 the International Primatological Society; (b) modified with permission from Nandini et al., 2018. Copyright © 2018 the International Society for Behavioral Ecology).

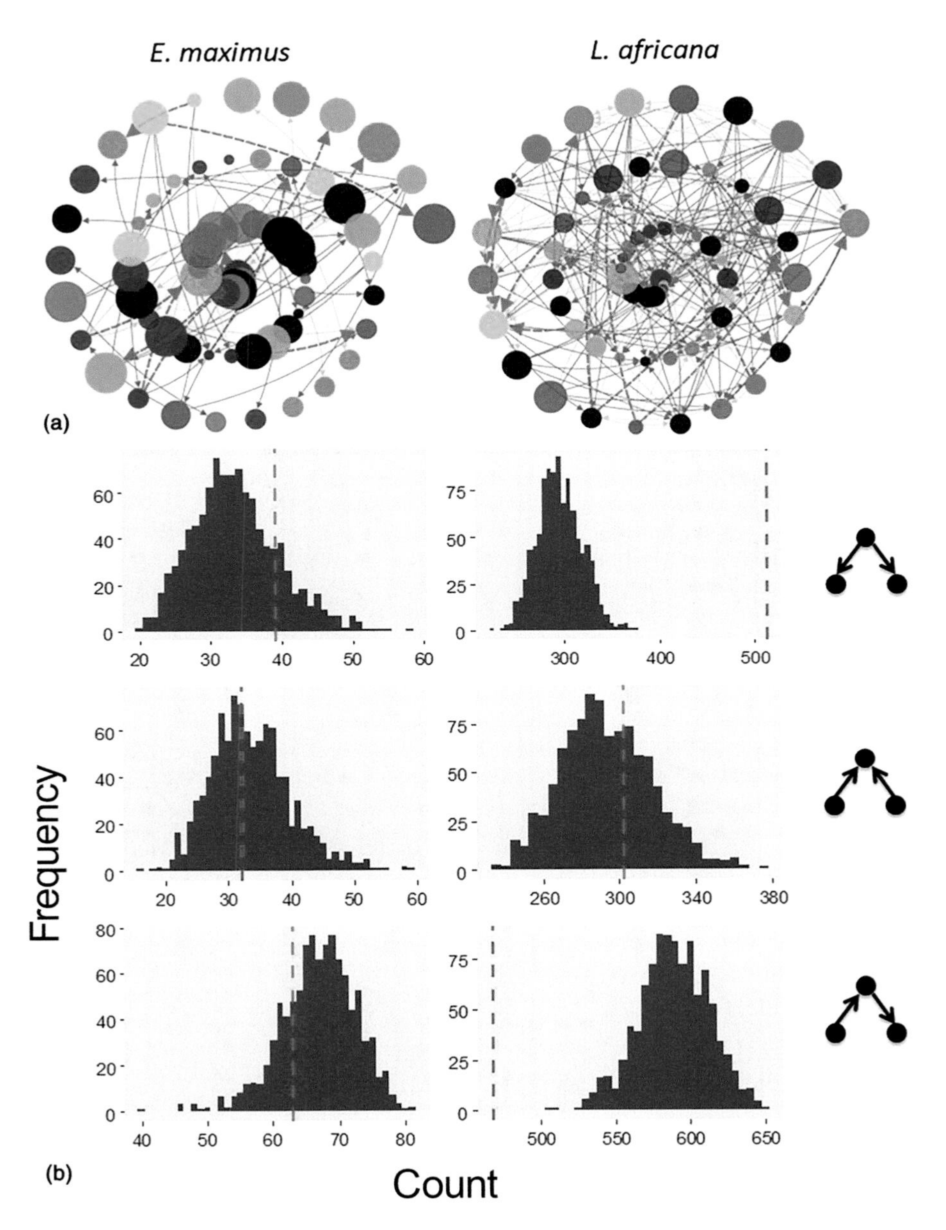

Figure 3.12 (a) Dominance networks among elephants at Udawalawe vs. Samburu, where nodes represent females >10 years old (larger = older) colored according to social unit membership. Peripheral individuals have more dominance "wins" relative to interior individuals. The Asian population has significantly fewer dominance interactions overall, and a proportionally greater number of age-reversed wins (red arrows). (b) Incomplete triad motifs exemplifying double-dominant (top), double-subordinate (middle), and pass-along (bottom) interactions. The first two would always result in a linear or transitive outcome, whereas the third could be either transitive (arrow going left to right) or circular (right to left). Many interactions of the first two types are required to form a linear hierarchy. Compared

Figure 3.13 A single adult female charges intruders while the rest of the herd makes their getaway. Casual observers might conclude that such an individual is the de facto "leader" and "matriarch" of this herd. However, given the fluidity of associations, there is likely no single individual that regularly takes this role within Asian elephant groups. Rather, such defensive behavior can be undertaken by any individual, frequently including subadults (photo by the author).

"win" confrontations in both species, age-reversed outcomes where younger (female) individuals actively harassed and dominated older individuals were not uncommon, and were proportionately more frequent in the Asian population. There was also little indication of any hierarchical order by social group. Savanna elephants, however, exhibited significant ordering both by age and by social group. We proposed that these differences are driven by the fission–fusion process itself. If one is changing companions relatively frequently, it is difficult to signal and assert dominance in the first place.

Matriarchs, who play leadership roles in African savanna elephant populations, are defined as the oldest and most dominant individuals within a family. While Asian elephants have biological matriarchs, these elephants do not necessarily play any leadership role, contrary to popular perception (Figure 3.13). Rather than moving together

---

Figure 3.12 caption (cont.)

with randomized outcomes (histograms), true outcomes (dashed lines) among the savanna elephants have significantly more transitivity and less potential for circularity, but not among Asian elephants (reproduced with permission from de Silva et al., 2017. Copyright © 2016, © The Author 2016. Published by Oxford University Press on behalf of the International Society for Behavioral Ecology).

guided by the strong, centralized leadership of an experienced individual, females and calves make more individualistic foraging decisions based on their own needs, as well as in avoidance of other individuals with whom they do not want to interact. I suggest that this is made possible by the fact that at least in UWNP elephants are not limited in their ability to disperse by either ecological constraints (water and forage are generally easy to find, if not always plentiful) or predation (large nonhuman predators are absent). The costs of exploring might therefore be lower compared to the African context, where resources are potentially both more scarce and dispersed and the threat of predation is higher. I termed this "ecological release" (de Silva et al., 2017: 249).

Elephants are classified as "generalist" herbivores, capable of consuming practically any vegetation, from fruits to grasses to bark and thornscrub. One expects dominance hierarchies to form when there is lots of competition so such versatile diets should mean that elephants do not need to compete very much for food. Classical theories attempting to relate the ecological conditions to social attributes of animal societies contended that dominance hierarchies should arise when resources are unevenly distributed, but not so scarce and scattered that group living becomes altogether impossible. This would increase competition between individuals within groups, but groups of individuals might still compete more effectively with other groups of individuals, offsetting the cost of group living (Chapman et al., 1995; Wittemyer et al., 2005b). Yet even horses, which eat grass – seemingly the opposite of a patchily distributed resource – form dominance hierarchies. Clearly, human intuitions of what constitute "patchy" or "uniform" may not always suffice because they are inherently biased by the commonplace (grasslands are not homogenous lawns), or it may simply be that the factors salient to these species, and thus driving competition, are imperfectly understood by us. The fact that savanna elephants form dominance hierarchies within and between families indicates that there is a high degree of competition among individuals, but grasslands are famously capable of supporting large numbers of herbivores (many of which are far more specialized than elephants), and so can support larger elephant groups. By contrast, in tropical Asia grasses are a highly seasonal, spatiotemporally patchily distributed resource. Large aggregations of grazers are quite rare. Elephants also forage selectively, depending on their own nutritional requirements. Thus despite its visual "greenness" Asian tropics may simply not have supported large groups of elephants (until relatively recently, with human modification of landscapes, see Chapter 8).

While the nature of dominance relationships among African forest elephants is not known, given their low rates of association it is likely that it would not be possible to construct linear hierarchies among individuals. Likewise, one might expect that savanna elephants in arid environments would simply not be able to create dominance hierarchies given their low frequency of interaction. Heavily forested environments as well as deserts both represent higher ecological constraints, not lower. Group fission is necessitated by the less abundant, highly dispersed, and perhaps unpredictable nature of available resources. Group sizes are a byproduct of some intrinsic tendency of individuals to be attracted to other conspecifics, which I call "social affinity." However, a larger affinity does not necessarily mean larger groups. As we have seen, group

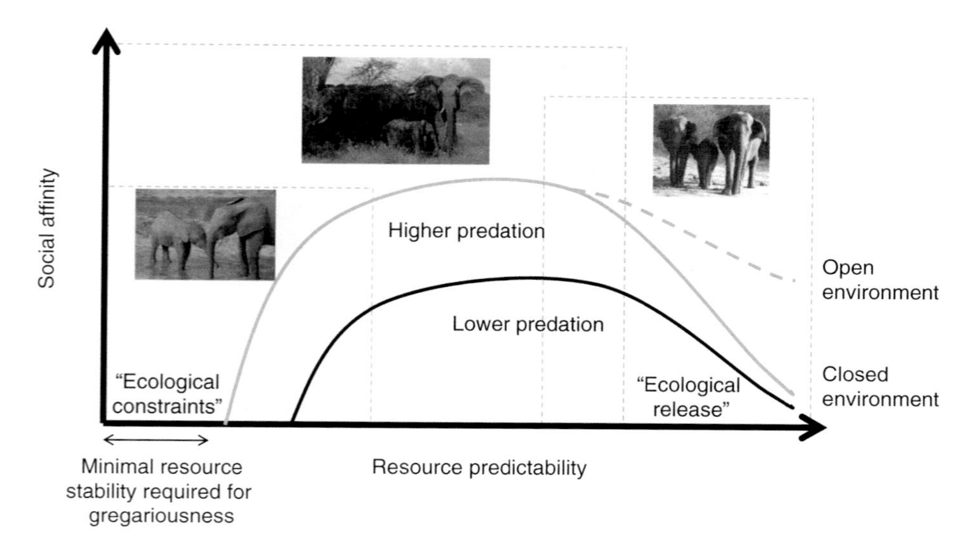

Figure 3.14 Conceptual model of elephant socioecology. Although elephants are thought of as "generalist" herbivores, African forest elephants are more frugivorous, while Asian elephants prefer grazing and African savanna elephants may graze or browse. All elephants require access to water. The vertical axis represents some inherent attraction to conspecifics (manifesting as larger and/or more cohesive groups) and the horizontal axis represents the predictability of critical resources (both food and water), that is some function of overall abundance, temporal variability, and spatial dispersion. A minimal degree of resource stability is required for social groups to be possible at all. The attractiveness of conspecifics is expected to increase with greater predictability and threat of predation up to a point. Gregariousness would still be beneficial in open environments with predators, but less so where there is cover, where crypticity would be favored instead. Greater social affinity may also outweigh within-group competition in intermediate regimes because of the value of older experienced/dominant individuals for resource acquisition, but not in more stable regimes. Thus, at the lower end of the spectrum, gregariousness is limited by resource stress whereas at the upper end of the spectrum it is inhibited by social stress (photos: Forest elephants by Victoria Fishlock, used with permission; savanna and Asian elephants by the author).

sizes can be misleading, masking larger social units through the process of fission and fusion in varying response to resource competition. Thus, it is not group sizes that should be predicted by the resource regime, but social affinities (Figure 3.14), which metrics such as group size and association index imperfectly approximate. Moreover, it may not be the particular configuration of resource abundance and dispersion that matter, but rather how they give rise to regimes of relatively low (low abundance/highly dispersed/ephemeral) versus high (high abundance/less dispersed/persistent) spatiotemporal predictability, reducing several possible independent variables onto a single axis.

Why should Asian elephants exhibit dominance signals at all, if signals are not used for the purpose of reinforcing some hierarchy? The question of what may have driven the evolution of a behavior is related but distinct from the matter of what maintains such behavior once it has evolved. Certainly, the latter may offer clues to the former, if circumstances in the present are comparable to those in the past. But it's also possible that behavior that evolved under one set of circumstances, much like any physical adaptation, gets put to some other use. Dominance signals may persist because they are holdovers from an earlier time, with individuals continuing to use them to assert themselves over others opportunistically, despite the absence of a stable hierarchy. But another interesting possibility is suggested by observations of a distantly related species: the social organization of some human societies. Many hunter-gatherer societies are famously egalitarian; however, Boehm (1999) develops the intriguing idea that this is a compromise situation. Humans, he contends, are naturally given to hierarchy just as the vast majority of primates are. Nevertheless, survival requires reciprocity in resource sharing and flexible band membership as individuals follow resources such as game over vast landscapes. Thus, we have a situation in which none can truly lead, and if one cannot lead, then the next best thing is to insist on equality. In consequence, resource distribution (in this case, *limitation* rather than predictability or abundance) drives fission–fusion, which in turn prevents the formation of hierarchies. The key lies in the flexible nature of relationships, that is, the degree of fission–fusion itself, not on the resource structure per se, though it is an underlying factor. A second piece of insight comes from Moffett (2013), who identifies a problem. As societies with fission–fusion dynamics grow in size, it becomes increasingly difficult to identify shared loyalties, especially if not all members of a social community know one another. Thus, he posits, humans have developed additional mechanisms of identifying social membership – labels of various forms.

Elephants potentially face a similar situation as these hunter-gatherer societies, given their wide-ranging movements, territorial overlap, and flexible social dynamics. But they stop short of doing as humans do, lacking the means to explicitly define egalitarian rules or identify social groups (as far as we know!). Even if elephants have the propensity to recognize a very large number of other individuals (possibly more than the average human), they are unlikely to infer social relationships among them without direct experience (unless this is somehow encoded in a communication signal, see Chapter 7). The absence of hierarchy merely results from individuals avoiding conflict through spatial segregation. Individuals must learn which conspecifics are members of one's own social unit and which are not, through familiarity. They cannot construct "anonymous societies," in Moffett's terms. Occasionally, encounters between individuals from different units may take place, but an absence of hierarchy does not imply complete tolerance. At such times, dominance interactions with or without aggression may be used to establish primacy of access to an area and who belongs within the social unit, and ultimately evict nonaffiliates from a desired resource.

These observations may help explain why Pinter-Wollman and colleagues found social segregation to be so prevalent between translocated and resident elephants. For savanna elephants, encounters with wholly unfamiliar social groups would be a rare and possibly very stressful experience. This may also explain the unfortunate outcome of elephant drives in Asia. Because the critical resources and carrying capacities for elephants are poorly appreciated throughout the range, it is not uncommon for wildlife managers to assume animals can simply be moved from one area to another without consequence. In preparation for development projects in southern Sri Lanka in 2004, elephant herds were forcibly driven by the hundreds into adjacent national parks and sanctuaries. Although no official estimates are available for the number displaced, and no formal monitoring of the consequence, tragically numerous females and calves later deteriorated and died on the visible boundaries of fenced protected areas. Why did they linger on the edges of their newly introduced habitat without pushing in to exploit the seemingly abundant and plentiful resources within? There was no apparent lack of forage or water. Instead, the cause may have been the invisible competitive exclusion between newcomers and resident elephants.

Once group living evolved, it likely had other benefits. Both savanna elephants and Asian elephants have in common the fact that females exhibit cooperative group defense, especially protection of calves. Younger females also show much the same inclination to exhibit alloparental care, even allowing infants to pseudo-nurse, which may be crucial practice for their own maternity. At Udawalawe my research group and I have observed anecdotally that some females will nurse offspring unrelated to them, with cases involving both wild and rehabilitated individuals. We have also observed that older orphaned calves rehabilitated and released into the wild have successfully integrated with wild herds, themselves exhibiting alloparental care later on as they matured into subadults. Interestingly, we have observed that cohorts of juveniles that are released together tend to stay together for many years, especially if they are females, distancing themselves from their wild counterparts once they become reproductive themselves. As with African elephants, these bonds likely represent substitutes for the relationships that would have originally held among relatives. Individuals in those populations that do not exhibit the tendency to aggregate may be even more vulnerable to social disruptions, such as those resulting from elephant drives or translocations. If networks are less dense and interconnected to begin with, individuals will have fewer social contacts to draw from in the event that they lose their own family members. The fitness benefits of social relationships among Asian elephants still remains a question wide open for investigation, with relevance for their survival little appreciated by either conservationists or wildlife managers.

4 Reproduction and Male Social Relationships

All elephant calves spend their early lives with their mothers and some are also surrounded by maternal relatives, as we have seen. But the social worlds of males and females begin to diverge as they approach sexual maturity, when the priorities of males become inseparable from the demands of reproduction. Females of all three species share similar physiological constraints on reproduction irrespective of their ecology, and thus the males of all three species have something in common thanks to the challenges of wooing members of the opposite sex. Males have unfortunately been studied far less than females outside the context of reproduction, owing to the perception that they are less social (though this is changing, as we shall see). This chapter will examine what we know of male social life as it begins to diverge from that of females, before turning to the reproductive behavior of both sexes, and finally taking a deeper look into the more uniquely proboscidean attributes of bulls. I've sandwiched topics in this manner because in order to make sense of the constraints and life history strategies of either sex, we have to unpack how their constraints influence one another.

4.1 Early Life and Dispersal

From birth to the age of 10 or so, an elephant's "calfhood" so to speak, males are mostly group living. Initially at least, they are just as gregarious as their mothers – which is to say that since mother and calf will spend a lot of time together until weaning, he has no choice but to be around her companions. Conversely, if she usually has few companions, so will the calf. However, the fission–fusion process does allow a buffer to some degree – when families mingle, calves have the opportunity to interact with new partners and indeed may actively seek them out. This is especially true of savanna elephants, which are quantifiably the most gregarious of the living species. It is not known how flexible the mother–calf relationship is among forest elephants, but our research team's own observations suggest that among Asian elephants the characteristic fission–fusion association pattern may start quite young, especially once calves are weaned, with both male and female calves spending time away from their mothers and congregating with groups of other calves. Such groups are usually under the oversight of a few adult females. Although Eisenberg and McKay remarked on their observations of such "nursery" groups as early as the 1970s, the phenomenon has not been systematically studied, largely owing to the difficulty of identifying individual juveniles when they are not with their mothers.

However, just as adults vary in gregariousness, so do calves. Some young males are extremely attached to their mothers and always appear as her closest companion, while others show much more independence. In Udawalawe, we saw one such close association between a mother and son. Although the two were seen with a social group until the calf was aged four or five, in subsequent years they grew more isolated. He lost his mother at the age of seven, and after a brief stint of aimless wandering around in poor health, he seemed to recover and was seen once again with his mother's former companions. He finally departed for parts unknown around the age of 13. Though this is only a single case, it suggests that even male calves can maintain some preference for the company of maternal relatives after an extended separation from those individuals, despite the evident fluidity of social relationships among Asian elephants. Moreover, it shows that such ties can be renewed even after some years, despite the absence of the mother. This is perhaps not so surprising, in light of the resilience in social restructuring demonstrated by heavily disturbed African savanna elephants (Goldenberg et al., 2016); however, the need for social companionship for young males has hitherto been very poorly studied and appreciated.

Which sex disperses and why is governed by both the evolutionary need to avoid inbreeding as well as the more proximate need to overcome breeding competition (Clutton-Brock, 1989; Clutton-Brock & Lukas, 2012; Greenwood, 1980; Pusey, 1987; Pusey & Wolf, 1996; Vehrencamp, 1983). The degree of social flexibility evident in different elephant societies suggests the costs of dispersing differ among systems, as already discussed. But, in general, any selective pressure on females to be gregarious (whatever this pressure may be) also places complementary pressure on males to disperse in order to avoid breeding with relatives. Given that both sexes have large area requirements, within which reproductive females are a scarce commodity (more on that later in Section 4.3.2), mating competition is intense and males are likely to encounter relatives in their ranging. In the Amboseli ecosystem, male savanna elephants are significantly less likely to exhibit sexual behaviors with relatives than nonrelatives and sire fewer offspring with them over their multidecade reproductive life, which suggests that the selective pressure against inbreeding is strong enough to influence mate choice and maintain kin recognition over a lengthy period of time (Archie et al., 2007). However, this is not the only reason to disperse. By leaving, males are able to make foraging decisions independently of female groups and also explore relationships with other males – potential friends or rivals. Males therefore have to both establish themselves among other males and also retain some recognition of who their relatives are, even though they may not encounter them frequently.

The dynamics of dispersal from the natal family has been documented most closely for African savanna elephants at Amboseli. Some males spent as much as 25% of their time away from family at the tender age of 5, whereas certain 18-year-old males spent up to 50% of time with their families. But males that left their families later tended to sever the connection more quickly than those that started the process earlier. The reasons behind dispersal decisions were difficult to pin down to a neat pattern, having to do with whether or not a male's mother was living, her social status, his family size, and the presence of other peer males. One thing that seemed clear though, was that it had little to do with ecological conditions (Moss et al., 2011). Thus, social factors appear to be the most important influences on when males decided to become independent. Being independent, however, is not the same as being asocial. Calves of both sexes like to be around age-mates, particularly members of the same sex. Early on, they will learn the identities and status of others, as well as begin to test how they measure up competitively against one another through play (Figure 4.1).

As discussed in Chapter 3, given the inherent competition for resources when foraging in groups, there must be a counterbalancing advantage to doing so. For younger males, association with older and more experienced individuals could provide opportunities for learning, but this doesn't necessitate group living. Once independent, an individual has much more control over his choice of companions as he explores his position in society and, at least among African savanna elephants, regional differences start to appear in the social inclinations of bulls. At Amboseli, bulls under the age of 20 were more frequently found in groups, but these groups were smaller (containing fewer than 25 individuals), compared to those of older bulls (Moss et al., 2011). These groups could be mixed sex, and contained more young males than expected by chance, though not as many as larger groups. It is thought that this might be to limit run-ins with potential competitors, but associating with age-mates (some of whom may be kin) also allows the opportunity for sparring practice (Chiyo et al., 2011b). In the Okavango Delta of Botswana, males between 10 and 20 years of age were found to be the *most* sociable among bulls (Evans & Harris, 2008). They engaged in more social interactions, tended to be in larger social groupings, and were in closer proximity to other elephants than other age classes (Figure 4.2). Evans and Harris (2008) suggest that just as older matriarchs function as social repositories among females, so too do mature bulls in male society. Older males have high centrality within their networks (Chiyo et al., 2011b). Complementing this view, observations at Etosha National Park in Namibia show that bull society can develop its own elaborate structure with a clearly defined hierarchy, especially when vying for scarce resources (O'Connell-Rodwell et al., 2011). Both in Namibia and Botswana, very large bull-only aggregations have been observed that simply do not seem to occur in Kenya (K. Evans, C. O'Connell-Rodwell, personal communications). In Botswana, it was also observed that the presence of older males curbed displays of aggression in younger individuals toward *targets other than elephants* (Allen et al., 2021). This is intriguing because it suggests that these social contacts help to maintain certain behavioral norms, including those that are less likely to escalate into negative situations – such as conflict with other species. One noteworthy long-term study made use of 16 years of data on male associations (sexually inactive state only, discussed further in Section 4.3.4) at a private nature reserve in South Africa to evaluate the stability of their relationships over time. In all, there were 39 males of varying age, all of whom were independent of their natal groups. Those aged 20–29 were considered "younger" and capable of being reproductively active, whereas those aged 30+ were considered "older" and likely to be more reproductively competitive. The researchers focused on two network

(b)

Figure 4.1 Young males at play. Teenaged male Asian elephants that have dispersed from their family groups nevertheless spend time sparring and socializing with one another. Here, a young tusker practices asserting dominance over another (non-tusked) male age-mate by playfully putting his trunk over (a) and even mounting him (b) (photos: Udawalawe Elephant Research Project).

metrics, namely *strength*, which indicates how frequently an individual associates with others, and *eigenvector centrality* (EC), which indicates how well connected they are to others in the network. They found that while the stability in EC of older

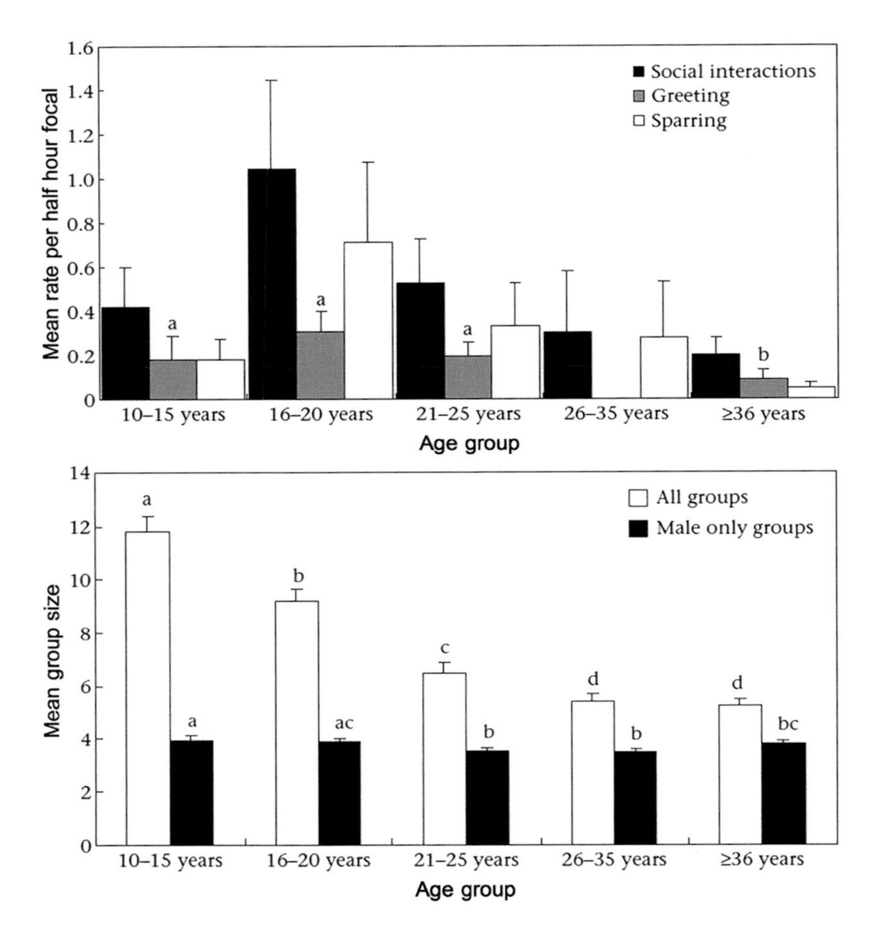

Figure 4.2 Young males are sociable. Top panel: Bulls in their late teens engage in more social interactions as well as sparring than any other age class. Time spent interacting with other bulls declines with age. Bottom panel: Young males also tend to be found in larger mixed-sex groups (figures reproduced from Evans & Harris, 2008).

and younger males did not differ significantly, the strength of associations for older males was significantly less variable than for younger males. This offers some indication that males might solidify their social relationships as they age, even if their social groupings are more flexible than those of females. Such familiarity may in turn be an important mechanism for maintaining behavioral norms and culture in male elephant communities. We will return to this in later sections.

These observations show that we are still in the early stages of understanding what bulls do when they part ways with their maternal relatives, what possible functions social contacts may serve, and why these patterns seem to vary geographically. Older males in Botswana, like older females in many populations, also seem more likely to lead collective movements (Allen et al., 2020), but perhaps this is more necessary under particular configurations of resources and risks on the landscape.

Figure 4.3 Bull groups. (a) A group of adult bulls of mixed ages graze amicably near one another at Samburu National Reserve, Kenya (photo by the author). (b) A male group at Boteti River in the western boundary of the Makgadikgadi Pans National Park, Botswana (photo by Elephants for Africa). (c) Two bulls of differing age share shade in Udawalawe National Park, Sri Lanka (photo by the Udawalawe Elephant Research Project). (d) A bull group pauses to get a drink at a communal reservoir belonging to Pokunutenna village, on the border of the protected area, before moving together toward agricultural fields (photo by the Udawalawe Elephant Research Project). Associations such as those in (a) and (c) appear to be relatively short-lived. It is not known whether relationships such as those in (b) and (d) are similarly temporary, or more stable.

Given that older bulls have been most heavily impacted by hunting pressure, the structure of male societies has likely been damaged as severely if not more so than that of females'. From a conservation standpoint, it is at least as important to understand how males are responding to these stresses as it is to understand the behavior of females. One reason for this gap in research is that bulls have traditionally been viewed as relatively asocial, given the temporary nature of most social encounters. But recent studies are changing that. Another major reason is the logistical challenge of studying animals that disperse over long distances, as behavioral studies are typically limited to particular sites. This is not unique to elephants – marine and freshwater biologists have long struggled to get a handle on what exactly happens during this phase of life for many aquatic organisms. This constraint is recently being partially overcome through the use of tracking devices, which also allow inferences of certain other behaviors besides movement (more on that in Chapter 6); however, such devices do not as yet substitute for human observers when it comes to interpreting interactions among individuals.

Dispersal is usually a risky period for any species, no less so in elephants, which must navigate a complex patchwork of different land uses and habitat types. Some of these land uses, such as protected areas or certain plantation forests, may be compatible with their presence, while others, such as agricultural fields, may not be. As many studies of elephant focus on adults, the fates of young dispersers are seldom known. While a male may explore and discover the lay of the land on his own, as we have seen from studies of females, social contacts may be a means of learning about resources in unfamiliar areas. Males may also get to know potential competitors or mentors in the relatively neutral context of foraging (Figure 4.3). It is possible, as hypothesized by Evans and colleagues, that young males learn from their older counterparts. There is some insight to be gained from a context in which male groups are seen regularly, even in East Africa; that is, relatively high-risk foraging – on croplands.

4.2 Crop Foraging

We will take a look at the effect of elephant behavior on natural ecosystems in Chapter 5. Here, I would like to focus on a foraging context in which social relationships may be particularly relevant for males. Mature bulls, having no calves to protect and little to fear from nonhuman predators, generally would not seem to benefit from groups. However, when foraging outside protected areas, an argument could be made that groups enable greater vigilance against people – especially true if a foray takes the animal directly into the midst of human activity. Foraging in croplands is extremely risky; therefore, the payoff must be worthwhile. Bulls are primarily implicated in crop "raiding" both in Asia and Africa, rather than groups of females and calves, likely because the latter are much more risk-averse (although in some landscapes dominated by large-scale plantations or where natural forage is of poor quality, female groups may not have much of a choice). Although both male and female elephants regularly use pathways for directed movements (discussed further in Chapter 6), bulls seem much more likely to use pathways to forage on crops whereas females prefer those that are further from human settlement (Von Gerhardt et al., 2014). Unless quoting text, I use the terminology of crop *foraging* rather than *raiding* regardless of which was used in the original study, because the latter is a loaded term (see Hill, 2018). Individuals that forage on crops seem to grow bigger relative to their counterparts that do not, even if the two cohorts do not start off with any appreciable difference in size (Chiyo et al., 2011a). This indicates that the strategy does indeed pay off, probably because crops often represent higher-quality and more concentrated nutrition than natural forage (more about this in Chapter 5).

In the Amboseli Basin ecosystem, the individuals that engage in crop foraging appear to be demographically similar to the rest of the population – that is, the age distribution of crop foragers was not significantly different from that of the overall population (Chiyo et al., 2012). However, it was found that the chances of being a crop forager increased as individuals got older, had close associates that were also crop foragers, and when those close associates tended to be older than the individual. Although this does not say that males are necessarily in groups while in the act of crop foraging, it is suggestive of it. Moreover, the strong community structure detected in the social networks of these bulls is an indicator that associations persist, although it is not known for how long. One of the observations most relevant to management is that not all crop foragers have the same profile. Some are habitual, others are occasional. Of those individuals that were identified as being part of the well-studied Amboseli population (about a third of total number estimated to be involved in crop foraging), just 12% were classified as habitual crop foragers, but they accounted for 56% of the crop losses (Chiyo et al., 2011c). Based on genetic samples left behind (i.e. dung), an estimated 247 individuals were involved in crop foraging across the three different overlapping localities that were studied.

There is a small but very interesting caveat though, revealed in the difference between estimates observed through direct sightings versus those obtained through genetic samples. Of those bulls that were regulars at Amboseli National Park, about twice as many were directly observed as the genetic sample size indicated (median estimates of 86 versus 41 respectively). But in the overall sample, the opposite was true – less than half were seen as opposed to sampled genetically (median estimates of 109 versus 247). The authors conclude that this indicates the Amboseli bulls were more easily observed than others from less habituated populations. Although Chiyo and colleagues stop short of extrapolating further from their findings, these discrepancies in the estimates seem to suggest something else as well. If one assumes that Amboseli bulls and non-Amboseli bulls defecate at the same rate (a reasonable assumption, since there is no reason to expect otherwise), even though they are *seen* at different frequencies, the same proportionality should apply to both subsets. This allows us to guess at the number of crop-foraging individuals that were detected *neither* through observation nor genetic sampling. If 86 bulls account for only 41 unique genetic samples, then 494 individuals are required to generate 247 unique samples. Therefore, the actual number of individuals participating in crop foraging may be closer to 500 (including both Amboseli and non-Amboseli bulls)! It is difficult to imagine what fraction of the total number of bulls this represents, but it has to be substantial.

These findings mirror those in Asia, where a minority, consisting predominantly of males, may be responsible for a disproportionate fraction of the damage; but there is likely to be a much larger pool of occasional opportunistic crop foragers (Sukumar, 1995). Given that the attempted removal of crop-foraging individuals through lethal or nonlethal means is a frequent response, these observations should give us pause. Translocation, which has been a favored strategy, potentially compounds these problems via social learning. When bulls are moved from one location to another, they do not simply leave their habits behind. An individual used to foraging on the provisions of humans, causing property damage, or breaking barriers will take these behaviors with them to the new area, where they then model these behaviors to new individuals (discussed more in Chapters 5 and 7). These issues also highlight that although it may certainly be possible to temporarily reduce crop damage incidence by taking out those "habitual raiders," so long as crops continue to present a highly condensed source of nutrition found within elephant range, there will continue to be those that take advantage of it despite the risk, because it pays off in the quest to gain strength and weight and thereby the reproductive edge. Removing any particular male may simply create a vacuum to be filled by others.

A study conducted in southern India found that the size of all-male groups tended to be largest in landscapes dominated by agriculture, as opposed to those that consisted of forest (Srinivasaiah et al., 2019). The former represent areas of higher risk, compared to the latter. Similar to observations in the African elephant context, teenaged males (ages 10–20) classified as "sexually mature but socially immature" were most likely to associate in all-male groups, whereas sexually immature males were more likely to be in mixed-sex groups and older bulls were more likely to be solitary. Based on informal long-term observations spanning at least 20 years, authors suggest that the occurrence of large, long-lasting (multiyear) associations among males seem to reflect a relatively recent and adaptive behavioral response to large-scale loss of natural habitat. The study does not address exactly what purpose such groups serve, but it was observed that of those individuals that did participate in all-male groups, only those in those of the intermediate age class showed influence of habitat type on their body condition. Tellingly, those that were in the best condition seem to be those that inhabited landscapes predominantly consisting of crops. The study notes the high prevalence of conflict in the area, including human and elephant deaths. Together, these observations are highly suggestive that at least one function of such larger groups may be as a safeguard against the risks of foraging on croplands. The authors speculate that the behavior of mature bulls may be emulated by younger individuals, with those that do not forage on crops modeling positive behavior; conversely the absence of such individuals in populations that have lost many older males may encourage riskier behavior among younger animals. Unfortunately, while there is ample evidence of the transmission of risk-taking behavior (examined further in Chapter 7), a similar transmission of such risk-avoidance behavior has not yet been documented, nor has it been shown that older individuals are any less likely to forage on crops. In heavily altered landscapes, it is entirely possible that elephants simply cannot subsist on the limited amounts of natural forage. The propensity of bulls to forage on crops is likely to be a function of the abundance of naturally available food as well as competitors for that food supply. Studies comparing landscapes of varying heterogeneity, representing distinct, noninteracting elephant populations, would shed light on whether there is a particular tipping point of cropland to natural forest and elephant density at which individuals are more likely to risk foraging on the tempting products of human labor in order to gain a competitive edge.

4.3 Reproduction

4.3.1 Mating System

In 1977, Emlen and Oring wrote a classic paper on mating systems that remains one of the most intuitive and widely applied theoretical constructs in behavioral ecology (Emlen & Oring, 1977), and indeed remains one of the most general explanatory frameworks the field has been able to achieve. It is helpful as a starting point for thinking about the constraints elephants face. They posit two key principles that govern mating systems. The first is the "ability of a portion of the population to control access of others to potential mates." The second is that "ecological constraints impose limits on the degree to which sexual selection can operate" (Emlen & Oring, 1977: 215). In other words, individuals may either be able to monopolize prospective mates, or control access to the resources that prospective mates might need, to various degrees.

Control of access is possible when females are desynchronized in their periods of sexual receptivity, especially if these are short, because then a male can defend a mate against potential rivals. If one male is able to mate with multiple females, the system is known as polygyny; hence the strategy is termed "female-defense polygyny." When females are more synchronized, or when periods of sexual receptivity are prolonged, a single male might find it challenging to ward off competitors. But he may not need to do so if he can control access to a key resource, such as food or water. The latter is possible only if the resources themselves are spatially defendable (i.e. more clumped), therefore this tactic is often referred to as "resource-defense polygyny." Of course, it is possible that neither females nor resources lend themselves easily to defense, in which case males may resort to displaying and competing with each other directly, similar to the formation of leks in species such as black grouse or Uganda cob.[1]

Another important consideration is the operational sex ratio (OSR): the number of individuals of either sex that are reproductively available at any given time. Emlen and Oring highlight a phenomenon that had been appreciated some time earlier, namely that there tends to be intense competition for mates among members of whichever sex is more reproductively abundant, with the result that fewer of these individuals actually wind up breeding successfully (Trivers, 1972). Therefore, the more male-biased the OSR, the greater the expectation of polygyny (where the most competitively fit male is able to mate with multiple females), while polyandry (one female mates with multiple males) is logically expected when the opposite is true. This is not a very intuitive result if one assumes, for instance, that the fewer females there are, the more likely each individual female might be to receive multiple suitors. However, members of the scarcer sex, especially if they can exert some degree of choice in the matter, are likely to select the best of the available options. Both polygyny and polyandry are forms of polygamy, which simply describes systems in which individuals have multiple mates.

[1] Leks are a phenomenon in which multiple individuals of one sex, usually males, display and compete for prospective mates in close proximity to one another.

A final consideration is the degree of parental investment required – the less investment required of one sex, the greater the potential for polygamy among members of that sex (see Clutton-Brock, 2017, for a more complete overview of the history of these ideas). Mammals typically tend to be polygynous because females must invest much more energy and time in parental care than males. In contrast, the majority of birds tend to be at least socially monogamous because both parents are required for incubating eggs and rearing offspring (although the genetic story may be more complicated). Polyandry is much more common in fish, reptiles, and amphibians than in either mammals or birds. Several species of spiders and insects fall on the other extreme, with polyandry being the norm, as males risk becoming a meal for their much larger and more formidable mates.

Male elephants offer no parental care and can (in principle, at least physically) breed throughout the year with more than one female. The OSR therefore is likely to be male-biased (with some exceptions, where males are disproportionately hunted); among males there is much competition. Like many mammals, elephants may therefore seem to typify polygyny, owing to the fact that males are likely to sire offspring with multiple females in any given year. In fact, elephants are technically "polygynandrous," meaning both males and females mate with more than one individual of the opposite sex (a form of polygamy), although of course ultimately only one male will end up siring the offspring of any particular female at any given time. Like some ungulates such as deer and sheep, both males and females exhibit heightened sexual activity at particular times of year and one may see more calves born in particular months, but neither sex could be said to have distinct breeding "seasons." However, elephants have evolved some uniquely proboscidean twists in their mating strategies owing to their life history traits.

Upon closer examination, elephants do not neatly conform to the framework initially outlined by Emlen and Oring. For one thing, it's difficult to classify the resources critical to such a generalist feeder, as this may vary depending on which species, resource, and spatial or temporal scale one means. For instance, grasses may be widely distributed in parts of sub-Saharan Africa at certain times of year, but are patchy in the tropical forests of either continent. Moreover, other types of resources such as fruit or water may only be available at certain locations at certain times of year, but plentifully at such times. Although one may therefore rule out resource-defense polygyny as an option for elephants, elephants fit none of the other four forms described (female/harem-defense polygyny, male-dominance polygyny, explosive breeding assemblages, and leks; see table 1 of Emlen & Oring, 1977: 217). We next examine the reasons why this is might be, but to do so we must first take a very deep dive into female reproductive considerations.

4.3.2 Female Reproductive Strategies and Constraints

Elephants are among the slowest breeders in the animal kingdom because of multiple constraints. First, the earliest age at which females can give birth (i.e. primiparity) is nine years old, although this is highly unusual (de Silva et al., 2013; Turkalo et al., 2017).

In the wild they typically have their first calf in the early teens (Aung, 1997; de Silva et al., 2013; Moss & Lee, 2011a; Sukumar, 2003), but forest elephants at Dzanga Bai currently hold the record for delayed reproduction with the median age of primiparity being 23 years (Turkalo et al., 2017). Second, gestation itself can take up to 22 months, one of the longest in the animal kingdom. For comparison, the much larger sperm whales have a gestation period of 19 months and blue whales have a gestation period of only about a year.

In addition to this lengthy gestation, calves are dependent on milk for a few years, and the nutritional demands alone may prevent a mother from being in good enough physical condition to have another baby. It will take at least a year before a female comes into oestrus again after giving birth owing to the nursing requirements of her calf, meaning that she cannot reproduce any faster than every three years, hence placing a lower limit on the interbirth interval. Savanna elephants seem more resilient than Asian elephants in this regard. In Kenya for instance, some calves manage to survive even when orphaned at the age of two years (Wittemyer, personal communication). However, in both Myanmar and Sri Lanka, calves that either lost their mothers or failed to receive adequate milk before the age of three did not survive (Lahdenperä et al., 2015; Mar et al., 2012). All of this means that a healthy birth interval between successive calves is typically closer to four years, but quite often even longer.

Longer birth intervals can arise for several reasons. One is that despite the longevity of elephants, there is at least one strong evolutionary case to be made for investing in early reproduction: simply because individuals are more likely to die as they get older (Williams, 1957). In other words, life is a gamble one is always at risk of losing. For females, which live just as long, if not longer than males, long-term data from a semicaptive Asian population suggests there is some selection pressure to reproduce earlier in life (Hayward et al., 2014; Robinson et al., 2012). The elephants employed by the timber industry of Myanmar have the distinction of having the most meticulously maintained long-term demographic data out of any elephant population in the world. Sexually mature females in this population show an increase in fecundity until the age of 19, followed by a decline. A high reproductive output before this peak age significantly reduced females' chances of survival later in life, but increased her lifetime reproductive success (Hayward et al., 2014).

Females therefore appear to slow down reproductively as they get older, producing fewer and fewer calves until eventually becoming senescent (Robinson et al., 2012), although the condition appears to be physically different from human menopause, a point to which I shall return. The survival of calves with very old mothers may also not be very good. In relatively large populations, where there is a mix of older and younger breeding females, the median birth interval might be around six years – this means that around *half* the population experiences even longer intervals (de Silva et al., 2013; Moss & Lee, 2011a; Turkalo et al., 2017). At Udawalawe, some older females were seen to produce a calf just once in 10 years (of course, it's possible that these females had calves that didn't survive long enough to be seen at all).

One might also expect reproduction to slow down when elephant populations are at high densities, simply because of the increased competition. But although we may observe that elephants at high densities reproduce slowly (de Silva et al., 2013), it's tricky to quantify how reproductive rates actually *change* with density (i.e. to show that variables such as birth intervals and calf survival are density-dependent). To do this would require keeping track of demographic rates within the same self-contained population at *different* densities, which could take a very, very long time in the case of elephants. An observer also has to catch a population just as it approaches "carrying capacity" but, so far, "carrying capacity" remains a theoretical construct that only applies to artificially restricted populations, if at all.

The only studies of elephants that have explicitly examined whether density-dependent population regulation is possible, through mathematical modeling, have been on African savanna elephants (de Silva, 2010a; Fowler & Smith, 1973). The empirical evidence, however, has been equivocal. In southern Africa, elephant populations that are able to move freely don't show any density-dependent effects (Addo Elephant National Park, South Africa: Gough & Kerley, 2006) while those that are limited in key resources found within a protected area, namely water, saw an overall decrease in population numbers. However, these short-term changes are again more likely owing to elephant movement than changes in reproductive rates (Hwange National Park, Zimbabwe: Chamaillé-Jammes et al., 2008). In Sri Lanka, the sudden and extreme compression of elephants into protected areas had catastrophic consequences, resulting in female and calf mortality rates as high as 50% (Kurt, 1974), but the effect of such artificially high densities on birth rates remains poorly understood.

Elephant populations, like that of many large animals, very likely don't remain static but instead are prone to fluctuate around some dynamic equilibrium (Bonenfant et al., 2009). In environments where resources are not constant themselves, it is can be difficult to differentiate the effect of population density itself from regular variation and observation errors (Freckleton et al., 2006). In Udawalawe, we observed that females that lost their calves within the first year were able to come into oestrus sooner, experiencing a reduction of about 15 months on average in the time between calves (de Silva et al., 2013). This adaptation could buffer elephant populations to some extent against particularly bad environmental events that result in high calf mortalities, but still runs up against the physiological limits mentioned. For these reasons, the reduction of elephant population sizes because of fragmentation and habitat loss presents an acute concern that is seldom acknowledged: It is not only the number of individuals that matter, but also their *age*.

Even though elephants can continue to breed well into their 60s, the engine that maintains the population growth rate consists of the younger females, as we have seen. It is as if the additional reproduction of older females were a bonus, but not one that populations (nor wildlife managers and conservationists) should rely on. If the age distribution gradually shifts toward older individuals past their reproductive peak, populations will very likely experience something known as the *Allee effect* (Allee et al., 1949). The Allee effect refers to any situation in which populations exhibit a negative growth rate once they fall below a certain size. Often, the culprits are things such as inbreeding depression, which may cause populations to collapse owing to the accumulation of harmful genetic mutations (Main et al., 1999; Rohland et al., 2010).

Alternatively, it may be because some species require groups in order to rear offspring and survive (e.g. African wild dogs: Woodroffe, 2011). In the case of elephants, it may additionally be brought on by the gradual aging of breeding females in a population, if reproductive rates and calf survival fail keep up with mortality in older age classes. This means that although elephant populations, being closer to the K-selected end of the spectrum (see Section 4.3.3), are expected to be more limited by ecological conditions than by reproductive rates, the latter must be maintained at near optimal levels precisely because they are already slow to begin with (de Silva & Leimgruber, 2019). Likewise, although populations that enjoy high reproductive rates and adult survival can tolerate surprisingly high rates of calf mortality (as much as 50%), even incremental negative shifts in these variables require dramatic improvements in calf survival as compensation (de Silva & Leimgruber, 2019; Jackson et al., 2019).

4.3.3 Reproductive Senescence

Classical life history theory distinguishes between organisms that are "r-selected" and those that are "K-selected" (Pianka, 1970). Species that have short lifespans are pushed by natural selection to produce many offspring as quickly as possible, and population sizes are largely limited by this reproductive rate – hence, the term *r-selected*. This is because, under normal circumstances, predation prevents populations from getting so large as to ever reach the environmental-carrying capacity, K. Such species are characterized by short gestation times along with short periods of parental dependence. Species that have longer lifespans are expected to have fewer offspring but invest more heavily in each offspring, which take longer to reach reproductive maturity. These species usually have larger body sizes and fewer predators, with population sizes more limited by resources (Peters, 1986) – hence, the term *K-selected*. Of course, as with so many things in biology, in reality these two sets represent the two ends of a spectrum rather than a clear dichotomy. In economic terms, species closer to the r-selected end discount future reproductive opportunities more than those near the K-selected end (Ramsey, 1928).

Elephants are usually presented as textbook examples of the K-selected species, given that they seem to embody the whole suite of requisite traits. But curiously, female elephants don't seem to go through menopause, the physiological shift after which females stop being reproductive but continue to live on for many years (while males continue being fertile well into old age). The ability of females to live beyond their reproductive years is actually not so uncommon among mammals (Cohen, 2004), though the term "menopause" seems to be popularly applied in describing humans and a few species of whales (specifically, killer whales, short-finned pilot whales, belugas, and narwhals; Ellis et al., 2018) because of the exceptional length of females' post-reproductive lifespans as observed in the wild as opposed to captivity. It is worth noting here that we are actually discussing two linked phenomena – the fact that females of some species stop being able to produce offspring at some point in life (reproductive senescence or menopause) and the fact that they go on living despite this (post-reproductive lifespan). In order to understand whether elephants truly represent an odd exception or merely a different manifestation of the same old rule, we must take a closer look at these phenomena.

While there would be no such thing as menopause in the first place if reproductive senescence was not decoupled from general senescence, the puzzle can nevertheless be presented from opposing viewpoints (Hawkes & Coxworth, 2013; Hawkes et al., 1998). One may ask: If females are capable of living for a certain number of years, why don't they reproduce for the entirety of it? This framing supposes that lifespan is fixed at its physiological upper bound and selection somehow pressures reproduction to terminate earlier than it otherwise might ("stopping early"). Alternatively, one may flip it around and ask: Given that females can only reproduce for a certain number of years, why do they go on living after this age? This framing supposes that the reproductive period is fixed at its upper bound and selection pushes for ever longer lifespan. It may be noted, for instance, that most primates live to around 50 years, with all aspects of health, including reproduction, failing at the same time toward the end of life, whereas humans can potentially live twice as long but have a similar number of reproductive years (Cohen, 2004; Hawkes et al., 1998). Whichever way one looks at it, these questions set up the need for adaptive explanations.

The so-called Grandmother hypothesis (Hawkes et al., 1998; Williams, 1957) has been one of the most studied and argued from both perspectives. It posits that post-reproductive contributions by individuals to enhance the survival of offspring (and grand-offspring) make up for the lost reproduction, and would encourage the persistence of both the trait and the lineage despite the fact that the individual herself does not continue to reproduce (Hawkes & Coxworth, 2013). This stems from the observation that pregnancy becomes increasingly risky with age in humans, therefore individuals should be better off taking care of their existing offspring than in producing new ones.[2] Implicit within this formulation is the expectation of an inherent tradeoff between current reproduction and future survival and reproduction (Kirkwood & Rose, 1991). Williams (1957: 408) states: "Menopause, although apparently a cessation of reproduction, may have arisen as a reproductive adaptation to a life-cycle already characterized by senescence, unusual hazards in pregnancy and childbirth, and a long period of juvenile dependence."

This hypothesis has been most extensively researched in humans and killer whales (also known as orcas), both in terms of the fitness effect itself and the mechanism of its operation. Some researchers have taken it in a somewhat different direction by emphasizing the latter. In human populations, grandmothers can enhance the ability of their daughters to rear more children (Lahdenperä et al., 2004), whereas in Southern Resident killer whales the effect is more on the survival of their own offspring, particularly males (Foster et al., 2012). Some form of food provisioning is implicated as the reason behind this boost to survival, either by handling food that is difficult for younger individuals to access or through guidance in finding unpredictable food

[2] As originally formulated, it was really more of a "Mother hypothesis." Indeed, many studies of nonhuman animals actually focus maternal effects, not on grandmothers, and there is a case to be made for distinguishing the two explicitly (see also Croft et al., 2015).

sources through accumulated ecological knowledge (Brent et al., 2015; Hawkes & Coxworth, 2013). The presence of a post-reproductive mother also appears to reduce the number of injuries males experience from others in the pod. One long-term study of Southern Resident killer whales in the waters off British Columbia and Washington revealed that post-reproductive females were more likely to lead pods in years when the preferred food source, salmon, were scarce (Brent et al., 2015). But all of these parental investments come at a steep cost – females with male offspring had drastically reduced reproductive success in subsequent years, nearly 70% annually (Weiss et al., 2023).

Are the benefits really sufficient to outweigh the cost of ending reproduction early? After all, the longest reproductive lives documented seem to occur in a species of baleen whales – fin whales seem able breed up to the age of 90 at least (Cohen, 2004), having clearly evolved a very different life history strategy. The crucial consideration is how long offspring remain with their mothers. Resident killer whale pods tend to be highly related and manage to avoid inbreeding by virtue of temporary trysts between members of different clans (Pilot et al., 2010). But this also sets up a potential reproductive conflict between mothers and their offspring (Cant & Johnstone, 2008). When a female and her daughter are both reproductive, they must compete for the same limited food supply and it appears that the calves of the older mothers in this situation have a higher risk of death (Croft et al., 2017). This, together with the preceding benefits of prolonged lifespan, is thought to explain the long post-reproductive period in females. However, killer whales are virtually unique in that neither sex disperses, whereas in other species, including humans, one or other sex leaves the natal group.

Cohen (2004) takes a wider perspective, reviewing whether post-reproductive females provide benefits in many other mammalian species, and finds the evidence wanting. Another possibility is simply that senescence of reproductive cells (oocytes) occurs on a different timescale than that of other cell types (somatic cells), owing to distinct selection pressures on each, which is a nonadaptive explanation for the dissociation between the two. This would not refute any of the adaptive explanations put forth for killer whales, but questions their generality. One of the issues with making cross-species comparisons, however, has been the absence of a standardized measure to make those comparisons with, which accounts for the probability that a female actually survives to a given age. The reasoning behind this is that in order for post-reproductive lifespan to be of any consequence to the population, a substantial fraction of females must exhibit it.

A recently proposed measure, termed "post-reproductive representation" (PrR), does this (Levitis & Lackey, 2011). The actual calculation takes a bit of time to wrap one's head around, but is to be understood as "the number of years an average newborn can expect to live as a post-reproductive adult … divided by the number of years an average newborn can expect to live as an adult" (Croft et al., 2015: 409). Humans hold the record for the longest PrR, with Japan leading the way with a value of 0.76 from data for the year 2002 and the rest of the world in the range of 0.60–0.64 (Croft et al., 2015). Chimpanzees in the wild have a value of 0.02, whereas those in captivity have a value of 0.224, comparable to that of wild resident killer whales (0.22) and short-finned pilot whales (0.28), the other two species readily admitted to experience "menopause." Semifree-ranging Asian elephants from the Myanmar timber camps have a value of 0.128, placing them squarely in the middle (Croft et al., 2015; Lahdenperä et al., 2014). Elephants in the Myanmar population continued to reproduce until the age of 65 and post-reproductive lifespan was 11–17 years.

If inclusive fitness benefits derived from the social and ecological knowledge of experienced leaders were sufficient to drive PrR, then one would expect African elephants to have values comparable at least to that of the toothed whales. And, given what we know about the differences in social organization between Asian and African elephants, one would predict PrR to be lower in Asian elephants, all else being equal. But this might be clouded by the high rates of mortality suffered by well-studied African populations, such as those in Kenya. It is important to remember that PrR is also function of the hazards that might claim older females – in African ecosystems, not only high rates of poaching but also events such as droughts. As a result, age structures can vary substantially over time, and can certainly differ from at least some wild populations in Asia, where females could live much longer (de Silva et al., 2017). The Myanmar timber camps also represent a highly unusual arrangement, where elephants are actively managed and routinely provided medical care that their wild counterparts would not receive, so one might expect the PrR value for the latter to be even lower. Low values in elephants would suggest that the mechanisms based on inclusive fitness alone are insufficient to explain the extended post-reproductive period of humans and limited subset of toothed whales. I suspect that the potential costs of late-life reproduction (or lack thereof) are likely to play significant roles in explaining the differences observed among species, not only the benefits of longevity. The various anthropogenic pressures elephant populations are currently under make it difficult to know whether current lifespans and reproductive rates accurately reflect those that they would have historically exhibited under undisturbed conditions. Until there are more data from wild elephant populations, the matter remains an important puzzle that is consequential to our understanding of the evolution of senescence and post-reproductive lifespan.

4.3.4 Male Reproductive Strategies

The preceding considerations have a range of consequences, irrespective of species. First, it means that females are limited to breeding within a brief oestrous window of three to five days once every four to six (or more) years and are more or less desynchronized, in that they can breed throughout the year. Although there may be certain years or certain times of year where more calves tend to be born, the pattern must vary geographically depending on local climatic specificities and it is virtually never the case that all potential breeders actually breed at the same time (unless the population is tiny). Second, females must range widely in order to meet their nutritional requirements; the poorer the environment in terms of food and water, the further they will have to go. Third, from a population perspective, the younger females are the most reproductively valuable but are likely to be a relatively small fraction of the breeding population. For all the foregoing reasons, reproductively available females are a going to be a scarce commodity even in a normal, healthy population. This limits the range of options available to bulls.

For a male attempting to maximize his own reproduction, the desynchrony and rarity of females in oestrus means he cannot efficiently guard a group as lions, horses, and mountain gorillas do. Moreover, given the fission–fusion tendencies of females, this would also not be very convenient even in situations where females are very gregarious, as with savanna elephants. Nor can he simply defend a resource or territory as some birds, ungulates, and primates do. This is because the chances of a reproductively available female passing through may be low while the resources (e.g. water sources) themselves may be distributed in such a manner (being too large, too small, or too scattered) as to make it impossible for him to guard them to the exclusion of other males. In other words, neither female groups nor the resources they require are economically defendable.

He can, however, try to maintain exclusive access to an individual female while she is in oestrus, which could be thought of as good old-fashioned female-defense polygyny, albeit a short-lived one, more like the consortships entered into by some primates. He must search for a female in oestrus, fend off rivals, and then guard her zealously. But in doing so, males employ a strategy originally omitted by Emlen and Oring: roving (a tactic also employed by white-tailed deer and cetaceans that have similar life histories) (Foley et al., 2018; Orbach, 2019). This is the favored tactic of mature bulls that are in good competitive condition, and is a primary reason for why bulls generally tend to traverse a vast range. But there is a catch.

A male cannot be in mate-searching mode all the time – mating and fighting is a costly business after all, and he has his own nutritional requirements to meet so that he can outcompete other males in the first place. Males therefore face a tradeoff between the time they invest in foraging versus the time they spend searching and contesting for mates (Foley et al., 2018; Pelletier et al., 2009). North American bison also face a similar tradeoff, as do many other ungulates, such as sheep and deer species. Because females breed synchronously in these species, the males are able to limit their reproductive efforts (and challenges to rivals) to those particular times of year and focus on feeding the rest of the time. We are probably all familiar with the grand battles between bull elk and the battering of bighorn rams. Many male ungulates decrease foraging effort during their reproductive period, known as the rut, which they appear to try and make up for during the rest of the year (Miquelle, 1990; Pelletier et al., 2009). Female elephants, on the other hand, can come into oestrus throughout the year, so it's not possible to partition reproduction and feeding into specific seasons in quite the same manner. Elephants have instead evolved a slightly different solution.

Bulls may become physically capable of reproducing as early as eight years of age (Lee et al., 2011; Sukumar, 2003) but they are not likely to be able to compete against larger males (nor be very appealing to females). They must therefore bide their time and focus on bulking up. Once a male becomes sexually active, possibly from the age of 15 onward, he will begin to differentiate between two distinct motivational states (Lee et al., 2011; Rasmussen et al., 2007). Because males appear to be seeking very different things when in these two conditions, it's important to keep the distinction in

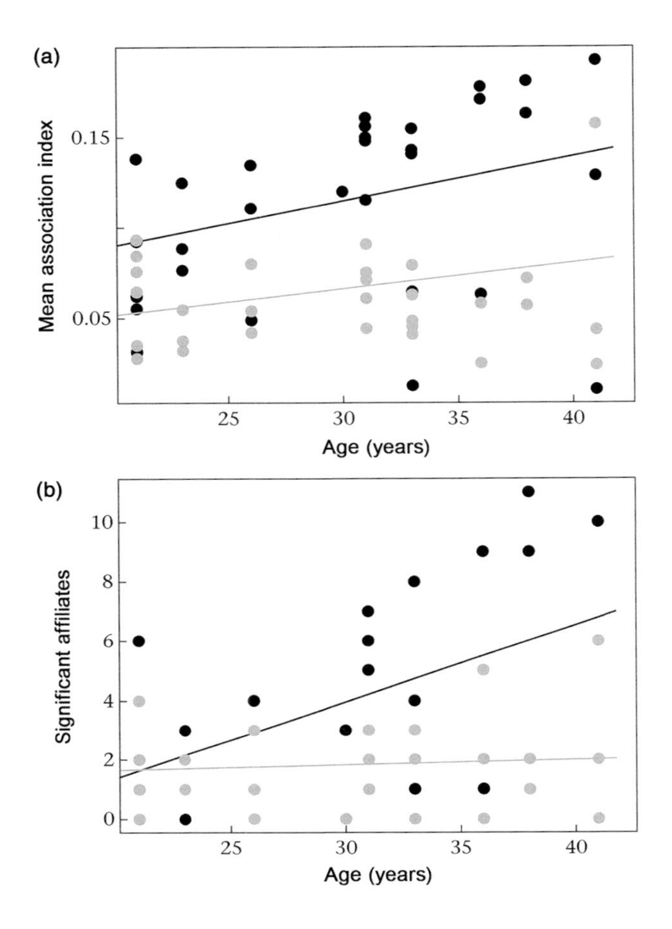

Figure 4.4 Bulls are more social when sexually inactive. Black circles represent males in the "sexually inactive" motivational state, while gray circles represent those that are in the "sexually active" state (with some overlap in the identities of individuals in these two sets). Males that are not sexually active have higher rates of association with other males (a) and more statistically significant contacts (b). These two trends increase with age (from Goldenberg et al., 2014, reproduced with permission from Elsevier Science & Technology Journals).

mind when thinking about behavior (Goldenberg et al., 2014). One state is focused on foraging and social relationships. But the other emphasizes mate search, male–male competition, and reproduction. In savanna elephants, bulls that are in "foraging" mode score higher, both in terms of the number of associates they have and the strength of those associations, on average relative to those are in the "reproductive" mode, a difference that grows more pronounced with age (Figure 4.4). Although similar studies in the other species are lacking, they are likely the same because of another attribute shared among elephants.

If a bull is in good physical condition, somewhere in his teens or 20s, he will undergo a physiological transformation termed *musth* (Figure 4.5). The term is taken from a condition described early on in Asian elephants by the *Rig Veda* (1500–1000 BCE,

Figure 4.5 Musth males in consortships. Secretions from the temporal glands on the sides of the head as well as urine dribbling are characteristic of males in musth. Among African elephants, both sexes can exhibit temporal streaming when excited (a), unlike Asian elephants where only males show distinct secretions with a reddish appearance (b). The phenomenon of musth in the former therefore went unrecognized for some time. Here, males are seen guarding females with whom they have mated (photos by the author).

present-day India), later also observed in African savanna elephants (Hall-Martin, 1987; Poole, 1987; Poole & Moss, 1981). Most recently, analyses of hormone traces in the tusks of woolly mammoths shows that they too underwent musth, with similar characteristics to those found in living elephants (Cherney et al., 2023). Thus, it's an ancestral feature of elephants. Although elephants can and do take the chance to mate whenever the opportunity presents itself, musth is the period in which a male visibly advertises to members of the same and opposite sex that he is sexually active and willing to be combative. In all elephants, musth is accompanied by intense chemical signaling, both through temporal secretions and urine dribbling (Figure 4.5). Though little is known about how this condition manifests in African forest elephants, it is safe to think that there are likely strong similarities based on their close relatedness to savanna elephants (see also Fishlock, 2010). Reminiscent of the rut, it is accompanied by heightened testosterone levels and aggression toward other reproductively mature bulls. Elephants also reduce their food intake and may stop feeding altogether in later stages of musth (Chave et al., 2019; LaDue et al., 2022). It doesn't appear that the bulls are particularly stressed by these physiological changes themselves though, insofar as their glucocorticoid hormone levels are concerned (Ganswindt et al., 2005). And unlike the other ungulates, individual males come into musth at different times of year, with musth periods lengthening with age. This makes the phenomenon of "musth" virtually unique in the animal kingdom.

The function of musth has usually been explained in game-theoretic terms, as a mechanism for honestly signaling condition to one another and females (Hollister-Smith et al., 2008; Poole, 1989a; Schulte & Rasmussen, 1999; Sukumar, 2003; Wyse et al., 2017). The signaling of condition to potential competitors can influence the latter's behavior and potentially reduce serious conflicts. In the absence of more mature males, younger individuals can exhibit musth as early as 17 years of age among savanna elephants (Poole et al., 2011; Slotow et al., 2000), while a captive Asian elephant was reported to have sired offspring by the age of 15 (Flower, 1943). For savanna elephants under natural conditions, musth onset more typically occurs around the age of 25 and can last from days to weeks, lengthening with age. At Amboseli, the median duration of musth for individuals aged 16–25 was just 2 days, while for ages 26–35 the median was 3 days, for ages 36–40 it was 52 days, for ages 41–45 it was 69 days, for ages 46–50 it was 81 days, and for ages 51–60 it was 54 days (note that age classes were binned at irregular intervals; Poole et al., 2011). Data are more scarce for wild Asian elephants, but our own limited observations at Udawalawe show that it similarly lengthens with age but can be highly individually variable (Madsen et al., 2022). In this population, the median duration was 24.5 days for ages 21–30, 32.5 days for ages 31–40, and 45 days for those > 40. Males increase in both size and dominance as they age. Among savanna elephants, the expression of musth by older bulls, which is accompanied by both vocal and chemical signaling, can suppress the musth of younger bulls (Poole, 1999; Slotow et al., 2000). The musth expression of young bulls is therefore usually shorter and less stable. If two musth bulls encounter one another and one is larger than the other, there will be no contest. But if they are evenly matched, a physical battle may ensue that lasts hours, or even days (Chelliah & Sukumar, 2013; de Silva et al., 2014).

Although bulls may congregate around oestrous females (reminiscent of "male-dominance polygyny" in the Emlen–Oring framework), the extent to which males form active coalitions against one another seems to vary not so much by species as by population. Asian elephants form coalitions whereby pairs of bulls may tag-team against a third (personal observations). Because this is a rare behavior, it has been challenging to study systematically and therefore the dynamics of such coalitions – for instance, whether they occur repeatedly among the same individuals – are not well known. In West Africa and Botswana, savanna elephant bulls have been observed forming larger aggregations than in East Africa, which may provide more opportunity for social-positioning tactics.

The musth signals may also serve to elicit the interest of females (Perrin & Rasmussen, 1994). When a bull manages to successfully mate with a female in oestrus, he hangs around her in a consortship, guarding against other prospective mating attempts (Figure 4.5). Bulls don't have it entirely up to them though. Far from passively waiting for males to approach them, females can actively solicit the attentions of suitors as well as try to play coy with those that don't strike their fancy (Poole, 1989b). The unusual architecture of the female reproductive tract in fact makes it difficult for a male to mate successfully without her cooperation. A study of the same Kaziranga population suggests that females exert some choice in their partnerships (Chelliah & Sukumar, 2015), though it is not known which males end up siring offspring. Avoiding the attention of amorous bulls can take a lot of energy, and at times even result in injury to the female, so it is still not clear how much control she ultimately has.

While the preceding explanations may indeed explain the *signaling* tactics of musth, it does not explain why bulls concurrently *reduce feeding*. The complete cessation of feeding during rut by prime-aged individuals is common in many ungulates (e.g. moose: Miquelle, 1990; reindeer: Barboza et al., 2004; white-tailed deer: Foley et al., 2018), but why they do so remains unclear, given that rutting is extremely energetically demanding. Proposed explanations are that males are trying to save energy because there is insufficient time for digestion through rumination (the energy-saving hypothesis), that they are resting instead of feeding in order to successfully compete against others (the physical rest hypothesis), or that is it a defense against parasite load owing to reduced immune function (the parasite hypothesis) and that appetites are somehow suppressed by chemical signaling (the appetite suppression hypothesis; see Miquelle, 1990; Mysterud et al., 2008). However, these explanations are largely based on ruminants and none appear very applicable to elephants, which are not ruminants. Why should a species, such as an elephant, which can in principle breed year-round, evolve this rut-like musth state in the first place? Given the rarity of reproductively available females, would it not make sense to try and breed as often as possible? Moreover, given that so-called capital-breeding strategies (i.e. reproduction using only stored energy reserves) are widespread among diverse taxa and both sexes (Soulsbury, 2019; Stephens et al., 2009), more general explanations may apply.

As discussed at the opening of this chapter, males are under heavy pressure to forage effectively *as well as* breed. There has been little attempt to situate these strategies within the context of individuals' space-use decisions in the presence of competitors and ecological risks (Laundré et al., 2001). It may have more to do with the tradeoffs between foraging and reproduction. If the physiological condition of musth evolved primarily as a means to resolve such tradeoffs inherent in the male-male scramble competition for oestrous females, it is possible that the signaling components of the musth state actually emerged *subsequently* in response to the asynchrony of breeding (and aggressive) periods in males, reversing the assumed direction of evolutionary causality. Thus, explanations solely focused on game-theoretic aspects of musth signals as driving its evolutionary origin may be conflating two distinct correlates of musth: the partitioning of feeding and reproduction periods (with suppression of appetite in the latter state) and signaling of condition. Indeed, this may suggest a far more general explanation for the apparent mystery of why some other male ungulates, such as moose, also broadcast chemical signals of their reproductive state concurrent with the suppression of feeding (Miquelle, 1990).

In contrast to females, the reproductive success of bulls continues to increase well into their 40s (Hollister-Smith et al., 2007), since they grow in size and thus become better at directly challenging other males, as well as more appealing to females. Somewhat counterintuitively, this can be explained in terms of the same tradeoff between current reproduction and future reproductive capacity/survival also faced by females. The difference is that for females the cost of reproduction comes from the physical resources she invests in her offspring, while the cost of reproduction for bulls comes from the physical resources invested in (risky) mate search and outright contests. Calf production happens to be more successful when a female is younger, while physical strength increases as a bull gets older. In other words, younger bulls, which are not very competitive against older bulls anyway, might be thought to gain less from effort invested in reproduction than their elders. Consequently, elephants would appear to be good candidates for exemplifying the "terminal investment hypothesis," which states that reproductive effort should increase with age so long as the costs remain constant (Clutton-Brock, 1984). The importance of older males as breeders has been one compelling argument to curb the selective hunting of these individuals, which are likely to be disproportionately targeted by both ivory traffickers and trophy hunters.

A nagging doubt may remain nevertheless – if life is inherently risky, it is even more so for males than females. Shouldn't this place a similar pressure on bulls to try and reproduce earlier? Indeed, it appears that in the Samburu and Buffalo Springs population, there is not as much reproductive skew as one might initially expect. Although older musth bulls sired most calves, nonmusth males sired 20% of calves and it was estimated that as much as 60% of bulls over 20 years old could have sired offspring over a 5-year period (Rasmussen et al., 2007). There is an alternative hypothesis, called the "mating strategy-effort hypothesis" (Yoccoz et al., 2002), characterized by a pattern of reproductive effort that peaks at some intermediate age and then declines. But how might one measure the reproductive effort of bulls independently of the number of offspring they sire or any visible behavioral indicators? One method is to measure weight loss, as has been done in red deer. Among red deer, males lose progressively more weight during the rut as they get older until they reach their prime age, after which point they lose less (Yoccoz et al., 2002). This argues that both effort and success decline after some point. A similar pattern is found among mountain goats (Mainguy & Cote, 2008), pointing to a phenomenon that is probably widespread among ungulates. Males in musth may forage halfheartedly, but because they are primarily on the move they do tend to lose body mass.

There is a problem with testing this idea nevertheless, not least because obtaining the weight of a wild elephant may prove a bit of a logistical challenge. Many savanna elephant populations lost their oldest males in the past thanks to the ivory trade. Needless to say, if the oldest bulls in a population tend to die before achieving their maximum lifespans, and near peak reproductive age, the two hypotheses become indistinguishable. In parts of Asia where bulls in the wild can still survive to toothless old age (personal observations), such bulls may eventually lose their competitive advantage to prime-aged bulls in better condition because they would simply fail to keep up with the nutritional demands of reproduction. In the Amboseli population, where older age classes have managed to persist, individuals sire a progressively greater proportion of offspring as they get older, until around 50 years old, when their share dips again (Hollister-Smith et al., 2007). This observation has even been used to suggest that sexual selection might have contributed to the evolution of longer lifespans in elephants. Irrespective of which hypothesis actually holds, from a practical perspective older individuals are still very important breeders.

Populations that have suffered extreme losses among their older age groups owing to poaching or culling show sociobehavioral disruptions extending beyond their loss as breeders. Perhaps the best-known example of this comes from Pilanesberg, South Africa, where juvenile orphans of both sexes were translocated in the early 1980s following culls at Kruger National Park and went on to mature without the presence of older bulls. Ten years later, in 1992–1997, the young bulls went on a "killing spree," resulting in the death of at least 49 white rhinoceros (Slotow & van Dyk, 2001). The victims were mostly also males, though elephants and rhinoceros generally have no problem with one another. However, the young males showed musth at an unusually early age (18 years), which was exceptionally prolonged (lasting up to five months), suggesting that the unfortunate rhinoceros became targets of inexperienced and unchecked aggression. The "delinquent" behavior ceased when older bulls (aged in their 20s) from the Kruger population were introduced, six of whom exhibited regular musth periods themselves (Slotow et al., 2000). The expression of musth in the older bulls drastically curbed the musth periods of younger bulls (Figure 4.6).

Typically, bigger body size should confer an obvious advantage during fighting contests. It's been suggested that musth might allow smaller males to sometimes outcompete larger, older males if the latter are not themselves in musth (Poole, 1999). Indeed, the intense signaling that accompanies musth is likely a way of avoiding costly conflicts, and individuals in the foraging state may wisely elect to give way to others hopped-up on testosterone. Nevertheless, if association with females is taken to be the main indicator of sexual activity (Figure 4.7), savanna elephants appear to have at least two if not three sexually active states (Ganswindt et al., 2005; Rasmussen, 2005), only one of which involves fully-fledged musth signaling.

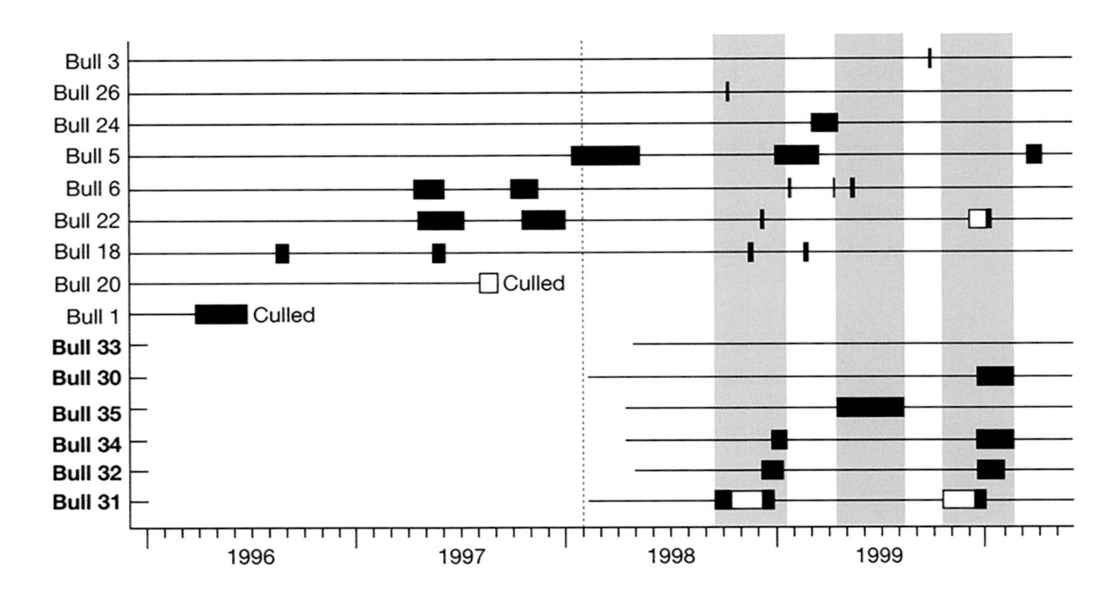

Figure 4.6 Musth suppression. Older bulls translocated from Kruger (boldface) limit the expression of musth in younger bulls of Pilanesberg when they are themselves in musth (gray bars), resulting in decreased aggression among the latter (from Slotow et al., 2000, used with permission of Springer Nature BV).

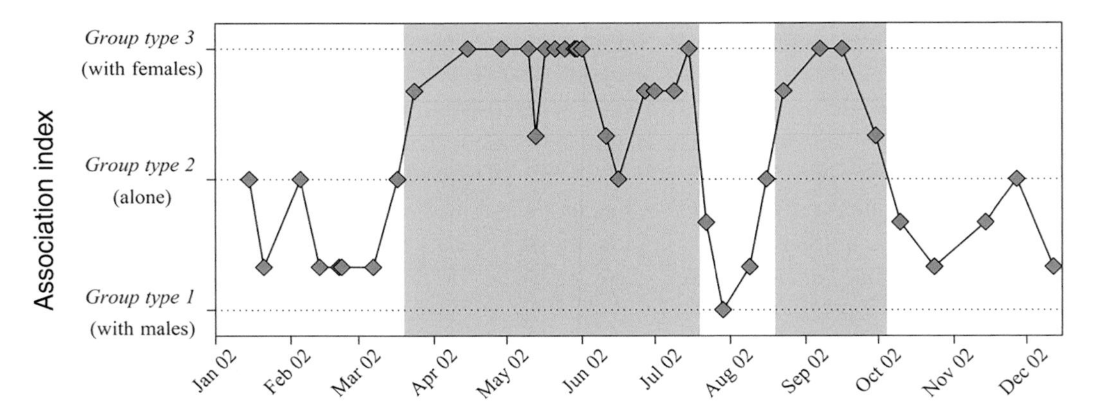

Figure 4.7 The association profile of a single bull, estimated to be 37 years old. Gray bars indicate sexually active periods, which are identified on the basis of his being with females. By contrast, inactive periods are characterized by associations with other males (reproduced from Ganswindt et al., 2005, with permission from Elsevier).

Temporal secretion may not always accompany urine dribbling; therefore, the latter is taken to be the more reliable indicator of musth in savanna elephants. However, sexual activity without signaling (termed "sexually active non-musth," Ganswindt et al., 2005), a sort of stealth mode, may be quite a frequent strategy employed by

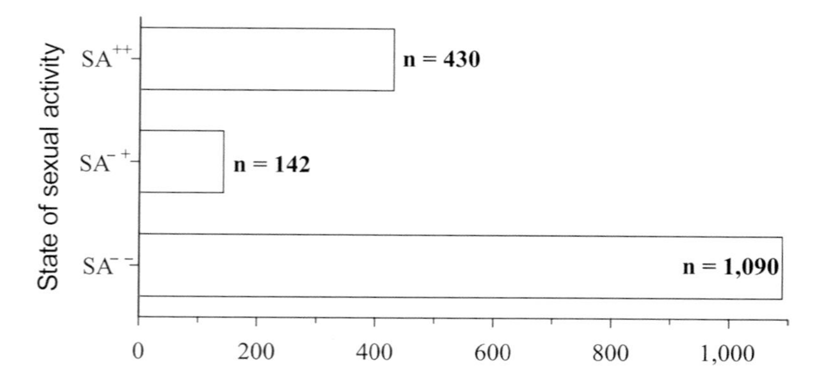

Figure 4.8 The frequencies at which bulls signal sexual activity. SA++ represents the presence of both temporal secretion and urine dribbling, SA−+ represents the presence of urine dribbling but weak temporal signaling, and SA−− represents the absence of any detectable signal (reproduced from Ganswindt et al., 2005, with permission from Elsevier).

savanna elephants (Figure 4.8). This might be because those that aren't likely to succeed in an open contest can employ an alternative tactic – attempting to mate sneakily while dominant bulls are distracted (Figure 4.9). It would behoove such individuals not to draw too much attention to themselves. It is not well understood which elephants employ this strategy and when, or whether it is distinct at all from the normal foraging state. Rather than a discrete sexual state, it may be that a bull will simply take advantage of any opportunity to mate if presented one, even if he is in "foraging" mode. This tactic may provide a partial answer to the apparent puzzle posed earlier of how much younger bulls might be expected to invest in reproduction versus what they seem to. It is possible that young bulls expend as much or more energy in reproduction as older bulls, but simply gain fewer offspring per unit of effort, making paternity a poor measure of that effort. Other measures of reproductive effort may be more illuminating.

The advantage of being in musth may be augmented by having a rare phenotype. From the year 2012 onward, a remarkable phenomenon was observed in Udawalawe: a wild, mature, dwarf male (de Silva et al., 2014; Wijesinha et al., 2013). Appearing in the park during his musth period, he was not afraid to pick fights with other musth bulls. His opponents were young bulls of ordinary stature and similar body mass, but the majority of males in the population are tuskless (discussed further in Section 4.4). In two out of three observed contests with different bulls across three different years, the dwarf male seemed to gain the upper hand, being more often the aggressor while opponents retreated. We infer that this is because of the fighting style of these tuskless bulls, which cannot gore each other. Instead, the strategy resembles sumo-wrestling, which involves ramming opponents and trying to push them back or knock them off their feet (see Figure 4.10). Packing bulk into a compact frame with a low center of gravity meant that the dwarf could charge his opponents in the legs head on, whereas they had to stoop awkwardly to meet this attack. The shorter stature actually helped against other males in this population, so bigger was not always better. However, it

Figure 4.9 The sneaky strategy. A young bull attempts to mate with a female in oestrus ((a), left) while her consort, the dominant bull ((a), right), is otherwise occupied. The younger bull does not show very obvious signs of musth. He is soon driven off (b) and the victor mates again with the same female (c) (photos by the author).

Figure 4.10 Sumo anyone? A dwarf male in musth beats back a male of average stature at Udawalawe National Park (de Silva et al., 2014; Copyright © 2014 de Silva et al).

Figure 4.11 A dwarf female photographed by trail cameras in southern Sri Lanka. Like the male in Figure 4.10, she seems to represent a case of disproportionate dwarfism, her stature being approximately half that of a regular adult female, but with more typical head size. She also exhibits reduced tusks, termed tushes (see also Figure 4.12) (photos by the Udawalawe Elephant Research Project).

can make mounting a female rather tricky. While the dwarf was seen in the company of female groups, mating was unfortunately not observed. A dwarf female was also observed in later years in the same population (Figure 4.11), suggesting the trait was not an isolated mutation but actively being passed on by one or other sex.

4.4 The Mystery of Tusks

4.4.1 Natural, Sexual, Frequency-Dependent, or Social Selection?

Tusks have appeared in one form or another in many proboscidean lineages, very likely including the common ancestor of all living elephants (Shoshani & Tassy, 1996). Among the African species, members of both sexes usually possess tusks (as in Figure 4.3, with rare exceptions), making it difficult to evaluate what benefits tusks may confer as there is little variation in the population. However, among Asian elephants, all females and a fraction of males do not have tusks (Figures 4.1, 4.3 and 4.5). Despite appearances, the incisors of these "tuskless" individuals are not completely absent; instead, they have slender teeth of reduced size known as "tushes" (though some are long enough to visibly protrude beyond the upper lip, Figure 4.12a). This is known as a dimorphism ("tusked" and "tuskless" being two different physical appearances or morphs within the same species). Because such dimorphisms are usually associated with sexually selected traits, this might seem to support the idea that tusks are under such a selection pressure. Alternatively, tusks may have been more useful as a defense against large predators historically found in the African environments, absent in the Asian. We do not as yet know how this dimorphism emerged in the Asian clade, or indeed whether the presence or absence of tusks in females is the ancestral feature.

The living elephants use their tusks for foraging – knocking over or debarking trees, even digging earth in some instances. It's therefore natural to think that tusks confer some basic survival advantage that has been selected for. But tushes are just as effective at foraging tasks such as debarking trees, and the existence of tuskless individuals on both continents (though the mechanisms behind their disappearance may be distinct, as discussed later in this chapter), with their varied ecologies, suggests that tusks are not strictly necessary for foraging either. However, tusks are also conspicuously used in aggression against conspecifics. From beetles to bovids, where there is stiff reproductive competition among members of the same sex, individuals of that sex tend to sport weaponry (even females, though this is rarer; bovids: Stankowich & Caro, 2009; beetles: Watson & Simmons, 2010). Any advantage conferred by tusks in the reproductive context, either through male–male competition or female choice, falls under the classical realm of sexual selection (Berglund et al., 1996).

The prevalence of tusked/tuskless dimorphism among male Asian elephants allows one to test whether musth, tusks, or other physical attributes such as body size actually confer the greater advantage in male–male competition. A study in Kaziranga National Park in northern India revealed that the state of musth trumped tusks as well as body size, conferring the greatest advantage in winning contests (Chelliah & Sukumar, 2013). Tusks not only failed on occasion to help to their owners, but could actually be a liability. Some of the more savvy tuskless bulls had learned to grab hold of the tusks and use their leverage to better wrestle against their opponents (Chelliah & Sukumar, 2013). The study went further, suggesting that while tusks are an ancient feature of proboscideans, perhaps musth is a more recently evolved physiological innovation in response to the competition among bulls. In populations where tuskers

Figure 4.12 (a) Tushes. Adult female "Wipuli" at shows a prominent miniature tusk known as a tush. Tushes can be used to debark trees as well as apply pressure on other individuals during dominance displays. Among Asian elephants that appear tuskless, including both males and females, the tush stays hidden at the lip line and is only visible when the trunk is raised. Length is often visibly asymmetric, the preferred side having been worn down. (b) Single-tusked male in Sri Lanka, where the second tusk appears completely absent rather than broken or worn down (photos: Udawalawe Elephant Research Project).

are more common such as Kaziranga, tuskless males may have gained enough experience by sparring together at younger ages to know effective tactics against them. But in populations where tuskers are extremely rare, such as in Sri Lanka, where they today comprise perhaps as few as 3–5% of all males, tusks could be analogous to the lefty advantage in sports. Confronted with these sharp objects only rarely, the majority of bulls may not know what to do, giving tuskers uncontested dominance. Fights between mature tuskers and tuskless bulls are seldom observed, though there have been news reports of adult males having been gored fatally during such contests. Thus, there may be some degree of frequency-dependent intrasexual selection in favor of tuskers when they are at low numbers, and against them when they are abundant. This hypothesis could also be tested with paternity data, but this has yet to be done.

Big, beautiful tusks may also be an attractive trait in a prospective mate as an indicator of good health and vigor. There is certainly a lot of variation in the appearance of tusks; while the tusks of savanna elephants tend to point forward, the tusks of forest elephants usually point down. Tusks can come in various shapes and sizes among Asian elephants, some being long, slender, and downward pointing, akin to those of African forest elephants, whereas others are broad and curved like those of savanna elephants. Yet while tusks seem to be formidable weapons, they are heavy and can even impede foraging by constraining movements of the trunk in some configurations, making them a potential handicap (Zahavi, 1975). The line between the benefits they provide and the costs they impose might therefore not be so clear. The Hamilton–Zuk hypothesis, originally proposed to explain the evolution of conspicuously bright coloration in (usually) male birds, posits that ornamental traits can signal resistance to parasites (Hamilton & Zuk, 1982). This has spurred substantial research into the role of parasites in mediating the interaction between sexual selection and species' ecology, although most of the evidence remains correlational rather than mechanistic (Balenger & Zuk, 2014). In Asian elephants, it has been found that males with longer tusks indeed had lower intestinal parasite loads (Watve & Sukumar, 1997). However, it has not been established whether females show firm preference for longer tusks (Chelliah & Sukumar, 2013, 2015), or whether they are able to exert much choice, given that musth condition seems to be more important in determining the outcome of male–male contests. If musth is more recently evolved than tusks, as I argued earlier, it may be that any evolutionarily antecedent preference for tusks might be obscured.

Weaponry also helps in intrasexual competition for access to resources, which raises another possibility that is perhaps more relevant for females than access to mates. Both tusks and tushes can be used in aggression and even mild assertions of dominance as individuals compete for better access to resources. This type of competition among individuals outside the narrowly circumscribed reproductive context falls within the domain of social rather than sexual selection (Tobias et al., 2012). One wonders whether the difference between the Asian and African genera might reflect more intense resource competition historically among females in African systems relative to those in Asian. The weaker dominance relationships among Asian elephants when compared to African savanna elephants (at least in the studied contexts) support such an explanation (de Silva et al., 2017). Unfortunately, there are not enough elephant species alive to test between these alternatives in a phylogenetically controlled manner (keeping in mind that female African forest elephants, like savanna elephants, also bear tusks). However, among bovids, which are a much more speciose clade that includes cattle and antelope, the presence of weaponry in females has been significantly associated with large, conspicuous species inhabiting open environments (Stankowich & Caro, 2009; Tobias et al., 2012).

It is commonly suggested that the ratio of tusked to tuskless males in Asian populations may be driven by the historic selective harvest of tusk-bearing males. This requires no new mutation, it only creates conditions favoring the prevalence of tushes in both sexes. True tusklessness – that is, the complete absence of tusks or tushes in either sex – seems to be a still more recent trait, rapidly spreading in populations where tusked individuals have been selectively hunted or captured (Kurt et al., 1995; Raubenheimer & Miniggio, 2016). However, the genetics of this phenotype in African elephants seems different than it is in Asia. Savanna elephants in Mozambique illustrate this possibility. Decades of civil war drove the loss of more than 90% of large herbivores, with elephants especially targeted for their ivory. The emergence of an X-chromosome-linked genetic mutation has resulted in the appearance of single-tusked and tuskless females, but seems to be lethal for males (Campbell-Staton et al., 2021). By contrast, some small populations, such as in Borneo, even seem to harbor bulls that are single-tusked, while entirely lacking tushes on the complementary side (N. Othman & S. de Silva, personal observations during collaring operation with animal under sedation). One-tusked males can also be seen in Sri Lanka (Figure 4.12b). These are likely distinct, independent, novel mutations. Ultimately, we may never know which selection pressures drive and continue to shape the evolution of tusks. They are multipurpose tools, which are likely shaped by multiple forces of selection working in concert. Most recently, humans have once again emerged one of the most dangerous selective agents.

4.4.2 Ivory, Demography, and the Future of Elephant Populations

In both African and Asian populations, individuals bearing tusks have been targeted by human hunters over many generations, resulting in the selective removal of individuals with the largest tusks. Typically, these were mature males, but, as noted, during the peak of poaching crises on the African continent they included many female groups and even calves (Goldenberg et al., 2016). The difference between Asian and African elephants in the prevalence of tusklessness has been consequential for these species both biologically and politically. On the one hand, in Asia, where offtake for domestic use has been going on over thousands of years and elephants have already lost vast portions of their historical range, elephant populations had already declined to critically low numbers by the turn of the millennium. Tusklessness (either with the presence of tushes or complete absence of any incisors), crucially, has allowed populations to persist in the face of humanity's appetite for ivory even if it has resulted in female-biased sex ratios at some locales. If tusks do indicate better resistance to parasites, there may have been other hidden costs. But if not for the existing dimorphism, elephants would already be extinct in most of Asia.

African elephants on the other hand, which had no such historical pressure and were more abundant at the start of the colonial era, were decimated by the subsequent hunting pressure specifically for the international ivory trade. Unlike the slow and long-term human pressure acting in Asia against tuskers, modern hunting pressure has been so rapid and devastating that it is difficult to imagine that the attribute of tusklessness could spread quickly enough to save African populations, especially given their lethality in males. The alarming rates of population decline documented as a result of the ivory trade have drawn much of the resources and attention of the conservation community on both continents (Maisels et al., 2013). This has yielded crucial conservation victories, most notably the landmark decisions to close ivory markets within the USA and China. But the story doesn't end there, and unfortunately narratives surrounding the ivory trade have largely eclipsed other threats to elephants, such as habitat loss and fragmentation, which are no less relevant to their long-term survival.

Animals are also removed from populations by lethal or nonlethal means – either by accident or intent – as a response to human–elephant conflict. This is a domain filled with unanswered questions, desperate for research. First, the terminology of "conflict" typically subsumes at least two major categories of disturbance to people – economic losses versus human injury or fatality. Although it is easy to assume that individual elephants responsible for causing hardship in the first instance are also those responsible for incidents involving the second, this need not be the case. Human responses, which are quite often punitive, can extend to target animals besides those directly involved, including females and calves. What effect do actions such as drives, translocations, or capture have on the individuals directly affected, if they survive? Do they drive changes in levels of aggression or habituation to human disturbance? How do death and loss through conflict affect other individuals in the population that are left behind? And, perhaps most crucially, do such removals actually solve the problems they are intended to solve? Available evidence suggests these measures are counterproductive (Fernando et al., 2012). Answering these questions requires researchers to collaborate with wildlife managers more closely than is typical, and to overcome the substantive social and political challenges these answers may present (Pooley et al., 2021).

As discussed earlier, elephants must reproduce at near-optimal reproductive rates and have relatively high calf survival in order to maintain stable populations when subjected to even modest levels of adult mortality. This is jeopardized by the ongoing habitat loss and fragmentation evidenced on both continents. Habitat degradation places more and more elephant populations in direct competition with people, not only driving up visible mortality through accident and conflict, but also eroding reproductive output. Nutritionally stressed mothers will have fewer calves, and of those that are produced, fewer may mature to adulthood. The confluence of these insults may cause elephant populations to cross demographic tipping points, where physiological constraints prevent populations from compensating for the mortality rates they experience, a leading to a slow-motion train wreck of population collapse (de Silva & Leimgruber, 2019). Yet these effects may hardly be noticed, because the reproductive health of elephant populations in the wild remains largely unstudied and undocumented, especially in Asia and in the forests of Africa. This is concerning because so many of these populations are already small and fragmented. The impact of human disturbance on demographic processes is therefore a largely hidden, poorly understood threat to elephants and numerous other species. Wildlife managers, policymakers, and conservation funders would do well to act on the premise that the greatest danger to future elephant populations will not be the felling of magnificent tusk-bearing adults; it will be the calves that are never born.

5 Elephants in Ecosystems

5.1 Elephants as Engineers

Elephants may be thought of as very large cogs in the machinery of nature, their multitude of interactions with other species as well as the abiotic environment meriting their title as "ecosystem engineers" (Campos-Arceiz, 2009; Haynes, 2012). It is worth stressing that a species does not have to be large to be classified as such – beavers are among the most industrious movers and shakers of the animal kingdom, with profound capacity for engineering habitats for themselves and countless other species. But large body size combined with a huge appetite, ability to move long distances, and the physical strength to mechanically alter landscapes make elephants very influential nodes in ecological webs. In this section, I explore what features of the environment matter to elephants or mattered to their extinct brethren and how elephants in turn affect (or were historically affected by) these environments.

5.1.1 Interactions with Plants

To sympathize with an elephant's primary motivation in life, one really has to understand its gut. Elephants have evolved their large body size with a digestive system that is relatively inefficient at breaking down cellulose. Like rabbits, horses, and rhinos, they are hindgut fermenters, meaning that food passes through a single-chambered stomach before being fermented by microbes in the large intestine, or cecum (Dumonceaux, 2006). As a consequence, elephants are able to extract nutrition from less than 50% of what they consume. In contrast, ruminants such as cattle subject their forage to a lengthy fermentation process within a multichambered stomach and thus are known as foregut fermenters. This fermentation process, together with specialized microbes, allows them to squeeze as much nutrition out of every mouthful of food as possible. Though hindgut fermentation might seem a wasteful contrast at first, ill-suited for a large animal that must anyway consume a lot of food, it potentially lets individuals exploit nutrient-poor environments because forage can be processed much more quickly, relying on quantity rather than quality. This also allows them to deal with the various toxins and secondary compounds plants have evolved as defense against herbivores and be generalists, critical to their ability to survive in such a diverse array of landscapes (Figure 5.1). The versatility of elephants can be breathtaking, from those in arid parts of Namibia that consume around 33 different

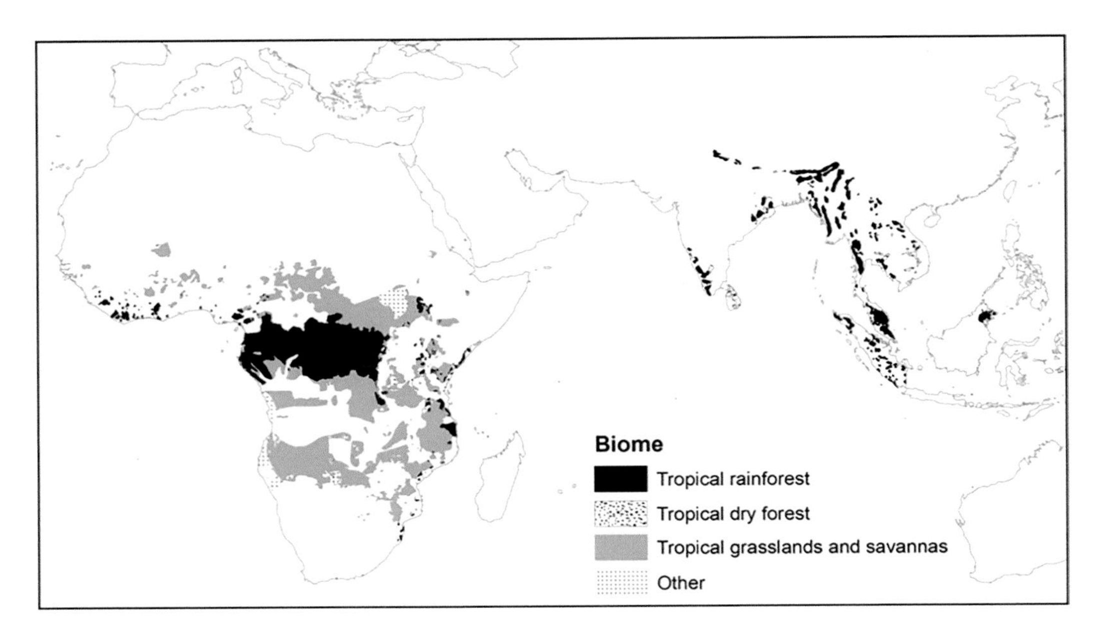

Figure 5.1 Distributions of elephant range according to the IUCN Red List and occupied habitat types as defined by Olson et al. (2004) (reproduced from Campos-Arceiz & Blake, 2011, with permission. Copyright © 2011. Published by Elsevier Masson SAS. All rights reserved).

plants, to those in populations in the Congolese rainforests that consume upwards of 500 species, while those in Asia can typically have access to at least 100 or so species (Campos-Arceiz & Blake, 2011, and references therein). Incidentally, elephants and rhinoceros have historically been able to attain much larger body sizes than their ruminating counterparts, as they have been able to achieve much faster rates of increase in body mass over evolutionary time, so it doesn't appear to have presented a hindrance to physical stature in the past (Evans et al., 2012).[1]

Seed dispersal is perhaps the most obvious beneficial effect that elephants can have on their ecosystems. As there have been numerous treatments of elephant diets in earlier literature (Campos-Arceiz & Blake, 2011; Sukumar, 1989, 2003), I will not dwell here on food preferences except with reference to the ecological role that elephants play and how different populations compare with one another. Thanks to their size and dietary diversity, elephants can move the seeds of many species quite some distance. The ability to swallow fruits whole without first crushing them also increases the chances that seeds will remain undamaged even if they are large (although some

[1] We must always be careful not to confuse correlation with causation. Equids have never grown to gargantuan sizes, being dwarfed by now-extinct bovids such as aurochs and cervids such as the Irish elk. Therefore, the evolutionary increases in size evidenced by some proboscideans, indricotheres (rhinos), and rodents might also have other possible explanations, for instance the suitability of their existing limb or bone structure and lifestyle to accommodate the added body weight, or sexual selection for large body size, and/or weaponry over speed and agility. Indeed, domesticated horses have been bred to be significantly larger than their wild ancestors, so it is curious to imagine what may have inhibited such body size increases in natural populations.

Figure 5.2 Elephants and trees. (a) Some tree species such as *Bauhinia racemosa* are able to withstand fire and foraging by sprouting many trunks if one is damaged. Their bean-like seed pods are particularly enjoyed by elephants (Photos: Udawalawe Elephant Research Project). (b) Seeds in elephant dung sprout in fertile conditions and are dispersed widely (photo by Ahimsa Campos-Arceiz used with permission). (c)–(d) Trees and vines consisting of different species with very similar compact growth habits are common in savanna-like environments in Africa and Asia, ideal for browsing as well as midday shade (photos by the author (c) and Udawalawe Elephant Research Project (d)).

species may have specifically evolved to require chewing and crushing, as discussed later). Elephants then deposit the seeds within a handy package of fertilizer in the form of dung (Figure 5.2). This makes them one of the most influential seed dispersers in any ecosystem. African forest elephants appear to have the greatest access to fruit, whereas savanna elephants and Asian elephants appear to consume fewer fruit species (Campos-Arceiz & Blake, 2011). This may be because fruit species accessible to elephants (i.e. those that fall to the ground upon ripening) may be less consistently available. Asian rainforests contain many dipterocarps, which are wind- rather than animal-dispersed, and many others bear fruit in synchrony during relatively rare "mast fruiting" events. As shown in Figure 5.3, the role of African forest elephants as seed dispersers is consequently more pronounced than that of the other two species (Campos-Arceiz & Blake, 2011). Savanna elephants are expected to disperse seeds over the greatest distances based on their movements, which can be up to 65 km between feeding and defecation (Bunney et al., 2017). Of the distances actually documented, African forest elephants hold the record at 57 km, whereas Asian elephants

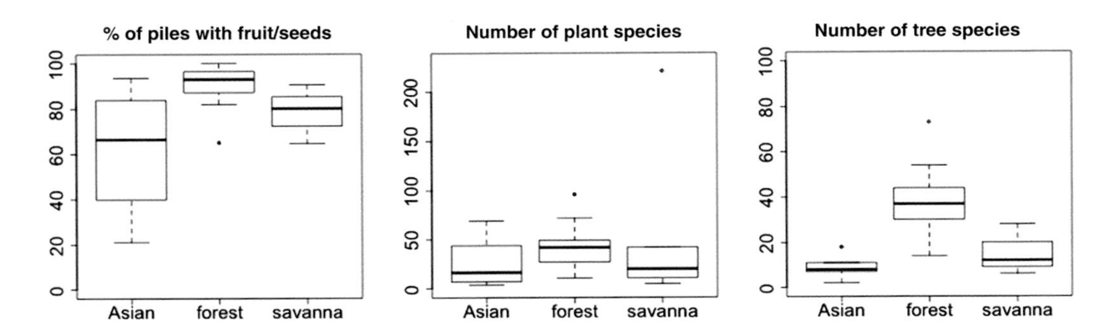

Figure 5.3 Seed content of elephant dung, based on N = 5 studies for Asian elephants, N=12 studies for African forest elephants, and N = 6 studies for African forest elephants (reproduced from Campos-Arceiz & Blake, 2011, with permission. Copyright © 2011. Published by Elsevier Masson SAS. All rights reserved).

may transport seeds anywhere from 6 to 20 km (Campos-Arceiz et al., 2008, Campos-Arceiz personal communication).

The impact of elephants on their ecosystems under crowded conditions has been a matter of particular concern and extensive debate in some of the drier ecosystems of Africa, where elephant browsing pressure on woody tree species can be intense, though often localized. The foraging behavior of large herbivores might be expected to have cascading effects not only on the plants themselves but also on other animal species. For instance, murid rodents (mice) prefer tall grass habitats because it provides greater cover and lowers the risk of predation. Grazing down grasses and maintaining more open habitat can reduce the suitability of such environments for small rodents (Hagenah et al., 2009). However, a meta-analysis of the impact of African savanna elephants on the landscape that reviewed 51 studies spanning 68 years across 5 countries in eastern and southern Africa found that while elephants do have an impact on the abundance and structure of trees, they did not overall appear to have negative effects on other co-occurring wildlife (Guldemond et al., 2017). Within the relatively dry ecosystems covered by the sampled literature, primary productivity doesn't explain much of the variance in elephants' impact on vegetation. Other important factors that structure these ecosystems include fire regimes (Holdo et al., 2009; MacFadyen et al., 2019). Some tree species even show adaptations that make them especially resilient to both browsers and fire (Figure 5.2). Moreover, in the much wetter elephant range in central Africa and most of Asia, the original distribution of natural forest cover clearly demonstrates that trees along the equatorial tropics have historically been better able to withstand and recover from herbivore foraging pressure.

Elephants are also usually not the only herbivores around, and studies in various African savanna systems have found that herbivores of different body sizes can have varied effects even on plant communities (Pringle et al., 2014). Often, herbivores even of similar body size have feeding strategies that differ not only in the types of plants consumed, but also the life history stage affected (owing to differences in plant size and available defenses) or even the specific parts that are preferred, enabling them to avoid direct competition with one another (Bakker et al., 2015). Elephants are capable of mowing their way through formidable plant defenses, such as thorns. They can strip bark off trees or topple them altogether in order to get at the sweet cambium or soft inner cores. Where they co-occur with other species that have similar diets, they may make more use of this ability to process coarse forage even if other vegetation is available. In Nepal, for instance, which contains some of the few remaining landscapes where Asian elephants still share habitat with the greater one-horned rhinoceros, nearly two-thirds of the elephant diet consisted of browse as opposed to grasses, whereas rhinos showed almost exactly the opposite pattern (Steinheim et al., 2005).

This would be an opportune time to remind ourselves that until the end of the Pleistocene, proboscideans were simply one among many megafaunal clades on land, and that they constituted key players in a host of ecological interactions, illustrated by Figure 5.4 (Galetti et al., 2018; Malhi et al., 2016). Reviewing the palaeoecological record as well as contemporary experiments in which herbivores of various body sizes were excluded from habitat patches in various environments, Bakker et al. (2015) cite evidence that megaherbivores (which authors define as those with an adult body mass of at least 45 kg) past and present are likely to have very dramatic impacts on plant communities, notably woody species (Table 5.1). Bakker and colleagues propose that many environments that hosted these large consumers may have been more open than they are today. One such example is the high-latitude grassland ecosystem of the Siberian steppe, which was evidently maintained by the activities of woolly mammoths. Being predominantly grazers, mammoths not only fed on but also facilitated the dispersal of grasses. Following their demise, these plains gave way to more wooded tundra (Bakker et al., 2015).

Meanwhile in North America, analyses of the teeth and gut contents of mastodons suggest that they likewise followed a mixed feeding strategy that included both grass and browse, possibly favoring cool habitats near water (Gobetz & Bozarth, 2001; Newsom & Mihlbachler, 2006; Teale & Miller, 2012). In the lowland forests of central America, relatives of the elephants known as gomphotheres dispersed large seeds from fruiting tree species (Janzen & Martin, 1982). Neotropical fruits dispersed by megafauna (in this case, species weighing over 1,000 kg) have a specific mix of characters which are hypothesized to be adaptations that specifically court the attention of these very large creatures. Such combinations of traits are generally known as "syndromes." For a more familiar example, flowers of plants that have evolved to attract hummingbirds as pollinators are generally red in color and tubular in shape, whereas those catering to bees may be white, or yellow with ultraviolet markings and flatter in shape. "Elephant fruits" are described as being either 4–10 cm in diameter with up to five large seeds or even larger in size with numerous small seeds (Guimarães et al., 2008). The plants themselves show features such as the ability to resprout vigorously when trampled or felled, which may have been originally evolved in response to browsing pressure.

It is difficult to say whether megafaunal extinctions have already directly caused the extinction of plants in the past, because the distributions of vanished plant species are difficult to discern, even with modern pollen analyses. Moreover, it can be

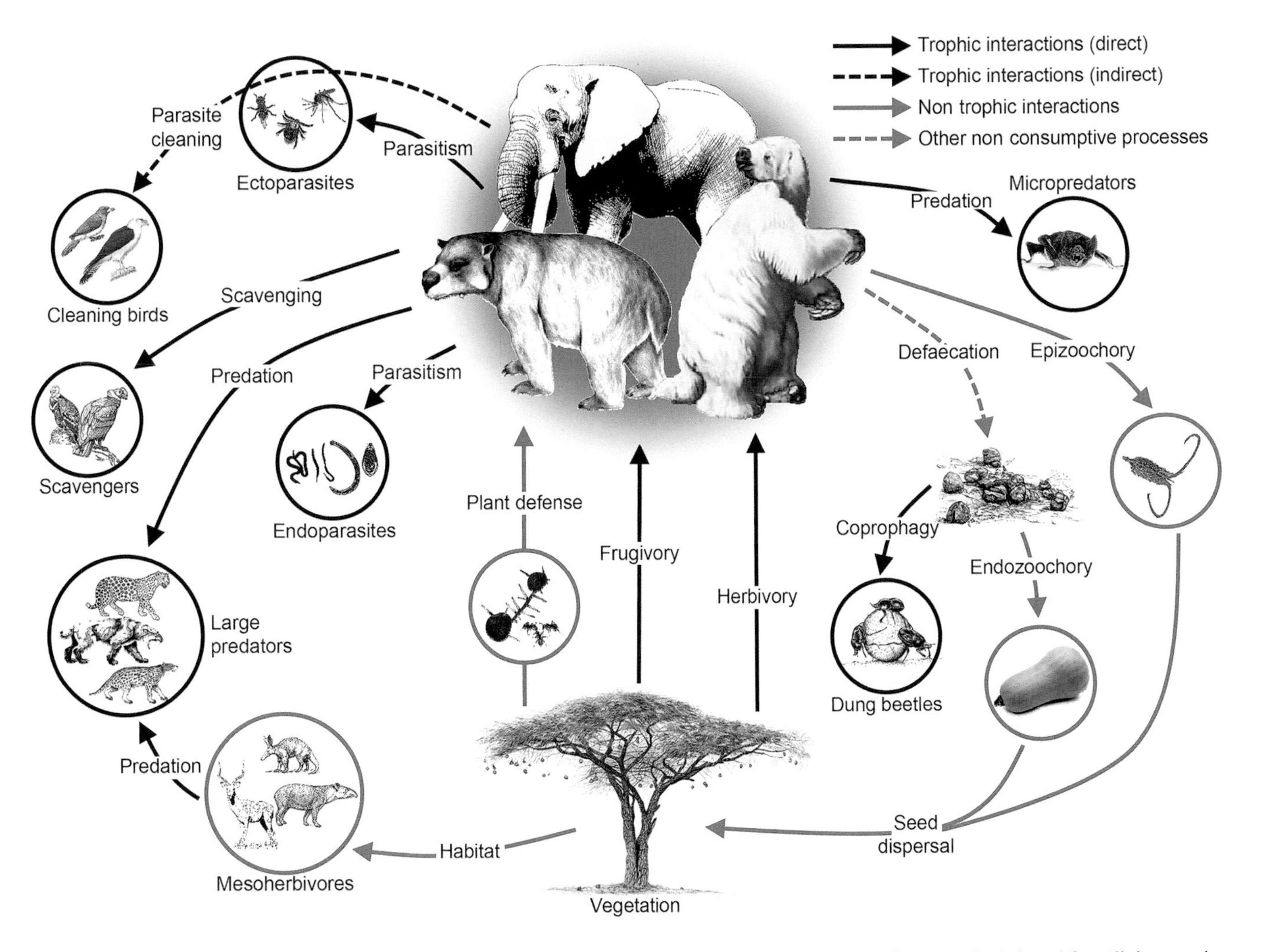

Figure 5.4 Ecological interactions involving extant and extinct terrestrial megafauna. While some interactions may be inferred from living species that exhibit so-called anachronisms that hint at past interactions, many coextinctions, such as those of parasites, may go unnoticed owing to the lack of any remaining trace or fossil record. Other species, such as pumpkins and gourds may have persisted thanks to cultivation by humans, but now may turn into an ecological trap for living elephants (reproduced from Galetti et al., 2018, with permission from Springer Nature BV).

Table 5.1 Herbivore impacts on woody plants

Process	Contemporary pattern	Palaeoecological record
Large herbivores reduce the abundance of woody plants.	Higher woody plant cover in exclosures and after removal of large herbivores.	Landscapes of previous interglacials seem to have been more open than after Pleistocene extinctions in the early Holocene. Moas may have maintained mosaics of open canopied woodland and scrub.
Large herbivores induce shifts in woody species composition.	Under intense browsing, unpalatable and thorny species thrive and palatable species are suppressed. Browsing may also promote browsing-tolerant species. Under intense herbivory, light-demanding trees and shrubs are promoted.	Increase in palatable and shade tolerant hardwoods immediately after the Pleistocene extinction in North America. Increase in unpalatable trees during historically high herbivore densities in European forest.
Large herbivore impact is mediated by soil fertility.	More thorny shrub species in fertile habitats may indicate higher browsing pressure. Higher elephant impact on treefall at fertile soils.	Vegetation openness was greater in fertile lowland areas than in less fertile upland areas.
Herbivores modify vegetation responses of woody plants to climate and soils.	In tundras, herbivores can inhibit shrub encroachment with climate warming, but this effect is site-dependent. In savannas, woody species cover frequently does not reach its abiotic potential owing to fire and herbivory.	Mosaic forest tundra in northeastern Siberia during the Last Interglacial, with browsing tolerant trees frequent – likely (at least partly) due to large herbivores. Large herbivore presence maintained the mammoth steppe in northeastern Siberia, which disappeared after Late Pleistocene extinctions. Higher openness of vegetation in last interglacial than expected based on climate and soil may be mediated by large herbivores.
Herbivores reduce fuel load for fires	Herbivores reduce herbaceous biomass and fire frequency, which benefits woody species, unless these woody plants are also browsed.	Increased fire activity immediately after the Pleistocene extinctions.

Large herbivores are likely to have a range of effects on plant communities in terms of their species composition, growth habits, and, as a result, on the structure of landscapes. Reproduced from Table 1 of Bakker et al. (2015), and references therein, with permission from the National Academy of Sciences of the United States of America.

a challenge to determine whether it was the lack of dispersal agents or some other factor, such as climate change, that dealt the fatal blow. Many species previously dispersed by megafauna still persist, albeit probably in a more restricted range. Their secondary dispersers include smaller animals (primates, pigs, birds) as well as environmental factors such as water runoff, particularly on floodplains (Guimarães et al., 2008). Guimarães and colleagues point out that the apparent resilience of these plant species to disturbances such as clear-cutting continues to serve them well even in the absence of megafauna.

The interplay of different actors can be observed today in systems where elephants still exist. A study at Buxa tiger reserve in India documented the dispersers of the large-fruited species *Dillenia indica* using camera traps, and found that while 64.3% of the seeds were taken by elephants, those that remained were also consumed by smaller taxa (Sekar & Sukumar, 2013). The seeds were subsequently able to germinate, demonstrating that elephants were not strictly necessary for this particular species, although they were possibly the most effective (Figure 5.5). It would seem evolutionarily sensible to have a back-up plan for seed dispersal in case your primary agent doesn't turn up (Sekar & Sukumar, 2013), but this sensible strategy is not followed by the numerous examples of highly specialized plant–animal mutualisms found in nature. Co-evolution, once it gets going, may be difficult to reverse. In southeast Asia and the neotropics, other medium-sized herbivores such as tapirs may substitute for elephants where certain plants are concerned. But they can also act as seed predators, and they do not disperse seeds as far as elephants (Campos-Arceiz et al., 2011).

A study conducted in the Democratic Republic of Congo offers far less optimistic results. Beaune and colleagues measured the impact of elephant declines on tree species over a 2.5 year period at the LuiKotale Max Planck research site (Beaune et al., 2013). There were 18 tree species exhibiting megafaunal syndromes, representing 4.5% of the tree species in the study area. They found that 14 of the species seemed to be entirely elephant-dependent, showing changes in both the number of saplings being recruited into their populations, as well as the spatial configurations of those saplings, as a function of elephant abundance. The smallest measured size class was saplings >50 cm in height. Twelve of the species exhibited no saplings of this size class, whereas saplings representing the remaining species were more clumped than other plant species. Clumping reflects lack of transport, with seedlings remaining close to parent trees. These observations prompt the conclusion that other animals are not compensating for the loss of elephants in this ecosystem. One therefore expects that the disappearance of elephants will have differential consequences in different systems, based on how many species appear to depend on elephants as their primary seed dispersers and whether other taxa might substitute for them.

The *Balanites* genus might offer particular insights, containing several species that are candidates for being elephant-dispersed and being found on both the Asian and African continents. *Balanites wilsoniana* seems to be heavily dependent on elephants, which appear to be their major consumers and without which seedlings show very poor chances of survival and germination (Babweteera et al., 2007; Cochrane, 2003). And although we have thus far been discussing relatively moist environments, elephants are also responsible for dispersing seeds in drier habitats. Elephant syndromes have been far less studied and characterized in these environments, with a few notable exceptions. One is *Balanites maughamii*, a species native to southern Africa widely used by various communities for medicinal purposes. This species is also suggested to be reliant on elephants for dispersal because its seeds actually germinate better

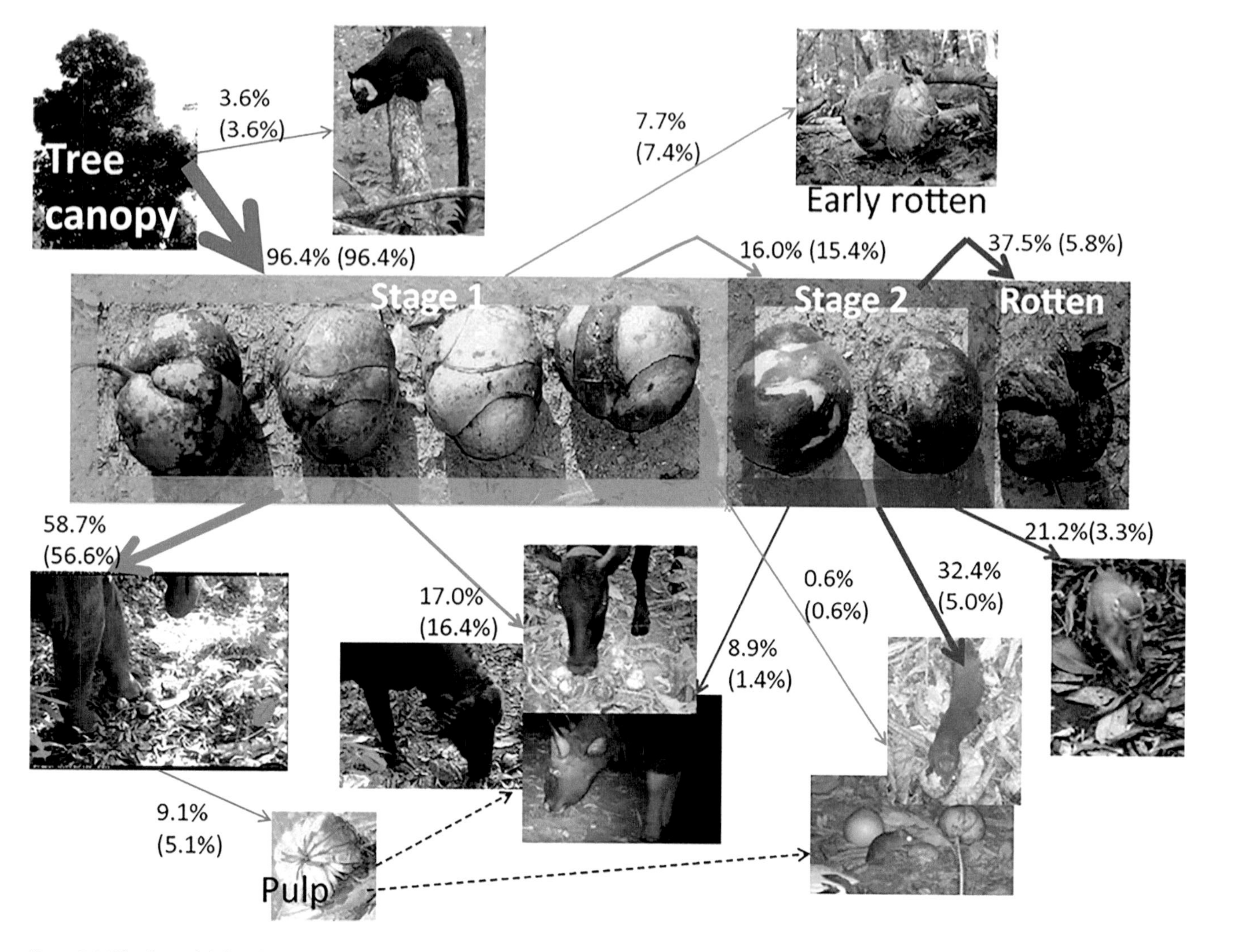

Figure 5.5 The fate of fallen fruit. Camera traps show which species consume the fruits of *Dillenia indica*, or chalta, a large-seeded tree species native to India. Elephants consumed most seeds, and they did so in earlier stages. But the fruit softened as it ripened, attracting smaller animals also capable of distributing seeds (reproduced from Sekar & Sukumar, 2013, with permission from John Wiley & Sons).

following crushing with a force comparable to what would be expected if an elephant were to chew on them (Midgley et al., 2015). The fruit are dull-colored, with tough seed coats, and fall to the ground upon ripening where they proceed to emit a strong scent. The species' Asian counterpart, *Balanites roxburghii*, is also suggested to be elephant-dispersed, though this has not been studied. Comparisons of the fates of species in this genus, on the two continents, may reveal informative parallels or contrasts.

It is strange to think that certain food plants that we take for granted may have originally evolved to lure the fancy of now-extinct behemoths. The glorious avocado, for instance, is endowed with the single large seed and green or dark coloration (when ripe) that seems suggestive of a color-blind, large-throated disperser. Might we have elephants, or more likely their brethren (i.e. the gomphotheres) to thank for this fruit of endless possibilities that improves any dish it graces, that beloved staple of café menus up and down the Western hemisphere? I would like to take the opportunity here to speculate about other prospective syndromes that might be associated with elephant dispersal that are less commonly considered to be so. These include tree or vine species with dull-colored elongated seed pods, as well as vines yielding large terrestrial fruiting bodies with thick skin. *Bauhinia racemosa*, already mentioned in Figure 5.2, is one example of the former. Elephants are the only species I have ever observed consuming its ripened seed pods, which resemble enormous flat beans, and the plant is remarkably tolerant to browsing even when trunks and bark are peeled multiple times as though it were a banana. Another candidate is *Tamarindus indica*, better known simply as Tamarind (Campos-Arceiz et al., 2008), which is a common ingredient used to add a sour tang in Asian cooking. Examples of vines growing at ground level include melons, squashes, pumpkins, and gourds. Pumpkins have origins in the Americas, whereas many species of gourds are native to Asia and Africa. Although humans have certainly bred species from these clades over several millennia to enhance their size and other attributes, it is possible that at least some varieties originally evolved their large size and hard coating millions of years ago in favor of megafaunal seed dispersers. Indeed, Galetti et al. suggest that domestication by humans and consumption by domesticated megafauna is one means by which some plant species may have avoided extinction (Galetti et al., 2018). If so, when humans cultivate these plants in areas with resident elephant populations, we inadvertently present elephants with an ecological trap. Ecological traps are defined as situations in which animals are attracted to cues that seem to indicate a high-quality resource (food, habitat, etc.) that might actually be detrimental or fatal to them. Such plants are not only palatable food for both humans and elephants, but also may have evolved a suite of traits specifically to attract elephants (Figure 5.6). This can end badly when animals consume crops laced with poisons, or, a more distressing recent development, explosives.

What is interesting about these observations is that they highlight synergistic effects between elephants and their preferred food plants that may be inadequately appreciated, though humans may have ultimately taken advantage of the results. These include adaptations specifically to attract elephants as dispersers (subsequently enhanced through human cultivation) and concurrent adaptations to withstand the

Figure 5.6 Squash vines sprouting from elephant dung in Sri Lanka. Since these varieties of pumpkins and squashes are not originally native to the country and do not easily grow wild in the dry environment without irrigation, the seeds would most likely have originated from cultivated plants (photo by the Udawalawe Elephant Research Project).

danger of attracting such attention from a potentially devastating forager. Positive feedback loops challenge notions of ecological carrying capacity at the heart of contemporary debates on elephant management, which are obviously concerned about the negative feedbacks. Expectations may rest on perceived changes to vegetation under conditions that have usually been human-modified, such as by constraining movements or introducing supplementary resources, such as water holes. Where ecosystem engineers are concerned, natural carrying capacity can neither be estimated nor predicted because it is fundamentally a moving target subject to dynamic interactions between the species and its environment. Certainly, overcrowding when space is constrained will probably have negative impacts on both elephants and other species. In practice, contemporary allusions to carrying capacity typically refer not to the capacity of the environment to support a given population of elephants, but its ability to support a population of elephants *while remaining unchanged*. The latter is clearly an artificial and unrealistic expectation since nature is never static and whatever equilibria may exist, they are dynamic.

5.1.2 Interactions with Parasites and Commensals

Elephants and other megafauna not only exploited their environment, but they were also themselves exploited by others, even before humans swept through. Parasites (species that are dependent on a host) are perhaps an overlooked ecological guild when it comes to conservation, mysteriously, given their obvious charisma. One may be tempted to think that the loss of host-specific parasites, despite technically also a blow to biodiversity, may not be too important in the grand scheme of things. Certainly, no one would try to motivate the conservation of megafauna by means of appealing to the public for concern over the fates of their parasites. The term "parasite" in and of itself carries negative connotations in everyday language. Be that as it may, the statement that "parasites represent the majority of species diversity on earth" (Nichols & Gómez, 2011) should give us pause, together with the observation that the conservation of this layer of biodiversity in its own right is much neglected within the discipline (Small, 2019). By regulating host populations and their distributions in natural food webs, parasites play an important role in maintaining diverse ecological communities – as important, if not more so, than the interactions among the free-living species within those webs (Nichols & Gómez, 2011). Moreover, they constitute reservoirs of potent bio-active compounds that have the potential to actually benefit human health. For instance, the anticoagulants and other specialized compounds evolved by blood-feeding parasites such as ticks and leeches are actively being researched for their applications in medical therapies (Small, 2019; Štibrániová et al., 2019). Parasite phylogeny and ecology can even offer insights into the biogeographic and evolutionary history of animal populations within a region, underpinning conservation arguments (Gardner & Campbell, 1992).

The interactions of large-bodied hosts with their parasites or mutualists constitute unique subsets of ecological interactions and evolutionary pathways, even if usually invisible to us. For instance, it appears that many megafauna served as hosts to species of botflies (Oestridae) that were highly specialized (Galetti et al., 2018). Their absence is only detectable through rare occasions when remains of extinct species have been preserved, such as those found within a frozen Siberian mammoth specimen, or observed in association with animals brought into captivity (Galetti et al., 2018). Such co-extinctions are quite probably even more difficult to detect than those of plants, as they leave so few traces. However, we might tend to notice on those occasions when parasites switch from their original hosts to species we care about – such as ourselves or our domesticated livestock. In South America, vampire bats (which are actually classified as "micro-predators" rather than parasites) also specialized in large bodied prey, and the disappearance of the megafaunal clades is thought to be related to the extinction of at least two giant species of vampire bat, *Desmodus draculae* and *Desmodus stocki*. But *Desmodus rotundus*, which specialized in feeding on large-bodied mammals, persists by dint of switching to introduced cattle, horses, and pigs in addition to the remaining native species such as tapir, deer, and capybaras, along with the occasional human (Galetti et al., 2016).

Where megafauna persist today, we might legitimately be concerned about parasites for a different, and more traditional reason that might seem at odds with any aim of conserving the parasites themselves. This is, of course, the fact that parasites by definition have a negative impact on their hosts, and if the latter are themselves of conservation concern, parasites represent an unwelcome cost that managers may actively seek to control or even eradicate. Drought conditions in the Laikipia-Samburu ecosystem in 2009 resulted in the deaths of savanna elephants that appeared starved and emaciated (Obanda et al., 2011). The dehydration and nutritional stress seems to have rendered individuals more vulnerable to the effects of infections by a type of nematode (*Grammocephalus clathratus*) and trematode (*Protofasciola robusta*). Forest elephants and Asian elephants as well all have their share of endemic parasites (Hing et al., 2013; Kinsella et al., 2004; Vidya & Sukumar, 2002). As we have briefly discussed, these species can be a consequential co-evolutionary partner, with their effects becoming ever more pronounced in a warming world where the mobility of wildlife is limited and resources are becoming ever more concentrated.

5.1.3 Interactions with Natural Resources

Aside from the obvious biological roles megafauna play as consumers and seed dispersers, it is now being appreciated that they also provide mechanisms by which nutrient cycling could occur on a global scale (Doughty et al., 2016; Sitters et al., 2020). At it happens, nutrients are rather like dust bunnies – they tend to accumulate in certain areas. For instance, the tendency of water to flow downhill results in nutrients running off easily enough from land to sea, where it proceeds to sink to the depths. In order for nutrients to cycle back into the system, they need to become mixed up again. But how does it get back to the interior of continents? In a word: poop. Animals and their waste are moving packets of nutrients. Doughty et al. (2016) point out that large marine vertebrates such as whales, by their feeding and movements, transport nutrients up from ocean depths to the surface. Anadromous fish such as salmon and sturgeon that move between freshwater and marine ecosystems as well as sea birds (and large predatory birds that feed on fish, such as bald eagles) transport nutrients inland from the sea. Large animals thus in general move nutrients away from places where they accumulate, redistributing them far and wide across land and water.

Elephants, for their part, also excavate and move soil, often creating microenvironments by their activities that are in turn exploited by other species. This highlights another class of resources the importance of which has thus far been greatly neglected within both the scholarly and conservation communities. Geophagy, or the consumption of soil, is a behavior exhibited by many herbivore species because of the need to supplement their diets with salts and minerals that may be scarce in regular forage. In many parts of the world animals repeatedly come to particular locations to obtain these trace elements, often with high fidelity over many years. Perhaps the most spectacular examples of this behavior are the so-called elephant caves of Mount Elgon National Park in Uganda where generations of elephants have excavated enormous cave systems in the process of salt "mining." These caves were found to be rich in sodium, calcium, and magnesium, among other things (Bowell et al., 2008). In the central African forests, the clearings known as *bais* (discussed in Chapter 3) may offer comparable resources and are sometimes also inundated with water. In drier ecosystems, elephants evidently avail themselves of earth from termite mounds (Holdo & McDowell, 2006; Kalumanga et al., 2017). In Asia, elephants visit very specific microsites that occur along exposed flood plains, seasonal riverbeds, or eroded hillsides (Figure 5.7). But do all these sites share the same characteristics?

Figure 5.7 (a), (b) Elephants on both continents excavate small wells along certain streams and riverbanks, carefully drinking water that seeps into the hollow. The water is very likely enriched with dissolved minerals that are somehow not available in the free-flowing water nearby (photos by the author). (c) They can also hollow out caves by eating the soil itself, some large enough to comfortably accommodate one or more people, like this one in Sri Lanka (photo by the author). (d) In the central African forests, elephants gather in large clearings that are a unique feature of these forests (photo by Victoria Fishlock used with permission).

While it is typical to refer to such locations as "salines" or "salt licks," this implication that elephants and other wildlife are primarily after salts is too limiting, therefore the term "mineral lick" is preferable. A different study that sampled locations in Sri Lanka compared geophagic sites and compared them with other random locations (Chandrajith et al., 2009). They found that the two types of locations did not differ in geochemistry, so the preference was not necessarily due to the salts that were present. Instead, the preferred clay substrate was significantly more enriched with minerals known as kaolinite and illite, as well as other rare compounds, relative to the soils found elsewhere in the landscape. These minerals were incidentally also primary components of pharmaceutical products marketed for human consumption in order to alleviate gastrointestinal problems, and the concentrations (50%) were similar. It is therefore suggested that the ingested compounds help to absorb the toxic secondary compounds plants have evolved as herbivore defenses, and essentially function as a

Table 5.2 Possible reasons for consuming earth, and associated physiological mechanisms

Biological effects	Mechanisms of realization
Mechanical effects in the digestive tract	Mechanical processing of food (in the gizzards of granivorous birds), enhancement of peristaltics, cleaning of the digestive tract
Supply of bioactive silicon compounds to the body	Due to the silicon compounds contained in mineral and organic substances
Regulation of ionic balance in biological electrolytes	Due to the absorption of ions and molecules as well as ion exchange on ion exchangers and gel
Detoxification function	Due to the absorption and inactivation of toxic substances
Supply of chemical elements and their associations required for the construction of biological tissues in the body	Due to the absorbed elements and their associations in clays and zeolites, as well as additional minerals, capable of dissociation in the gastrointestinal tract
Participation in the enzymatic reactions, primarily hydrolytic Prolongation	Prolongation of the action of enzymes, direct involvement in biochemical reactions (catalysis)
Binding of excess water in the intestines	Due to the hydration of ions, molecules, and water sorption on silica gel, clays and zeolites
Effect on the symbiotic microflora	Activation of fermenting forms, inhibition of fungi and actinomycetes, binding atypical microorganisms and their removal from the digestive tract
Enhancing the antioxidant defences of the body	Due to the antioxidant properties of minerals
Training the immune system	Due to the constant influx of antigenic material to the body in the composition of mineral adjuvants

Reproduced from Panichev et al. (2013) (originally Table 3) with permission of Springer Verlag-Dordrecht.

digestive aid (Chandrajith et al., 2009; Houston et al., 2001). Reviews of this behavior in humans and other ungulates reveal many more possible functions of soil consumption, depending on the particulars of circumstance, listed in Table 5.2 (Abrahams, 2012; Panichev et al., 2013).

But soil is not the only thing that elephants are after at such locations. As it turns out, elephants are also rather fond of mineral water, which is accessed by means of digging wells or "sinkholes" at particular locations (Fishlock et al., 2016). These sites can occur alongside mineral licks, and require a fair bit of practice to make use of, described by Fishlock et al. (2016: 431–432):

> Elephants using sinkholes do so with a specific repertoire beginning with the removal of the upper substrate of sand or mud using a forefoot or tusk before inserting the trunk into the hole, blowing out forcefully and then sucking in water from the base of the hole (Wrege, 2015;[2] Fishlock pers. obs.). Once they have a trunkful of water they drink, discarding the last sediment-filled portion. Holes below the water level rapidly backfill with upper sediment, requiring frequent re-excavation to access the base.

[2] This refers to P. Wrege (2015). Why elephants come to bais. In: V. Fishlock & T. Breuer (Eds.) *Studying forest elephants*, pp. 88–89. Neuer Sportverlag.

The interactions between elephants and termite mounds, which may appear to be more spatially diffuse resources, are a good example of how a physiological need for rare nutrients can influence the ecology of landscapes through the actions of two ecosystem engineers. A study conducted at Ugalla River Game Reserve in Western Tanzania suggests that termite mounds that are preferred are located on floodplains close to watercourses rather than in surrounding Miombo woodlands (Kalumanga et al., 2017), but not all were mineral-enriched compared with the background, which seems to explain why not all mounds located near watercourses are consumed (Figure 5.8). However, the study could not pinpoint which minerals in particular were important because they found elevated levels of many different elements, including iron, zinc, copper, magnesium, potassium, sodium, calcium, and manganese. Another study in Zimbabwe came up with similar results, and, moreover, found that trees growing on termite mounds had higher concentrations of nutrients and were consequently browsed more intensely by elephants as well (Holdo & McDowell, 2006).

What, exactly, are elephants gaining? Curiously, observations of human behavior might provide some clues. In the 1930s and 1940s, researchers working in present-day Malawi documented that pregnant and nursing women as well as growing children consumed soil collected from ant-hills (Abrahams, 2012, and references therein). These lime-rich soils compensated for dietary deficiencies in calcium, with a similar practice observed among indigenous people who do not consume dairy. Other studies link human soil consumption with the presence of iron. The lack of systematic studies makes these few observations far from conclusive, however. Reviewing various studies of geophagic sites used by elephants in Africa and Asia, Fishlock et al. (2016) report that they seem to represent several different types of resources. But what is interesting is that in some cases studies do not find an appreciable difference between locations actually used by elephants and other parts of the landscape (i.e. the background). So why the preference? Fishlock and colleagues propose that the sites become frequented over time through a process that one might call "tradition." We will discuss this more in Chapter 7. How such traditions become established is a mystery, but once they do, they can be faithfully transmitted through social learning. This can be a double-edged sword, as we shall see in Chapter 6. For the vast majority of herbivores, elephants included, very little is understood about how important minerals and nutrients are obtained. Whether or not these sites are somehow special, they potentially represent highly valuable and rare resources that are by and large ignored in conservation planning. In the absence of studies to understand how these locations influence animals' movements and physical health, it would be all too easy to lose them.

Elephants are sometimes referred to as "ungulates," which lacks taxonomic coherence, but like hooved mammals, elephants do need to drink water on a regular basis. Elephants use regular surface water both for drinking and bathing, both being vital to thermoregulation (Dunkin et al., 2013). In captivity, elephants are variously reported to drink between 60 and 155 liters of water per day (Seaworld, 2019), yet this offers little insight into what is required under natural circumstances for the populations occupying varied environments. Elephant home range sizes tend to decline with

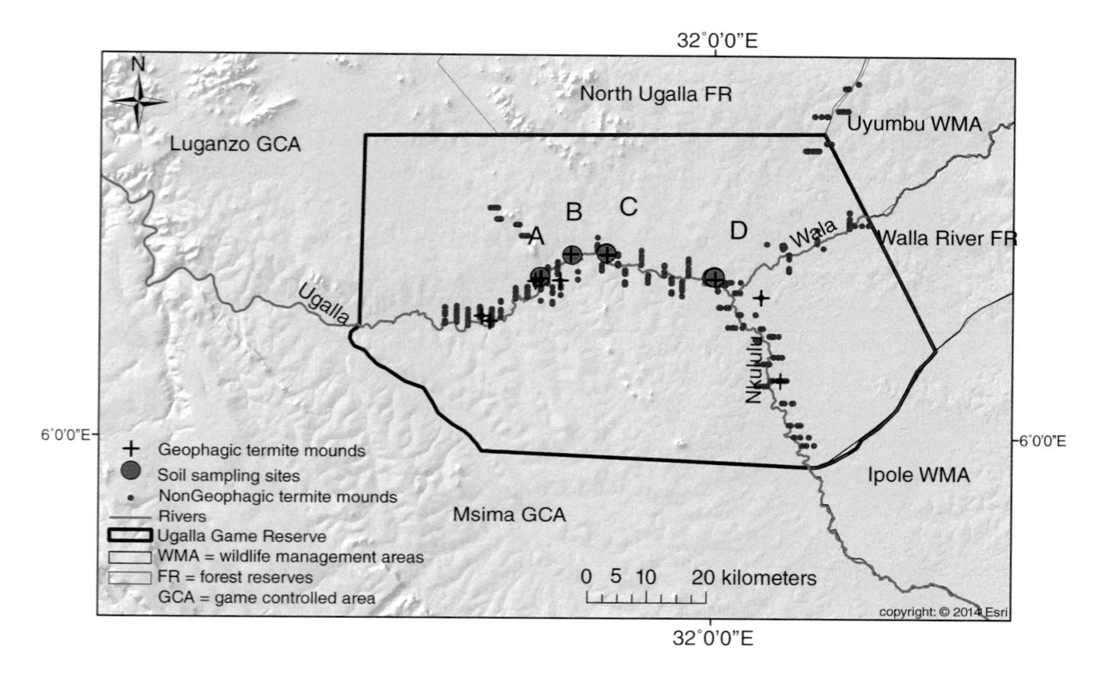

Figure 5.8 Geophagic termite mounds. All termite mounds that are consumed are found close to a watercourse, but not all termite mounds located near watercourses are consumed (reproduced from Kalumanga et al., 2017, with permission from John Wiley & Sons).

increasing rainfall, a pattern that is seen most strongly among African elephants and to a lesser degree among Asian elephants (reviewed extensively in Sukumar, 2003 and discussed further in Chapter 6). Rainfall of course determines vegetative productivity and hence the distribution of food. But it also, critically, determines the amount of available surface water. The relationship between the amount of rainfall received and elephant area requirements is far from straightforward, because it is also influenced by the heterogeneity of the landscape, whether water is accessible at the surface, and how disturbed it is, as we shall see. Temporary or permanent surface water sources are essential features of the landscapes that elephants occupy on both continents, though water sources may be more widely dispersed is some places than in others. While it may be guessed that elephants in hotter and drier conditions would require more water than those in cooler, wetter conditions, there are surprisingly few studies on the water requirements of elephants apart from certain savanna populations where water availability may be intentionally manipulated for management purposes (Chamaillé-Jammes et al., 2007).

5.1.4 Interactions with Artificial Resources

In the present era, elephants are exploiting several resources that are a byproduct of human labors. These include crop lands and engineered water sources, which elephants will have likely encountered to varying degrees throughout history. These will be discussed more in coming sections. However, elephants are also increasingly being exposed to novel items that they would not have had access to before but can learn to exploit. Artificial salt licks are sometimes created intentionally within protected areas (PAs) to draw wildlife for tourism viewing purposes and may provide a benefit to the animals as ranging constraints and land-use changes limit access to natural mineral deposits. But the ecological effects of doing so have not been studied. In populated areas, certain bold individuals, typically bulls, try to assuage their taste for salt by breaking into household kitchens. Numerous anecdotal accounts of home damage by elephants indicate that elephants seek out grain stores, salt, and even alcohol.

In what might appear to be a contrasting situation, people sometimes intentionally attempt to feed elephants. Where wild elephants have become habituated to the presence of people, the latter tend to mistake the former for docile and even domesticated creatures. Humans, of course, have a general tendency to want to offer food to wild animals if it will afford the opportunity for a close encounter, a photograph, or well-intended sense of communion with nature. However, feeding wildlife is widely discouraged for numerous reasons. First, people tend to feed animals foods that are rich, sweet, and, if not artificial, at least uncommon in the regular diet. A constant diet of fruit, for instance, in an herbivore whose physiology is evolved for a diet of mixed greens and roughage, can create a diabetic. Additionally, it encourages otherwise highly mobile animals to become more or less sedentary – which we can all appreciate as being associated with poor health outcomes. Finally, animals are exposed to novel items that they may later be tempted to seek out intentionally, putting both themselves and human communities at risk. Certain individuals have learned to beg from motorists along roadsides, some of them harassing passers-by by exacting a "toll." Such scenarios are very unlikely to end well.

Two case studies illustrate the long-term effects of intentionally feeding wild elephants. One concerns a male known as "Rambo" at Udawalawe National Park who had perfected the art of standing innocently by the electric fence running along its southern boundary so that passers-by would provide him with hand-outs. He had been at it for at least twenty years, by which point other males also surmised (through simple observation) that this might be a rewarding foraging tactic and emulated the behavior. We will return to this case study and another example from India in Chapter 7.

Feeding is not always intentional. A growing problem accompanying poorly planned development and urbanization globally is the proliferation of garbage dumps (Plaza & Lambertucci, 2017). Illicit dumping grounds, especially on the edges of wildlife habitats, can attract numerous species with the promise of easy, calorie-rich food and novel tastes (Katlam et al., 2018). Elephants, who would normally range over different habitats across the year to keep up with seasonally shifting food resources, may find it far more convenient to stay put at a dump and have the food brought to them (Figure 5.9a). Bulls seem particularly happy to occupy particular sites, though females and calves also use them (Figure 5.9b–c). In Sri Lanka there have been repeated attempts to translocate bulls from garbage dumps into PAs. However, removal of some individuals only resulted in others taking their place (SdS personal

Figure 5.9 Garbage consumption. (a) This unplanned dump close to Minneriya National Park in Sri Lanka is a consequence of the poor waste management policies and infrastructure that currently represent the norm for habitat edges in many parts of range countries. The attraction of wildlife as well as domestic animals such as stray dogs and cattle provides ample opportunities for pathogen transfer. (b) Bulls are most often seen at the site during daytime; nevertheless this group of females (c), photographed inside the National Park itself, shows visible evidence of having ingested garbage, with one individual having plastic waste dangling partially from her anus (arrow). The youngest member of the group is a calf that is barely a few days old (box), seen beneath its mother. (d) Elephant dung littered with plastic bags is commonly found in and around PAs: (i) inside Minneriya National Park (ii) inside Udawalawe National Park (photos by the Udawalawe Elephant Research Project).

observations). Moreover, tracking these translocated individuals showed that at least some were strongly motivated to return (Fernando et al., 2012). One notably tragic example was that of Homey:

> Homey after his first and second translocations over 48.2 and 46.2 km homed back in 5 and 41 days respectively. Homey on his third translocation of 161.7 km showed homing movement for 62.0 km in 4 days but entered a town causing conflict. Chased back to the release location, he settled at the perimeter of the park, raided surrounding villages, was shot repeatedly and died 15 months after from gunshot injuries. (Fernando et al., 2012)

Bulls are normally expected to range widely. Nonetheless such homing behavior, clearly stressful and ultimately fatal for the individual, indicates remarkably strong site fidelity. Theories concerning of male motivations in this classically nonterritorial polygynous species would suggest that Homey should have been satisfied to roam anywhere with food and females, yet this was clearly not the case. Why try so hard to return to familiar stomping grounds? One possibility is that he may have been the king of his particular hill; perhaps the ease with which he could forage within his known range, among familiar individuals, was unmatched by his new surroundings filled with unfamiliar competitors and challengers. This hypothesis gains support from a study of 216 males observed over a period of nine years at Udawalawe National Park (Madsen et al., 2022). Males over the age of 40 were the most likely to be seen returning to the PA during their musth periods with regularity across years, whereas younger males tended to be more transient. This suggests that males require decades of exploration before finally "settling" into an established routine, therefore mature bulls might be especially unwilling to start anew in some unfamiliar location. There is still much to be learned about the social relationships and personalities of bulls, an area that is relatively understudied as we have already seen. From a management standpoint, this case begs additional questions – on what basis should an animal be defined as problematic? An elephant consuming garbage but otherwise causing no disturbance would seem to present less of a problem than one that leaves a trail of destruction in its attempt to return to its home. We will return to the issue of translocations in Chapter 6.

But feeding at dumps *is problematic in itself*. The sites may contain items that can physically harm animals, such as sharp edges and plastics. One would expect such objects to cause external or internal injuries, yet cases are so poorly documented that fatality rates cannot usually be determined (at least, with data that is publicly available). However, in East Africa, there have been at least three documented cases of deaths linked to garbage consumption, where a post-mortem revealed the gut to be packed with plastic, glass, and other debris (P. Lee, personal communication). Animals that are exposed to contaminants at young ages, if not injured immediately, may nevertheless develop pathologies later in life from exposure to contaminants (Figure 5.9c). Material that is not expelled can obstruct the gut, whereas that which is expelled contaminates the landscape. Garbage dumps effectively hijack the dispersal power of elephants to redistribute plastics and other pollutants rather than seeds and compost (Figure 5.9d).

Another serious concern is the risk of disease transmission (Plaza & Lambertucci, 2017). Tuberculosis (TB) is one example. While many species have their own endemic strains of pathogens such as TB, cross-species transmission can be fatal. In the 1980s Robert Sapolsky documented the disintegration of a troop of olive baboons after they started regularly salvaging human food waste, with several of its members contracting bovine TB from contaminated meat (Sapolsky & Else, 1987). He notes that it was the more aggressive individuals that were most susceptible to TB, perhaps because they were able to forcefully exploit a larger share of the spoils. Initially, the death of the more aggressive individuals had the effect of resetting the social culture of the troop into a more peaceful and relaxed state, which seemed to persist thanks to the expectations enforced by the remaining members (Sapolsky, 2007). In later stages, however, he observed that the entire social structure of the baboon troop simply eroded. The elaborate rituals and relationships in animal societies are maintained through interactions primarily centered on two types of things: resources and other individuals. Social contacts often mediate access to the resources (Sapolsky, 2005). Not having to work for the resources, perhaps the typical machinations of baboon society simply ceased to be relevant.

Interestingly, a study of banded mongooses shows another possible mechanism by which TB can spread. It was found that troops of banded mongooses feeding at dump sites had much higher levels of aggression than those foraging elsewhere (Flint et al., 2016). Higher aggression was in turn associated with a higher rate of injuries. Because TB is transmitted either nasally or through cuts in the skin, the researchers suggested that the increased levels of aggression might be directly facilitating exposure to the pathogen. Whether animals incur injuries as a result of aggression or from coming into contact with dangerous objects, the potential for disease transmission seems substantial. The human strain of TB was documented in wild banded mongooses in the early 2000s, a first among free-ranging wildlife (Alexander et al., 2002). When wildlife contract diseases through anthropogenic disturbances such as these, resultant mortalities may nevertheless pass for "natural" since the causes will rarely be in evidence. Is there an uptick in such instances as human activities push ever further into formerly wild spaces? The degree to which parasite loads and disease in wildlife are being altered as a result of anthropogenic disturbance is an emerging area of research, as this has obvious implications for human health as well. The observations of baboons and banded mongoose also demonstrate behavioral shifts among social species. We have as yet no understanding of what effects refuse sites have on interactions among elephants.

The issues surrounding elephants have many analogues to the situation for bears in North America, though some of the early history concerning bears might not be as well known or remembered today. At the beginning of the twentieth century, bears were actively fed in national parks in the United States both by managers and tourists for the spectacle it created. This changed eventually, as both bear populations and their exhibited aggression increased, alongside incidence of injury and dangerous encounters with humans. Bears that injure people are killed, even today. Now, the National Park Service has a clear policy against feeding wildlife, along with infrastructure such as bearproof bins to minimize the possibility. However, bears also feed in and around urban areas, which continues to be a concern (Baruch-Mordo et al., 2014). Capture and translocation of animals in such situations have been tried and found to be only temporarily effective (Gunther et al., 2004a). One study of black bears in Sequoia National Park found that aversive conditioning (such as a sprayed or projected deterrent) was successful at preventing non-food conditioned bears from becoming so, but equivocal once they were already habituated (Mazur, 2010). Run-ins with people seem to coincide with a lack of availability of natural food for some populations of grizzly bears as well as black bears (Baruch-Mordo et al., 2014; Gunther et al., 2004a). An important insight gained from tracking black bears in Aspen, Colorado, is that the bears seem to switch between wild and urban forage depending on natural food availability, using the latter during poor food years (Baruch-Mordo et al., 2014). Therefore, the individuals using the forests much of the time are the very same ones as those using urban settings some of the time. Similar patterns may hold for elephants that are crop-foraging. Since these are years when survival is anyway more difficult, the researchers point out that subjecting animals to additional stressors such as translocation can have an overall negative impact on the population. At least in some places, management has shifted toward proactively preventing problematic bear behavior by seeking to change the human behavior that encourages it (Gniadek & Kendall, 1998).

In the case of elephants, there is currently no understanding of what fraction of a population makes use of human provisioning and refuse, although as we have seen there is already evidence that some individuals habitually rely on such resources (Chiyo et al., 2011a, 2011b). Clearly, changing the behavior of the animal will be a never-ending battle if the human behavior is not addressed. Sorting waste is a partial solution, in that it can at least minimize wildlife exposure to toxic contaminants such as heavy metals but does nothing to prevent disease transmission. There can be no substitute for responsible urban planning and waste management. Where there is a strong cultural component to food provisioning, such as those associated with religious customs or societal beliefs, changing mindsets can be extremely challenging. Yet it is no less important to devote resources to doing so than to spend them on managing the species itself.

5.2 Risk and Fear

5.2.1 Ecology and Landscape of Fear

A categorically distinct consideration governing where elephants roam is safety. There is a growing literature on how the threat of being eaten shapes animal movement and foraging patterns, evocatively termed the "landscape of fear" (Hernández & Laundré, 2007; Laundré et al., 2001), which has framed many studies of habitat selection by large herbivores in North America and Europe.[3] The rising popularity of this term has led to its application in various ways, but is most simply defined as "the spatial variation in prey *perception* of predation risk" (Gaynor et al., 2019; emphasis added). But the more general topic of how predation risk shapes foraging behavior (or "ecology of fear") has long been a subject of study (see Gaynor et al., 2019 for a brief discussion of that history). The prototypical and most famous, if contentious, example of this is the effect of the presence of wolves on the browsing of elk and bison (Laundré et al., 2001), which in turn is proposed to alter a series of other ecological interactions and habitat variables. When wolves were eradicated from Yellowstone National Park and

[3] What is revolutionary about this terminology invoking "fear" is that it explicitly acknowledges the ecological consequences of emotion as well as nonhuman perception of emotion. Ecology, as indeed much of biology, has generally tended to shy away from using internal states as explanatory variables, out of a virtually Cartesian dualism separating the mental realm from any actual behavior that has a measurable impact on the material world. Thus one may traditionally speak of predation risk as dictating behavior, but not the fear that an organism experiences because of it. "Predation risk" merely deals with probabilities, a domain comfortably familiar to the scientist, whereas "fear" consists of subjective experiences that may as well be voodoo. Hence much of the work on this topic continues to reference the behavior rather than the emotional state driving the behavior. The landscape of fear nevertheless challenges this reticence by inviting us to imagine fear itself as an invisible force that is as real as any in the physical world, shaping and sculpting landscapes. As Gaynor et al. (2019) discuss, the term itself has been borrowed from anthropological literature describing landscapes that evoke terror in human communities, and the nod to emotions of nonhuman species with its risks of anthropomorphism still has its critics.

its surroundings, herbivore foraging behavior went relatively unchecked, resulting in the stripping of much vegetation within feeding height. The reintroduction of wolves not only controlled herbivore numbers but also where the latter chose to forage, which is argued by some to have more profound effects on the ecosystem than the number of animals that the wolves actually managed to eat. Female elk and bison showed increased vigilance and avoidance of watercourses and habitats more likely to conceal predators, though only elk seemed to have changed their feeding habits (Hernández & Laundré, 2007). These habitat shifts are thought to have allowed recovery of vegetation in those areas, which in turn is suggested to have had broader effects on other species (Laundré et al., 2001). The process is more formally called a "behaviorally mediated trophic cascade." However, studies have reached mixed conclusions about the actual impact of wolves, possibly because of the time lags required to discern certain ecological responses as well as the complexity of these multispecies interactions in a completely open ecosystem that is subject to other environmental effects, such as climate change (Kauffman et al., 2010; Ripple & Beschta, 2012).

Though I have so far emphasized the role of humans as predators, and we are accustomed to thinking of ourselves as such, we can also have quite the opposite effect. Proximity to human settlement and other activities can also serve as a refuge for herbivores who may otherwise be hunted by carnivores. This "predation shelter" or "human shield" effect is the indirect result of carnivores being deterred by human presence – seemingly an example of the adage that "the enemy of my enemy is my friend," at least from the point of view of the prey. This is especially evident in cases where human activity is relatively innocuous, such as the tourist visitation of PAs. In Grand Teton National Park, elk and pronghorn antelope show behavioral changes such as reduced vigilance and smaller group sizes when within 500 m of a road with heavier traffic compared with an adjacent road with less traffic (Shannon et al., 2014). However, animals choose habitats at multiple spatial and temporal scales. The sheltering effect, which occurs at relatively fine scales, is constrained by coarser spatial and temporal scales such as seasonality, which govern where there is food or cover at particular times of year. A study of roe deer in Germany showed that at the coarser scale, species may be obligated to trade off the potential risk of predation (in this case, lynx) in favor of habitat where there is adequate food (Dupke et al., 2017). One must eat after all, and the associated risks appear to be a lesser concern. As we shall see in Chapter 6 in the case of elephants, these trade-offs can be very consequential.

There is also some concern that the behavioral changes (namely a reduction of fearfulness and vigilance) brought about by human-mediated predation shelters may have unintended consequences. There are at least two possible avenues by which it could potentially wind up increasing vulnerability to predation (Geffroy et al., 2015). One is indirect, whereby if an activity is sufficiently disruptive over a large area, there may be an overall reduction in the presence of predators that results in reducing antipredator behavior in prey. This could then make individuals more vulnerable in other areas, for instance when they disperse or move seasonally. The other route is direct, where human habituation may favor the propagation of certain phenotypes, such as individuals that are especially bold or docile. If these individuals transfer these behaviors from the nonthreatening human context to one in which there is an actual predator, this so-called *behavioral spillover* may have adverse impacts on these individuals. Extrapolating to the population level, these individual effects could potentially alter population demography. Another consequence might be the loss of behaviors such as migration. In different studies involving two wild populations of elk in which individuals where characterized along a shy–bold spectrum of personalities, it was found that the more "shy" individuals were three times more likely to initiate migration whereas the "bolder" were just as likely to remain resident year-round (Found & St. Clair, 2016).

Experimental studies from controlled microcosms offer more striking evidence of how fear itself can alter even the abiotic components of ecosystems, the size of its key players notwithstanding. Grasshoppers are tiny herbivores that live with the threat of being eaten by spiders. In one study, a subset of grasshoppers were exposed to spiders while the others were not (Schmitz et al., 1999). Those that were exposed to spiders were further divided into those that had to contend with spiders that could actually eat them (predatory) versus spiders that had their mouth parts glued shut, and thus only posed an apparent threat (risk only). It was found that the grasshoppers that were exposed to spiders suffered higher mortality than those not exposed to spiders, irrespective of whether the spiders were actually capable of eating them. This is because the risk alone was enough to reduce the amount of time they spent feeding, resulting in starvation. This striking observation supports the idea that the effect that predators have on their prey goes beyond the number of animals actually hunted – and although large ungulates may be less prone to starving themselves than grasshoppers, there may still be a reduction in nutritional intake depending on the severity of the risk. Still more astonishingly, another study illustrates that this signature of fear can penetrate into the soil community itself (Hawlena et al., 2012). As it turns out, grasshoppers that are stressed by the presence of spiders have a different balance of carbon to nitrogen in their bodies relative to those that are unstressed. When the grasshoppers died, their bodies got buried in leaf litter. It was found that the leaf litter harboring the bodies of grasshoppers that were stressed subsequently decomposed more slowly, as a result of these elemental differences! One cannot but wonder whether larger prey species could have analogous impacts on larger spatial scales.

5.2.2 Stress

As the largest terrestrial herbivores, elephants' body sizes have evolved to outgrow the risk of predation (as well as outcompete other foragers; Head et al., 2012), and this strategy must have been mostly successful until the rise of our own ancestors. And yet, as we have already discussed, elephants in different types of environments may employ different strategies to manage risks. Those in more open environments may use group foraging to their advantage, whereas those with more shelter may resort to crypticity. Even within those broadly different environments, there will be microenvironments and habitats that are either preferred or not. To what extent are these preferences governed by the resources themselves as opposed to the threat of predation that may be lurking within? Do elephants have much of a choice?

To the first approximation, we may use physiological stress as one indicator of how (un)comfortable elephants are in different landscapes. Frequently, this is assessed by means of the levels of glucocorticoid metabolites detected in faecal samples (fGCMs). The voluminous literature on this subject could constitute a chapter in itself, but I pick out a few studies that offer intriguing or surprising observations. But first, the caveats. There may be an unknown time lag between a particular stressful event and the defecation event (which is not the same thing as gut passage time, as this is the time between feeding and defecation), which may take place at an entirely different and unrelated location. The fGCM levels in the dung may represent an average of the physiological state over the previous 30 hours (Wasser et al., 2000). Therefore, it will be difficult to relate hormone profiles to specific incidents or locations. Moreover, the chemical molecules involved are much more sensitive to environmental conditions than genetic material and can degrade more readily. In the tropics, exposure to sunlight can alter apparent concentrations and sampling may need to be done within eight hours or less to get a relatively accurate measurement (Wong et al., 2016). As a final confound, excreted fGCM can consist of both active hormones and those that have been used in a glucagon response. Distinguishing levels in response to "stress" is difficult if not impossible (Romero & Beattie, 2022).

Given these constraints, we are limited to relatively gross population-level assays of association between fGCMs and habitat type. For instance, we may predict very crudely that animals experiencing nutritional deficits or anthropogenic disturbances should exhibit more signs of physiological stress. However, this simple statement appears to be bedeviled by details. A study in Kenya compared fGCM levels in dung found in or near crop fields against samples taken from individuals in two of the nearby PAs, Maasai Mara game reserve and Amboseli National Park (Ahlering et al., 2011). Using molecular sexing, it was determined that the samples from the crop fields were all males whereas those sampled in the PAs intentionally included members of both sexes from different family groups. The samples from putatively crop-foraging males showed significantly higher levels of fGCMs relative to the other two sampling locations, but this seems to be largely due to differences with the females. When only males were included, there was a significant difference between the crop-foragers and the Amboseli samples but not between them and Maasai Mara. One is immediately inclined to wonder whether this is owing to differences in the propensity of males in the two PAs to forage on crops, differences in how people respond to the elephants within their local areas, or something else entirely. The authors point out, for instance, elephants in Amboseli may have lower stress levels due to tourism, as vehicular traffic is more tightly regulated than in Maasai Mara.

Making matters more confusing, a study of Asian elephants in Karnataka, India, found results surprisingly contrary to intuition. The authors compared fGCM levels from samples collected in the PAs of Nagarahole and Bandipur national parks with those from the Hassan district, which contains a patchwork of forests scattered throughout agricultural production landscapes. Given the high levels of conflict and antagonism against elephants in the latter, they expected stress levels and associated fGCMs to be greater. The study also included an additional critical dimension:

habitat quality measured in terms of vegetation greenness (Normalized Difference Vegetation Index), and forage nutritional content assessed directly through measurements of the carbon to nitrogen ratio in faecal samples. Higher carbon:nitrogen ratios were taken to indicate lower diet quality. What they found was that fGCMs were significantly higher in the PAs compared with the production landscapes, for both males and females. In the case of females, both seasonality (i.e. wet or dry) and habitat quality had a significant effect on fGCM concentrations, but for males only habitat quality was significant, and habitat quality was markedly better in human production landscapes by both measures. Curiously, this suggests that having access to higher quality forage, despite whatever risks are presented by the activities of people, represents a less "stressful" situation than being in a protected environment with lower quality forage. In a separate study by the same authors, it was found that fGCM levels were indeed related to elephants' seasonal body condition scores, especially for females, thus nutritional "stress" appears to translate into actual physiological stress (Pokharel et al., 2017). Given the contrasting observations from Kenya, this pair of studies reinforce the context-specificity of elephants' responses to potential stressors.

Species can reduce the odds of stressful human encounters by shifting their activity patterns, a strategy readily employed by many mammals (Gaynor et al., 2018b). In Asia, wild elephants are virtually impossible to observe during the daytime outside PAs, and this is especially true of females and calves. In the Laikipia-Samburu ecosystem of northern Kenya, elephants traveled more at night during periods when there was increased poaching activity (Ihwagi et al., 2018). This was more pronounced in the movements of females than males. Ihwagi and colleagues therefore propose that changes in the ratio of daytime to nighttime movements could be used as an indicator of poaching activity, complementing patrols and other monitoring efforts. Likewise, elephants shifted to showing more nocturnal activity when in the vicinity of *bomas*, which are shelters constructed by nomadic pastoralists in order to safeguard livestock at night (Duporge et al., 2022). Though the occupants of *bomas* are ostensibly less threatening than poachers, the behavioral response is analogous. This has consequences, such as limiting elephants' ability to access water supplies and forage in relatively open habitat during the daytime. Interestingly though, *bomas* can act as predation shelters for other species, such as giraffe. A study of Maasai giraffes in neighboring Tanzania found that predation risk was the most important consideration for groups with calves, which used bushlands more than open areas and ranged closer to *bomas* though they stayed further from towns (Bond et al., 2019). The calves of giraffes may be more vulnerable to large predators in these environments than those of elephants, or humans may react differently to elephants than to giraffes; one megaherbivore is not the same as another.

The quality of the available habitat, whether protected or not, remains an important outstanding question for elephants and other wildlife today. Whereas elephant herds may have once moved with relative ease across borders and continents, the forces of change unleashed since the colonial era have dramatically altered the amount and configuration of habitat available for elephants and other wildlife in Asia and Africa (more on this in Chapter 8). These range from the wholesale destruction of certain habitat types, such as cloud forests and lowland rainforest, to the ongoing development of roads, canals, and other infrastructure that carve up remaining habitat. Such developments may either impede movements directly, or presage settlement, cultivation, and other human activity. They also provide ready access for prospective hunters and poachers as much as tourists, which represent different kinds of disturbances from the perspective of an elephant. Besides potentially influencing the physical presence of elephants on landscapes, how might these processes be altering the very biochemistry of these systems with each birth and death of an elephantine ecosystem engineer? Although these large-scale systems may defy neatly controlled experimental manipulation, the question presents a realm of research that has been barely explored for those with the curiosity and ingenuity to pursue it.

6 Space Use

This chapter is organized by ecological gradient rather than species or geography because it highlights similarities and dissimilarities in how elephants use varied landscapes, which is useful for identifying what generalities, if any, might be extracted. Quite simply put, the more abundant or densely packed the resources are per unit area, the more individuals it can support, and/or the less area each individual requires to meet its needs. Obvious? Not quite. The "ideal free distribution" (IFD) sounds like a concept that belongs in physics rather than ecology, but is the proposal that in an idealized world animals are expected to distribute themselves in the environment in proportion to the concentration of resources available in it (Fretwell & Lucas, 1969). It is a disarmingly simple extension of optimal foraging theory, knowingly or unknowingly tested by young children feeding ducks the world over.[1] But behind it lie many assumptions; for instance, that animals can accurately perceive the quality and discrepancies among patches, go wherever they want both in sampling and settling with negligible travelling costs, and that they need not worry about anything besides the single resource they need. These assumptions might all be met at the small scale of a duck pond (barring occasional squabbles over mates or territory among the ducks, if not the children). Nevertheless, nature is of course rather messier. Indeed, there is a well-known bias by which organisms tend to overuse poorer sites and underuse richer ones (Kennedy & Gray, 1993). With a moment's reflection, I'm sure anyone can come up with reasons for why any of the underlying assumptions may be unrealistic. Theoretical studies show that violation of these assumptions can result in departures from the IFD (Matsumura et al., 2010).

For many species, constraints on movement represent one of the most fundamental violations of IFD assumptions. This may be due to intrinsic reasons, such as intraspecific competition and competitive exclusion; or extrinsic factors, such as avoidance of predators or anthropogenic impediments. As discussed in Chapter 5, the notion of a landscape of fear (Laundré et al., 2010) offers a major insight that avoidance of predation and other threats might explain why animals may use areas that might not initially appear well-suited for them. Additionally, animals do not automatically have perfect knowledge of their environments – the configuration of resources must be learned, all

[1] If you have not already done so, you may try this with a friend, in the company of willing ducks or fish. Position a person on either side of the pond, and have one toss out bread crumbs at twice the rate of the other. After some time, modify the ratio. You will observe that the number of ducks (or fish) at either end equilibrates to match the ratio at which crumbs are dispensed.

the more so in environments that might be unstable or disturbed. These issues, together with time lags between habitat modifications (e.g. owing to land use change) and populations' behavioral responses, may result in very suboptimal habitat use. It cannot always be assumed that the mere presence of elephants in an area deems it suitable for them (especially when such areas are heavily human-dominated). Clearly, a population must not only be present but also thriving for this to be the case. In this chapter, I discuss examples illustrating more specific considerations that govern elephant space use in different types of ecosystems. These situations underscore challenges for predicting elephant foraging behavior, and, by extension, adequately appreciating and accommodating their survival requirements.

6.1 Deserts and Their Edges

6.1.1 Western Africa

Only a few elephant populations traverse landscapes that are naturally arid enough to qualify as deserts, all of them occurring today in Africa. As one might expect based on the dispersion of resources in such harsh environments, these elephants have the largest home ranges on record among the extant species. In these populations, more so than any other, the dependence of elephants on surface water comes into stark relief. The Gourma region of Mali, which covers part of the Sahel along the southern edge of the Sahara and receives on average 300 mm of precipitation per year, hosts a population of just over 250 elephants according to the most recent estimate (Chase et al., 2016). One of the striking features of the landscape is the presence of a tight bottleneck – a narrow passage between two steep rocky outcroppings known as "la Porte des Eléphants," through which elephants must seasonally pass to maintain access to a substantial portion of their range (Figure 6.1). One collared female was reported to have circled an area of 24,265 km^2 over a 17–month period based on the minimum convex polygon (MCP) method of estimation (Blake et al., 2003). Another female and male were recorded to range over 19,338 km^2 and 11,651 km^2 respectively, over a comparable period. All movements essentially concentrated around waterholes or were movements between them (Figure 6.1). A later study found no statistically significant difference in range size among the sexes, but females did on average appear to have more distributed ranges than males, and the largest MCP range recorded, at 32,062 km^2, belonged to a female (Wall et al., 2013). A more detailed look at the areas elephants use shows that elephants are, as one might expect, also tracking areas of higher plant productivity as measured through satellite imagery by a measurement referred to as the Normalized Difference Vegetation Index, or NDVI (Wall et al., 2013).

The Gourma region receives on average the same amount of rainfall as Tsavo East, yet in the latter, clearly not classified as "desert," the home range sizes reported in older studies were smaller. Female ranges at Tsavo were between 1,009 and 2,975 km^2 and that of males between 516 and 1,756 km^2 (Leuthold, 1977). Blake et al. (2003) propose that this is because in the Gourma region rainfall is highly concentrated in

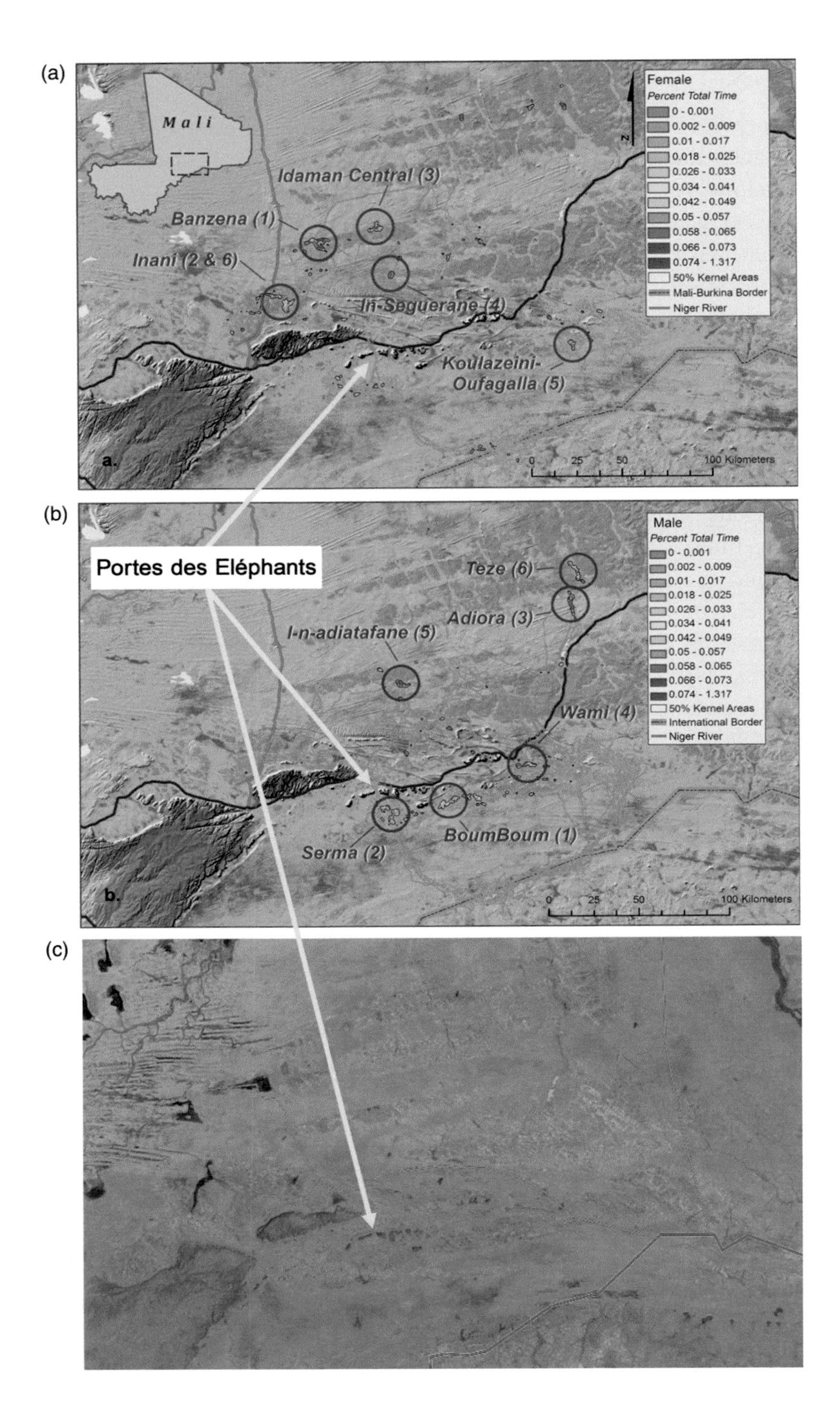

Figure 6.1 Movement paths of Gourma elephants. The named “hotspots” indicate areas of higher use and routes are bisected by the RN16 highway, shown as a brown line. Hotspots concentrate on small water sources that are virtually invisible when viewed aerially ((a) and (b) reproduced from Wall et al., 2013, with permission from Elsevier; (c) from NASA EarthData, https://earthdata.nasa.gov).

time, 70% of it falling within just two months of the year. This limits not only water but also the growth period for vegetation, forcing animals to move over a larger area than in a regime where rainfall might be more distributed throughout the year, as in the two wet periods in Tsavo. This does not necessarily mean that they are travelling over longer distances, on average, than elephants in other arid regions. Comparison of the linear distances travelled by the Gourma elephants and those in Kenya showed that the average path distance covered in a year was between 3,500 and 4,000 km in both populations, with no statistically significant difference. The Gourma ranges also appear to be about two and a half times larger than those documented among the desert elephants of Namibia (see below) and nearly a third larger than in Botswana (Wall et al., 2013). This is an odd pattern in itself, because although Botswana and Namibia share a common border and therefore have contiguous ecosystems, the drier portions belong to Namibia and hence one would expect elephant home ranges to be larger in the latter. Some possible explanations are considered in Section 6.1.2, but it would be interesting to similarly compare linear path distances for other populations, which requires more studies to report these values in addition to home range area estimates. The linear distance traveled may nevertheless be tuned within a narrow range across the different populations despite their very diverse local circumstances.

There is another possibility behind the exceptional size of the Gourma home ranges. The data for Tsavo came from older studies and included an unsettled protected area (PA). In contrast, the elephants of Mali are ranging largely outside any PA, alongside many human settlements. This places them in the minority, because according to an aerial survey of African savanna elephants conducted between 2014 and 2015 by Chase et al. (2016), which encompassed 18 countries, 84% of elephants were counted inside PAs.[2] One wonders if the huge circuit covered by elephants in Mali might also be the result of increased settlement around lakes in recent years, along with heavier competition for water from both people and their cattle (Blake et al., 2003; Canney & Ganame, 2012). Unfortunately, there are no earlier studies of elephant movements in Mali for comparison, although verbal accounts suggest that elephants have indeed had to change their movements in response to human and livestock pressure over a 30-year period (Canney & Ganame, 2012). During particularly difficult years, the additional travel distance required to find water that outlasts human consumption can increase mortality. Canney and Ganame (2012) record an incident in 2010 in which 180 animals left one lake, under heavy use by livestock, for another 70 km away. In total, 21 animals died en route, predominantly young or subadult males, an occurrence that prompted investigation and conservation action in this particular case. However, the impact of these chronic pressures on a broader scale remains poorly studied and appreciated. A further mystery is why the elephants do not seem to make use of more abundant water resources further north. Is there some physical impediment to travel

[2] This should *not* be taken to mean that elephants spend most of their time inside PAs, as much may have to do with the timing of aerial surveys. The rest of this chapter will explore the complicated relationship between elephants, resources, risks, and PAs.

or is it that exploration of new routes in such an environment is so risky that elephants just stick to what they know?

There is now evidence from various species of ungulates that long-distance migration routes must be learned (Festa-Bianchet, 2018; Milner-Gulland, 2011). Migration is therefore a socially transmitted, culturally preserved behavior that builds on the existing species-specific predisposition to be sensitive to particular environmental cues. It is this social learning that gives rise to the mass movements of large herds. In one study, researchers compared the migratory behavior of well-established populations of moose and bighorn sheep to those that were newly reintroduced (Jesmer et al., 2018). While 65–100% of animals in the established populations migrated, it took up to 90 years (12–13 generations) for the majority of individuals in translocated populations to become migratory. This corresponded to increasing knowledge of how to track the availability of fresh forage, colorfully termed "surfing the green wave," in the face of impediments including predation and anthropogenic barriers such as fences. The experience an individual has acquired over its lifetime, stored away in memory, is likely to be an important resource both for itself and its groupmates (Bracis & Mueller, 2017; Polansky et al., 2015). Likewise, the loss of such knowledge might be catastrophic (Brakes et al., 2019). We return to this in Chapter 7.

How then do animals learn and establish new routes in particularly harsh environments, where errors might be costly? A somewhat disconcerting answer is that perhaps they don't. For instance, if the distance between water holes is sufficiently great that animals risk death on the way during a bad year, it is a good bet that this will discourage exploration of the unknown. One could imagine a scenario in which populations initially established their movements at a time when resources were more plentiful – this may have been in the distant past, many generations ago. As conditions gradually become less favorable, populations may maintain their tried-and-tested routes by relying on the memory of experienced individuals; however, they may become averse to exploring novel routes because the risk of doing so is simply too great. Animal populations could then become locked into their habits by this gradual ratcheting effect, becoming vulnerable to extinction if there is sufficient disruption in those established habits. At the same time, such a population, once extinguished, is unlikely to be reestablished.

6.1.2 Southwestern Africa

A study from 1980–1983 highlights the importance for elephants of habitat associated with rivers in what was referred to as the Kaokoveld region on the edge of the Namib desert, where annual rainfall could be as little as 19–150 mm (Viljoen, 1989). During the dry season, elephants relied on browsing the trees and shrubs growing along dry river courses, whereas in the wet season they preferred grazing on grasses and forbs on the flood plains. They were somewhat limited in their movements by expanses of rugged terrain, but even these seemed to be preferred over rocky or sandy plains and dunes as they evidently contained more forage. Elephants occasionally also availed themselves of a curious skin treatment: They spread chewed leaves of the species

Euclea pseudebenus on their ears and heads, possibly as a sunscreen or means of cooling down. Mud, which is the preferred substance elsewhere, is presumably in very short supply. This rare behavior represents a highly localized innovation that I have not heard described anywhere else. It would be of great interest to know what other tricks and novelties might be hidden within such populations. Unfortunately, at the moment the Kaokoveld elephants appear to be among the most critically threatened of all savanna elephant populations, having declined precipitously from nearly 360 counted in 1983, to 80 or fewer by 2015 (Desert Elephant Conservation Report, 2015).

A subsequent study tracked collared elephants from six areas located within the Kunene region of Namibia, which is part of the area just described, where the mean precipitation ranged from 13 to 305 mm (Leggett, 2006). The ranges of adult males were found to vary from 2,168 to 12,800 km^2, while those of adult female and immature males ranged from 871 to 5,900 km^2, using the MCP method (Leggett, 2006). However, the size of the area constituting 95% of the space actually used ranged between just 628 and 3,251 km^2 for the former and between 256 and 2,224 km^2 for the latter, when estimated under a different statistical procedure known as a fixed kernel density estimator. In absolute terms, the latter likely provides a better characterization of the area requirements of the subjects because it de-emphasizes parts of the home range that are seldom, if ever, used. Whether the MCP method provides a useful basis for comparison among different populations, especially given that they are the only estimates available from prior studies, remains to be evaluated. There are other home range estimators that improve upon these, but in the interest of maintaining focus on elephant behavior rather than dwelling on methodologies, which are continuously being updated, I do not discuss them here.

The importance of water resources is nevertheless once again strikingly evident. Elephants in this region engaged in water-associated activities to a greater degree than reported for other savanna elephant populations (Leggett, 2008), and prefer to remain close to water sources, which can include artificial boreholes (Tsalyuk et al., 2019). They even demonstrate a remarkable ability that offers insight into how elephants may be finding the best foraging locations at any given time throughout their entire range. By relating the movements of nine collared animals to local rainfall patterns, it was found that individuals independently exhibited pronounced and nearly simultaneous changes in their movements when rainfall was occurring over 100 km away (Garstang et al., 2014). Another study conducted at Etosha National Park found that while elephants did make movement choices based on localized conditions, they were strongly responsive to long-term patterns of productivity at familiar locations (Tsalyuk et al., 2019). While the authors describe the resultant movement strategies as contributing to an "ideal free distribution," they observed that elephants showed *increasing* preference for resources that were *declining*, which suggests otherwise. Because rain in this environment is so limited, the green flush that follows is very short-lived. Elephants are therefore under especially strong selection pressure to detect cues that help them detect and even anticipate plant productivity, perhaps much more so than in environments that are less extreme. What cues might these be? The possibilities are discussed in Chapter 7.

The need to range widely, and conserve energy for doing so, is likely the reason why females in this population appear to spend so little time socializing, as noted in Chapter 3. Competition within and among groups is likely to be high for both forage and water, which are circumstances under which one would predict the formation of very strict dominance hierarchies as well as spatial segregation among dominant and subordinate social groups as they jockey for priority of access to key resources at the optimal times of year. These kinds of interactions are yet to be studied, but similar processes are seen among male elephants at waterholes in Etosha, described earlier (O'Connell-Rodwell et al., 2011). These preliminary observations point to another challenge for preserving crucial knowledge in the face of resource scarcity, which is that animals may simply not have the time and energy to invest in maintaining the social relationships they might need to learn new behaviors and knowledge of the landscape.

6.2 Savanna Grasslands, Woodlands, and Deciduous Forests

What constitutes a savanna biome turns out to be surprisingly difficult to define (Ratnam et al., 2011), with different disciplines seemingly using the term as they wish. Most often, the term is used to refer to a landscape dominated by open grassland (Veldman et al., 2015), in which trees may be widely scattered. For instance, this was the common conception behind the early "savanna hypothesis" of human evolution, which contended that early hominins acquired bipedal posture and other specializations as a consequence of colonizing more open environments (Bender et al., 2012; Lamarck, 1809). For many, relatively open grasslands are also likely to be the image invoked in reference to African "savanna" elephants. These environments, which can also sometimes be described as arid or semi-arid, are maintained by seasonal rainfall regimes together with the interaction of fire and herbivory (Staver et al., 2009; Staver et al., 2011a, 2011b), elephants being a particularly influential agent of the latter (Holdo et al., 2009; Norton-Griffiths, 1979). As such, these biomes have the potential to be altered by changes in any of their components, such as the frequency of fires and the density and movements of herbivores (Holdo et al., 2009, 2010).

The dynamism of savannas also has the consequence that the term can also be used in reference to woodlands (Andersson et al., 2004), not to mention the much wetter regions that border wetlands and deltas. It now appears that the earliest hominins were actually also rather comfortable climbing trees, having inhabited a much wetter and more wooded environment than previously supposed (Ashley et al., 2010). It appears that ~650 mm of rain might be the magic threshold at which more open grass-dominated landscapes can convert to tree-dominated landscapes in the absence of fire and herbivory (Sankaran et al., 2005), and depending on how the rain is distributed annually. For our purposes, it suffices to think of savannas as representing a broad continuum between grassland interspersed with trees and open-canopy woodland interspersed with grasses, subject to seasonal rainfall of varying durations (Smith, 2016). In continental Asia, such habitats grade into scrubland and deciduous monsoon-watered forests. Savannas, meaning a gradient of grasslands and woodlands, therefore exist in both Africa and Asia, although in Asia there has been a struggle to recognize them as such (Murphy et al., 2016; Nerlekar et al., 2022; Ratnam et al., 2011, 2016). Many savannas are also the consequence of (or perhaps maintained by) direct human activity, both recent and past (see Chapter 8).

6.2.1 Eastern and Southern Africa

Elephants in these environments may either graze or browse, depending on the specific vegetation available, but they too try to surf the green wave whenever they can, following the fresh foliage and especially grasses that sprout after rain (Cerling et al., 2006). The researchers de Beer and Van Aarde (2008) investigated the relationship between home range size and landscape heterogeneity along a rainfall gradient ranging from the drier Etosha National Park (~200 mm of annual rain) and Khaudum Game Reserve, to Ngamiland District 11 (NG11) along the Okavango Panhandle (~650 mm annual rainfall). Whereas elephant numbers were reportedly relatively stable in Etosha National Park, the population at Khaudum had increased from just 80 in 1976 to more than 3,000 by 2008. At NG11 the population fluctuated between over 3,500 in the dry season and just over 1,000 in the wet season. Sixteen adult females, belonging to each of the three populations, were collared and followed. In Etosha, their dry season home range size decreased exponentially with increasing water point density in one out of two studied years. In Khaudum they found a similar relationship, but in the wet season and over both years (Figure 6.2). Elephants also chose patchier habitat (i.e. more heterogeneous) during the wet season in Etosha and NG11, whereas in Khaudum this was true of dry seasons. Conversely, elephants appeared to avoid more homogeneous areas. In all three locations elephants for the most part seemed to prefer areas characterized as "edges," which contain transitions between different habitat types, where the authors posit there to be more diversity in forage. However, there was no clear pattern among the three elephant populations in the way that home ranges were distributed relative to quantitative measures of patch diversity. Moreover, although it was expected that elephants would have more dietary choice and could afford to be more selective during wet seasons, when water availability is less constraining, the mixed results did not show this. Other studies show that although elephants may use more habitat types in wet seasons than dry seasons, as expected, specific fine-scale habitat choices vary individually, even in populations with relatively small home ranges that are on the order of a few hundred square kilometers (e.g. Maputo Elephant Reserve, Mozambique: Ntumi et al., 2005). One of the conundrums for a researcher studying a generalist herbivore is that if the species is not overly constrained by diet, it is at liberty to make decisions on the basis of other considerations (such as historical knowledge or intergenerationally acquired dietary preferences) that may be difficult to observe and perhaps too subtle to pin down.

Studies that track some subset of individuals may be able to relate movements to environmental variables, but the picture is likely to be incomplete because there is

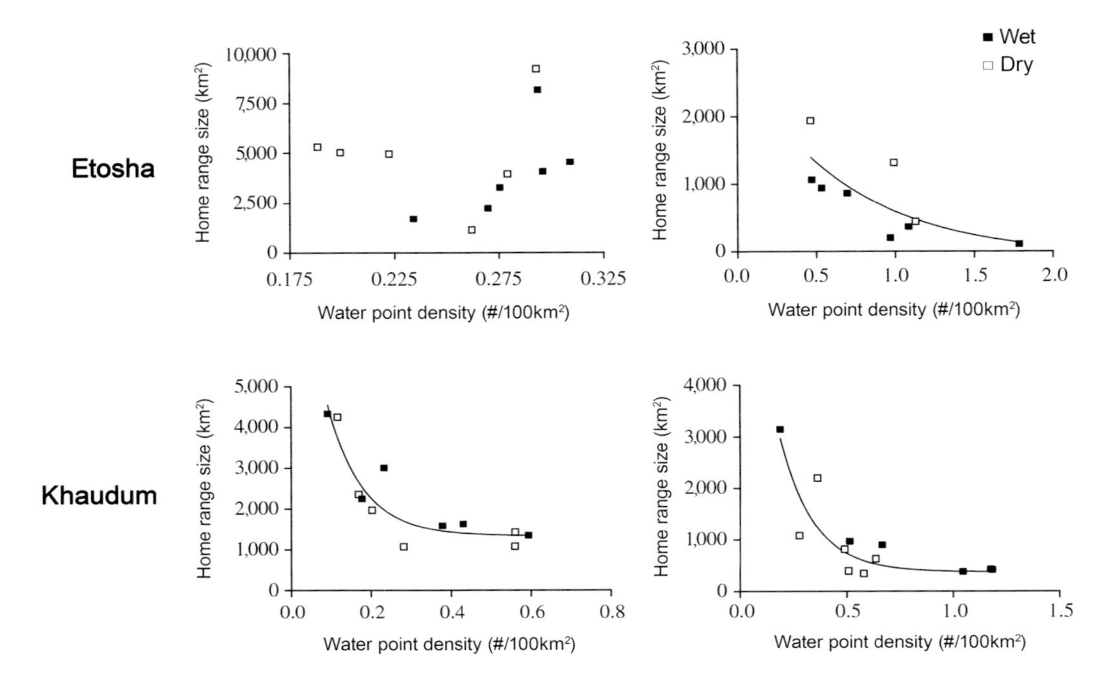

Figure 6.2 Relationship between *L. africana* home range size and water point density in arid/semi-arid areas. Each column is a separate wet–dry season pairing. Curves show significant relationships (modified from de Beer & Van Aarde, 2008, with permission of Elsevier).

another crucial yet invisible dimension to animal movement decisions: social relationships. Studies of well-studied family groups in East Africa show how animals juggle their needs for food, water, and safety while contending with competitors who are trying to do the same. In the Samburu-Buffalo Springs ecosystem, movements during wet seasons appear more random than in dry seasons, which is in keeping with expectations that they are more reliant on certain areas with key resources during the drier periods (Wittemyer et al., 2008). Movements inside PAs more closely matched measures of plant productivity than movements outside PAs, indicating that elephants have to trade-off their access to higher quality forage against the potential competition and disturbances presented by human or livestock activity (Boettiger et al., 2011). As one would expect, more dominant individuals (and, by extension their families) are able to maintain better access to areas in which all three sets of needs (safety, high quality food, water) are met, namely the core of the PA which is closer to water sources (Wittemyer et al., 2007, 2008), as is clear from Figure 6.3. In fact, the lowest-ranked family in the study actually showed significant avoidance of the PA, suggesting competitive exclusion by the higher-ranked families. Although home ranges in dry seasons were smaller than home ranges in wet seasons for all individuals irrespective of rank, those ranked lower nevertheless walked greater distances in dry seasons (Wittemyer et al., 2007). As discussed elsewhere, families that rely on water sources outside PAs not only face possible physical harm not only through poaching, but also competition from people and livestock.

Thus it would appear that lower rank carries a distinct energetic cost for elephants. Since familial rank is closely determined by that of the matriarch, and matriarch rank tends to be determined by her age and size (though there may also be personality effects), these observations illustrate a clear mechanism through which families benefit from the presence of the older individuals among them besides her knowledge of the physical and social landscape or ability to assess predatory threats (McComb et al., 2001, 2011; Mutinda et al., 2011). Do these effects also translate into fitness gains? This remains a very important yet outstanding question. A female's fitness derives from multiple components – her age at primiparity (i.e. first reproduction), her rate of reproduction, and her ability to rear calves successfully. Calf survivorship is strongly influenced by an individual's own age and environmental conditions, then to a lesser degree by the presence of other social resources, such as allomothers (Moss & Lee, 2011a). Among primates, there is a substantial body of evidence that social relationships have consequences for an individual's fitness, but the effect of rank in itself is not necessarily as important as the quantity and quality of those relationships (McFarland et al., 2017; Silk, 2007). Since dominance rank does appear to have very straightforward energetic consequences for savanna elephants, however, it might be possible to directly test whether and how the distances an individual has to travel affect reproductive success – provided that one can longitudinally monitor a population that is not affected by other confounding stressors, such as poaching. This has not yet been done.

Zimbabwe and Botswana, which share a common border, today contain some of the largest elephant populations in Africa (Chase et al., 2016), offering a glimpse of how elephants may have once behaved when they were more numerous. Northern Botswana contains the Okavango delta along with Chobe National Park, Moremi Game Reserve, and Nxai Pan National Park. Elephant populations cross into neighboring Zimbabwe and Namibia. It was found that home range size increased along with greater distance to dry season surface water (Verlinden & Gavor, 1998). Studies in Botswana as well as Zimbabwe show that some herds can be characterized as resident, while others migrate to varying degrees (Tshipa et al., 2017; Verlinden & Gavor, 1998). These two groups partitioned the habitat differently, with residents favoring different woodland tree species than migrants (Verlinden & Gavor, 1998). Migrants also traveled much greater distances to surface water, up to 200 km in the dry season (Verlinden & Gavor, 1998). One naturally wonders why individuals choose one strategy or another. Based on the studies in East Africa, it might be expected that residents dominate migrants, with migration being the result of competitive exclusion. But a closer look at elephant movements in and around Hwange National Park throws up a hitch in this hypothesis. Tshipa et al. (2017) report that during dry seasons elephants congregate at artificial water points located within Hwange, overlapping their ranges substantially as a result (Figure 6.3). As soon as it rains, individuals return to their wet season areas, which contain more ephemeral water sources. This suggests that the scarcity of water in this environment is such that the draw of the man-made water sources is stronger than the social intimidation that individuals might face. It's also possible that there is some attribute of these water sources (e.g. size, dispersion,

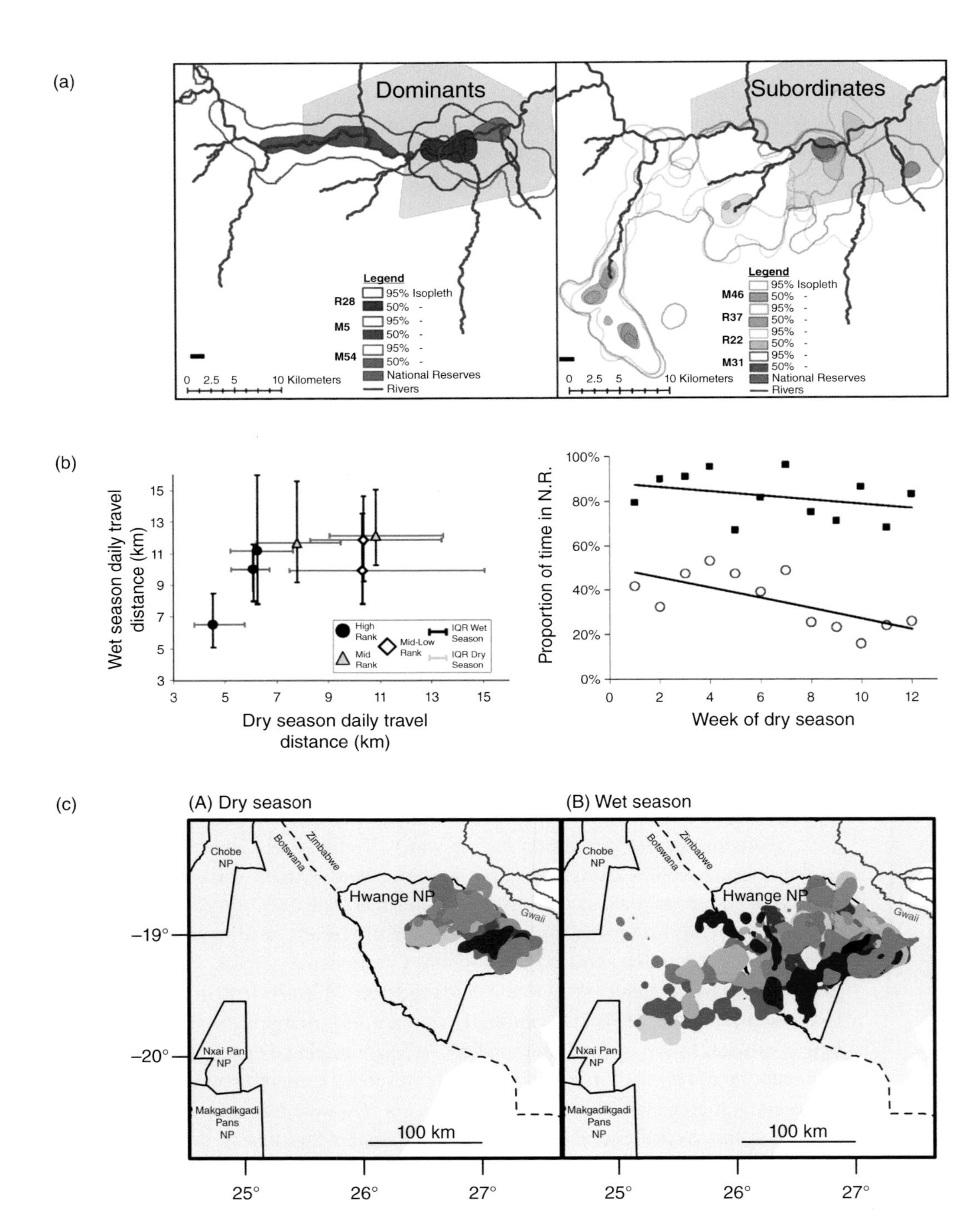

Figure 6.3 Spatial relationships among social groups. (a) At Samburu National Reserve, Kenya, higher-ranked matriarchs and their families are able to stay closer to water in dry seasons. (b) As a consequence, they need to travel shorter distances than lower ranked individuals, therefore lower ranked families (open circles) thus show a significant reduction

numerical abundance) and their local environment (dispersion of forage or cover) that makes monopolization by dominants more difficult.

A study at Sengwa Wildlife Research Area in northwest Zimbabwe adds a genetic dimension to the story. Individuals that share similar dry season ranges were defined as "clans" in this study (Charif et al., 2004). It turned out that clans consisted of two or more mtDNA haplotypes, or maternal lineages, and the same haplotypes could be found in more than one clan. Because the divergence time of mtDNA haplotypes is quite long, it's possible to say that individuals with different haplotypes originate from different matrilines, but individuals sharing the same haplotype need not be from the same matrilineal family. Individuals sharing the same haplotype therefore were not necessarily family nor were they more similar in their home ranges. On the other hand, at least four pairs of individuals from *different* families appeared to coordinate their movements by remaining within hearing range. Three of these pairs consisted of individuals with different mtDNA haplotypes. So it would appear that elephants in the Sengwa Wildlife Research Area population not only share dry season range rather amicably, but they even do so with apparently distantly related individuals. Given the observation in Samburu that unrelated individuals can become so closely bonded as to be indistinguishable from family members (Goldenberg et al., 2016; Wittemyer et al., 2009), these space-use patterns beg to be understood not only in terms of what elephants are doing in the moment, but also their history.

Kruger National Park contains elephant census data for the period spanning 1985–2012, which served as the basis for examining what influenced habitat selection by bulls versus female groups (MacFadyen et al., 2019). Although elephants in general preferred to be near rivers, bulls tended to use areas with lower rainfall that were susceptible to a higher frequency of fires, whereas female groups did the opposite. Toward the end of this time span, the population of elephants just about doubled from around 8,000 to 16,000. The segregation between bulls and females grew less pronounced as density increased, but female groups intensified their use of different areas more than bulls did. However, these fine-scale sex differences might be less important when one considers a larger spatial scale in relating elephant movements directly to plant phenology. In another study that tracked 68 individuals from 12 sites spanning 7 countries representing a strong rainfall gradient, it was found that elephants in general choose areas that are "greener" than average (Loarie et al., 2009). In dry seasons, these tended to be well-wooded habitat, which exhibited little seasonal variation in productivity. In wet seasons, elephants were more likely to use open woodlands, shrublands, and grasslands. This pattern held irrespective of the typical amount of rainfall each landscape received and in spite of local constraints on surface water availability.

Figure 6.3 caption (cont.)

in their use of the National Reserve whereas higher ranked families (closed circles) do not ((a) and (b) reproduced from Wittemyer et al., 2007, with permission from SNCSC). (c) By contrast at Hwange National Park in Zimbabwe, elephants converge around water sources in dry seasons and overlap their movements much more than they do in wet seasons (reproduced from Tshipa et al., 2017, with permission from Elsevier).

The movement of elephants can be influenced by the provision of resources by people, as one might well expect. Often, this resource is water. The impact of doing so has been much debated, and I do not here attempt to present an exhaustive examination of this topic (Chamaillé-Jammes et al., 2007; Guldemond et al., 2017; Tshipa et al., 2017). But in keeping with the preceding discussion is another study at Kruger, which documented the movements of 26 females and their families to and from natural as well as artificial water sources over a period of nearly two years (Purdon & van Aarde, 2017). The natural sources consisted mainly of rivers, whereas the artificial water holes were point resources scattered at varying distances away from the rivers. Although elephants predominantly used rivers, the presence of artificial water sources appeared to allow them to move nearly twice the distance away from rivers than they would have if reliant only on the river water. The preference for rivers might have to do with the availability of both forage and shade, but the water holes allow them to use areas that they might not otherwise (or at least do so at a lower intensity). This might reduce their impact on areas closer to the natural water sources, but conversely, it could increase their impact on other areas that are more distant. Purdon and van Aarde (2017) express concern that such water sources de-couple elephant populations from natural processes, including those that might suppress or regulate their populations, which might have even heavier, long-term consequences for local ecosystems. One might wonder, as well, whether there is a flipside to this. If elephants are sufficiently well-provisioned artificially at a particular location such that they need not leave it, might this gradually reduce their familiarity with seasonally varying resources, thus eroding their overall resilience as a population? As discussed in preceding sections as well as Chapter 5, seasonal foraging requires learned behaviors – especially those that are long-range migrations. Elephants that rely on rare resources undertake seasonal movements to access them. If they cease to do so thanks to the convenience offered by human provisions, they may not regain this ability and thus remain human-dependent. This could then convert a resource provided with well-meaning short-term intentions into a long-term ecological trap. On the other hand, provisioning may be deemed necessary to replace a natural resource that has been disrupted through human activity. The aims and approaches must be carefully evaluated, with similar concerns arising for both continents.

Elephant populations do have the capacity for adjusting their behavior as conditions change, as we have seen. The Laikipia plateau of central Kenya contains the country's second largest population of elephants, many of which range outside PAs (Graham et al., 2009). Interestingly, it would appear that this is quite a recent phenomenon, as researchers note that elephants were not observed in the area by nineteenth-century colonial explorers, but numbers seem to have increased since the 1960s (Graham et al., 2009, and references therein). It was strongly suspected that there may have been immigration into the area in the 1970s from the nearby Samburu district, which had experienced much more intense poaching pressure. Following the second wave of poaching during the 2000s, another study was conducted in the neighboring Laikipia-Samburu ecosystem where several known families were tracked over a period of 17 years, a longer window of observation than is typical for movement studies. It demonstrated that ranging patterns of a family can indeed change from one generation to the next (Goldenberg et al., 2018). First, it was observed that there was a general shift northward that seemed to be associated with increases in primary productivity. Range sizes increased for some families and decreased for others, resulting in no difference overall among all families, but younger individuals tended to expand their ranges relative to older individuals. Noticeably, certain families showed movement away from heavily poached areas whereas others did not. Range expansions were associated with areas showing increased productivity and a lower density of poached carcasses, not necessarily located within designated sanctuaries. However, the authors note that such areas did include long-established communally managed lands that benefited from ecotourism. One family did begin to use a PA located quite a distance from their original range, but unfortunately there were no data on poaching intensity to determine whether this actually represented a safer location.

The propensity of carcasses to be found in particular areas allows us to dig a little deeper into the interplay of elephant and human behavior. Analyzing patterns of carcasses detected through aerial surveys over Ruaha-Rungwa National Park in Tanzania, researchers set out to investigate whether elephants as well as poachers were changing their activity over time and whether the presence of ranger posts had any deterring effect on illegal hunting (Beale et al., 2018). The elephant population using this landscape had reportedly plummeted by more than 50% since 2009, possibly indicating what has been described as poaching at "a near industrial scale," amounting to thousands of animals per year (Beale et al., 2018). These are staggering figures, all the more so because the decline seemed to be happening despite active ranger patrolling efforts. Beale et al. propose that at least some of the apparent declines might have been due to temporary emigration from the landscape in response to poaching pressure as well as chance undercounting during one of the surveys. But elephants are tethered to areas that contain critical resources, and aside from a slight dry-season shift between 2013 and 2015 to favor more open areas, the types of habitats used by elephants did not change much. During dry seasons, which is when aerial surveys were conducted, live elephants were seen in higher densities toward the southeastern periphery of the landscape near the Ruaha river. A greater proportion of the ranger posts also occurred along these areas (Figure 6.4). On the other hand, carcasses were concentrated centrally, in areas elephants were presumably using during wet seasons, when water availability was less constraining. The locations of carcasses shifted more toward the interior over the three-year period but did not significantly change in terms of the environmental conditions associated with them. However, the association between carcasses and individual ranger posts varied quite a bit. The picture that emerges is that poachers appear to have been concentrating their activities around wet season water points toward the center of the PA, away from other human activities (which include tourists as much as patrols), possibly in areas that elephants simply could not avoid.

As already mentioned, another prominent process is the ongoing land-use change that has fragmented elephant habitats throughout the world. As the bulk of these studies imply, elephants obviously can and do range outside the confines of PAs,

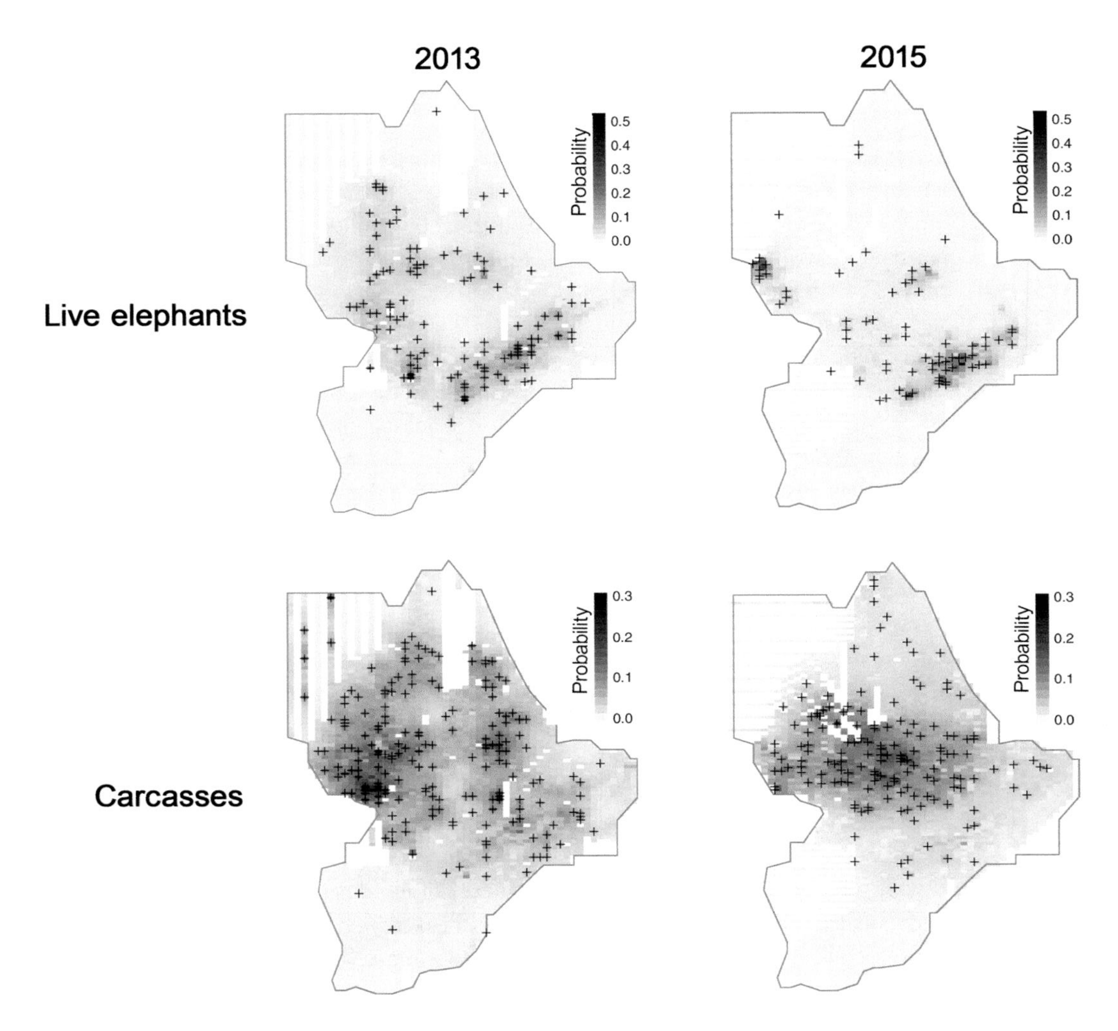

Figure 6.4 Locations of live elephants, carcasses, and patrolling stations at Ruaha National Park, Tanzania. Aerial surveys performed during dry seasons between 2013 and 2015 record a slight shift in sightings of live elephants toward more open and peripheral areas of the park but carcasses are mostly recorded in the interior. Dots show the locations of ranger posts, most of which are positioned near a river course (modified from Beale et al., 2018, with permission from Elsevier. © 2017 The authors. Published by Elsevier Ltd.).

encountering cattle, settlements, and croplands. But they are usually a lot more difficult to observe in such settings. This is an issue that tracking devices and trail cameras can help to shed some light on. Several studies have investigated how elephants deal with the risks associated with human use of these areas. Often, wildlife tend to avoid humans through temporal partitioning of the habitat, which is a fancy way of saying they use the landscape at different times of day than people do (Gaynor et al., 2018b). This has been shown in multiple species, notably carnivores (Carter et al., 2012; Gaynor et al., 2018b; Weaver et al., 1996). Unsurprisingly, this is also a strategy employed by elephants (Adams et al., 2022; Buchholtz et al., 2019; Galanti et al., 2006; Gaynor et al., 2018a; Graham et al., 2009). Graham and colleagues tracked 13 animals of both sexes over a period of two years in a mixed-use landscape on the Laikipia plateau, which contained a mix of smallholder farms, pastoralist areas, and large private ranches as well as some reserves. People in some areas were more tolerant of elephants, as they benefited from tourism revenue. They found that elephants were more nocturnal in areas where they were less tolerated (defined as areas in which they were at risk of injury or death upon encountering people), than in areas such as large-scale ranches where they were more tolerated (Graham et al., 2009). Surprisingly, they seemed to actually prefer large-scale ranches the most, followed by forest reserves. There was variation among individuals as well, with some that were habitual crop foragers spending more time in smallholder areas than others, as might be expected.

Another study that included multiple populations in Kenya and bordering Tanzania found that both males and females used PAs as the core of their home ranges, tending to move more quickly when passing between different reserves than when they were inside them (Douglas-Hamilton et al., 2005). Such rapid movement is sometimes referred to as "streaking." Yet another study conducted in Tanzania focused exclusively on females, all of whom were initially collared inside Tarangire National Park (Galanti et al., 2006). However, the study identified what appeared to be two distinct clans. Those that occupied the north-central part of the national park stayed within the PA most of the time, whereas those in the southern part undertook migrations that resulted in home ranges that were approximately five times larger than the northern home ranges. Clearly, elephants are not all alike in the ways they use space, even among subsets of the same population; therefore the inferences that can be made from any particular study of movement will rest on which individuals happened to be tracked, and where. If there is a single generalization to be made, it is that elephants are very aware of the risks they face outside PAs; but they manage these risks in distinct ways depending on sex, dominance rank, and foraging strategy.

These studies collectively illustrate the complexity of elephant movement decisions. Elephants move either to access water directly or to access forage, some of which is itself water-reliant. Grasses and other herbaceous vegetation are preferred if they are available and accessible, but coarser or woodier species serve as fall-back food. Depending on which is more limiting, elephants constrain their movements in space and time, but in some cases their ability to exploit these resources may also be influenced by the presence or absence of other elephants, whether of the same or opposite sex. Another general pattern is that PAs are attractive to elephants for several reasons. First, they often tend to contain prominent water features that are either naturally or artificially maintained. Many PAs also exclude livestock, thereby eliminating a source of competition for both food and water. But what about predation? Where PAs contain robust populations of large predators, one might expect this to negate some of their benefits. However, in many landscapes elephants face a greater danger from humans, and PAs should in principle be a refuge from illegal hunting. As mentioned earlier, a large-scale aerial survey across Africa found that most elephants were indeed seen inside PAs (Chase et al., 2016). But surveys were also largely focused on PAs, where most elephants were expected to be found – such surveys necessarily being performed during the daytime. Unfortunately, the study also reported that the number of carcasses detected within PAs did not differ significantly from those detected outside them, which indicates that PAs are not necessarily shielding elephants from hunting pressure, and that elephants are not exclusively reliant on them. The weight of these studies seems to indicate that space use is primarily governed by resource availability, with animals having to enter certain areas despite their risks. The challenge for conservationists and wildlife managers, then, is to find ways of reducing these risks, both to elephants and to people.

6.2.2 South Asia

Asian elephants switch between graze and browse opportunistically, with the relative proportions of each varying quite substantially even in adjacent areas (Choudhury et al., 2008, and references therein). Whereas elephants may be sharing landscapes with many other herbivores when viewed at a coarse scale, space use among different species can be partitioned more finely among habitat types even within a single PA. A study in Royal Bardia National Park in Nepal, where elephants and rhinoceros cohabitate, found that elephants made extensive use of two different forest types as well as tall grass floodplain (Steinheim et al., 2005). Rhinoceros used the floodplain as well, but only one of the two forest types. The differences might reflect dietary choices, discussed earlier, while overlap very likely reflects common needs – notably, water. The researchers speculate that as populations increase within the limited habitat, elephants could outcompete rhinoceros, given what appears to be their broader niche breadth. Constraints of available habitat and artificial barriers that attempt to limit dispersal can lead to the overexploitation of habitat within PAs by any number of resident species, with largely unstudied consequences.

In comparing the movement habits of Asian elephants with their African counterparts, a striking difference is how much smaller home ranges in Asia appear to be. Asian elephants do not appear to undertake regular long-range seasonal movements from one foraging area to another, and generally maintain strong fidelity to their ranges, although they may use different areas preferentially during wet as opposed to dry periods (Fernando et al., 2008). However, given that they do seasonally move across ecological and political boundaries, they are often termed "migratory" and have been recognized as such under the Convention on the Conservation of Migratory Species of Animals. Nevertheless, PAs in South Asia do not serve elephants in quite the same way as they do in the African context. One major reason is that PAs in South Asia tend to be much smaller (Chowdhury et al., 2022). In India, PAs average around 300 km^2 and cover less than 5% of the total land area (Karanth & DeFries, 2011). Sri Lanka has managed to offer some level of protection to something like 15,617 km^2 or ~22% of the land area, at least on paper (Silva, 2011). By contrast, in Namibia Etosha National Park alone covers 22,270 km^2; Kruger National Park in South Africa has an extent of 19,480 km^2; Tsavo East in Kenya covers 12,750 km^2; Hwange National

Park in Zimbabwe covers around 15,000 km^2, and the Moremi Game reserve in the Okavango delta of Botswana is just under 5,000 km^2. Although smaller PAs of course do exist, and it is difficult to come up with an average figure owing to the existence of numerous conservancies and private reserves, it is clear that they add up to be much larger in extent in Africa than in Asia.

Just half of the elephant range in Asia can be considered "wildlands" and only around 8% of the total range had any form of legal protection by the turn of the millennium (Leimgruber et al., 2003). Elephants in South Asia (India, Sri Lanka, Nepal, Bangladesh, and Bhutan) range in what now appear as deciduous woodlands, scrub jungle, and grasslands, as well as mosaics consisting of agricultural fields and plantations including timber, tea, rubber, various palms, and coffee. Several studies have by now documented that even the designation of PA status does not alter land use conversion rates, with PAs in South Asia being largely indistinguishable from unprotected areas, and up to a quarter being human-modified (Clark et al., 2013; Neupane et al., 2020; Ram et al., 2021; Rathnayake et al., 2022a). As a consequence, it would appear that elephants in South Asia have to make a living alongside people to a great degree. This has become an increasingly difficult proposition. A study in Nepal found that 60% of cases in which "elephant attacks" on humans occurred in an area where forest was converted into agriculture or settlements between 1930 and 2020 (Ram et al., 2021). In India, several hundred human fatalities occur annually owing to elephants, resulting in both psychological and financial hardship (Barua et al., 2013; Gulati et al., 2021).

The largest elephant populations occur in India, being the largest of the range states, where there were still wildlands in this century with an extent of at least 2,000–4,000 km^2 (Leimgruber et al., 2003). It is therefore not surprising that elephants in India have some of the largest home ranges documented in Asia. The notable exception to this was a female in Malaysia, tracked following a translocation attempt, whose apparent range was over 6,000 km^2 (we will discuss southeast Asia in a later section). In parts of India there are landscapes that are sufficiently large as to contain different forest and other natural habitat types sometimes in relatively close proximity, allowing us to assess habitat preferences. At least one set of observations from the Brahmagiri, Nilgiri-Eastern Ghats landscape suggests that when multiple social groups have the option to choose among different habitat types, their area use may be dictated not only by habitat types but also by social factors (Nagarajan et al., 2018). Researchers tracked two males and two females across four different forest types (dry deciduous, moist, mixed deciduous, and thorn). Of the females, both initially used dry deciduous forest to a greater extent than any of the others, but one of the two seemed to use the less productive thorn forest disproportionately more in later seasons. The authors speculate that this shift occurs owing to competitive exclusion as a result of dominance hierarchies, bringing about an "ideal despotic distribution." As discussed in Chapter 3, there is as yet no evidence for the existence of clear dominance hierarchies in Asian elephants, with spatial avoidance being a mechanism that prevents both conflict and hierarchy formation. Based on the small sample and limited behavioral observations, this study cannot offer supporting evidence for the

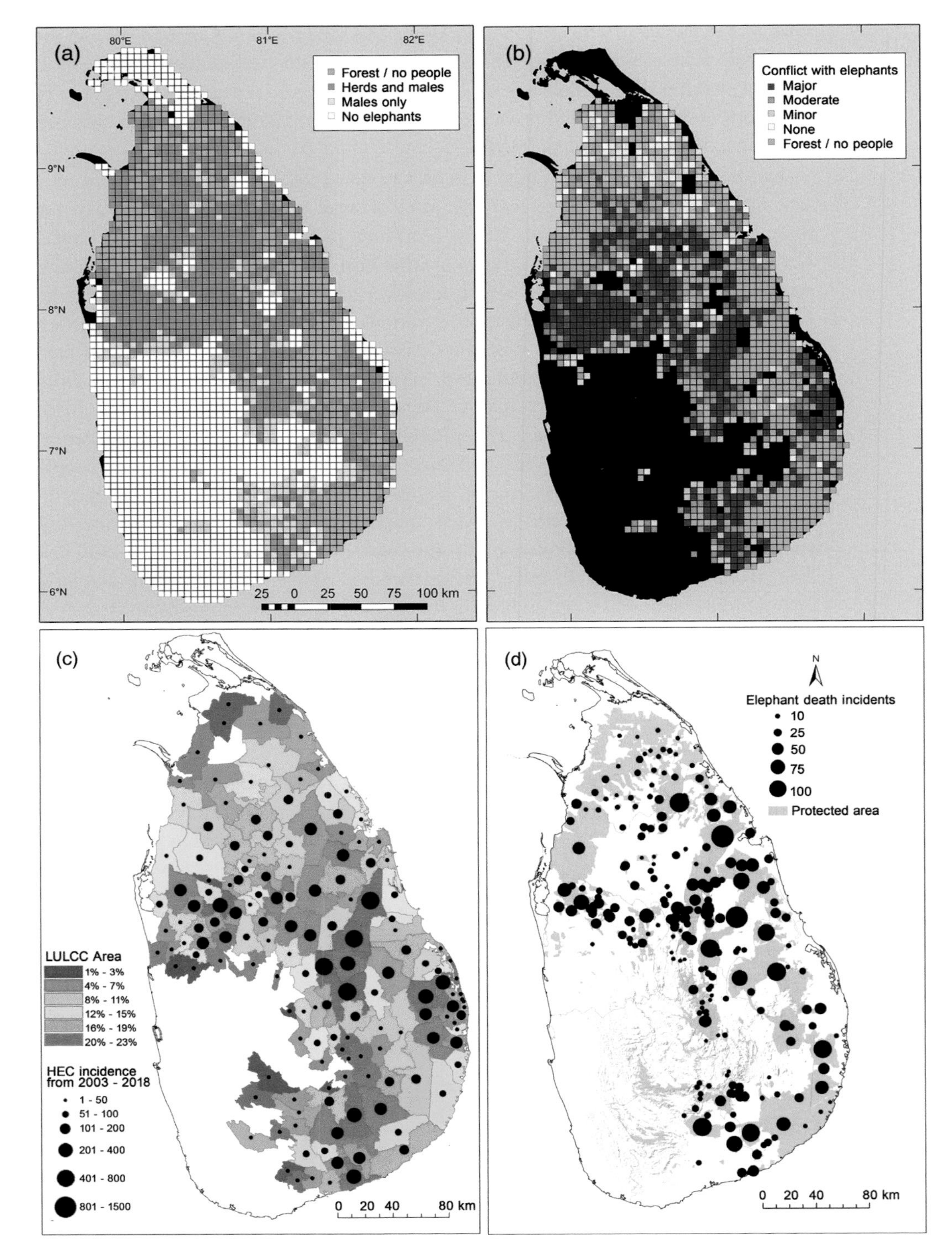

Figure 6.5 Land-use, PAs, and conflict in Sri Lanka. (a–b) Upper panel: Grids are 5 x 5 km cells in which elephant occurrence and conflict levels were surveyed using questionnaires (reproduced

claim, but does raise the question of why some individuals would use what appear to be suboptimal habitats, if not because of some form of competitive exclusion. Possibly individuals or social groups that are well-established in an area might show aggression toward others that are newer and less well-established, resulting in their exclusion.

Elephants in Sri Lanka appear to have relatively small home ranges, compared with elsewhere in South Asia. An island-wide survey claimed that nearly 70% of the area elephants use was outside the existing PAs, and nearly two-thirds of this consisted of areas with substantial human activity (Fernando et al., 2021). But the PAs considered were not complete, as many of the smaller areas of lesser status were without historical data, so it is impossible to say whether the smaller ranges represent a long-established norm, or are the result of more recent habitat loss and resultant space constraints. There may be some behavioral adjustments to risk exposure. Females and calves appear to be a bit more restricted to the interior of the existing range (which is *not* to say that they are restricted to PAs), whereas males also persist on the edges (Figure 6.5a). This may be because males are much bolder and less risk-averse than females, and in Sri Lanka are far more likely to venture into croplands. Moreover, bulls may be forming longer-lasting social groups in response to these recent land conversions to allow them to better exploit these risky resources, as discussed in Chapter 4. However, it is also possible the apparent absence of female groups in certain grids adjacent to those that are known to have them (e.g. PAs) is because of their tendency to be active almost exclusively at night, and thus be less frequently seen by people (Figure 6.6c). Corroborating this, we found that more than three-quarters of the elephant population of Udawalawe National Park, including both sexes, was likely to be ranging outside the PA (Madsen et al., 2022). In keeping with the fluidity of female associations observed in this population (and in contrast to African savanna elephants), female social groups did not disperse as a unit; while larger social units had more residential individuals, all social groups had individuals that were less residential. In fact, only half of the 130 adult females observed within the first two years of the study were seen in all nine years of the dataset.

If this pattern is broadly reflective of the many landscapes in south Asia with substantial human populations, it would indicate a remarkable feat of behavioral

---

Figure 6.5 caption (cont.)
from Fernando et al., 2021 © Fauna & Flora International 2019. Some of the designated PAs are shown in green (but many small PAs are omitted). (a) People report seeing males only (light blue) along range edges whereas both males and groups of females are seen throughout most of the range (dark blue); nevertheless, elephant herds are present along range edges (see Figure 6.6). (b) Intensity of conflict as reported in the survey. (c) Colors represent percentage of administrative divisions undergoing land-use and land-cover change (LULCC) between 1993 and 2018 (green 1–3%, red 20–23%) and total incidence of HEC between 2003 and 2018 scaling from between 1–50 and >800 (reproduced from Rathnayake et al., 2022b with permission from Elsevier Science and Technology Journals). There is a clear association between LULCC and the number of incidents (see also Figure 6.6a). (d) Elephant deaths (circles scaled 10–100) number of elephant deaths occurring in or within 5 km of PAs (light green) between 2003 and 2018; reproduced from Appendix 2 of Rathnayake et al., 2022a with permission from Elsevier.

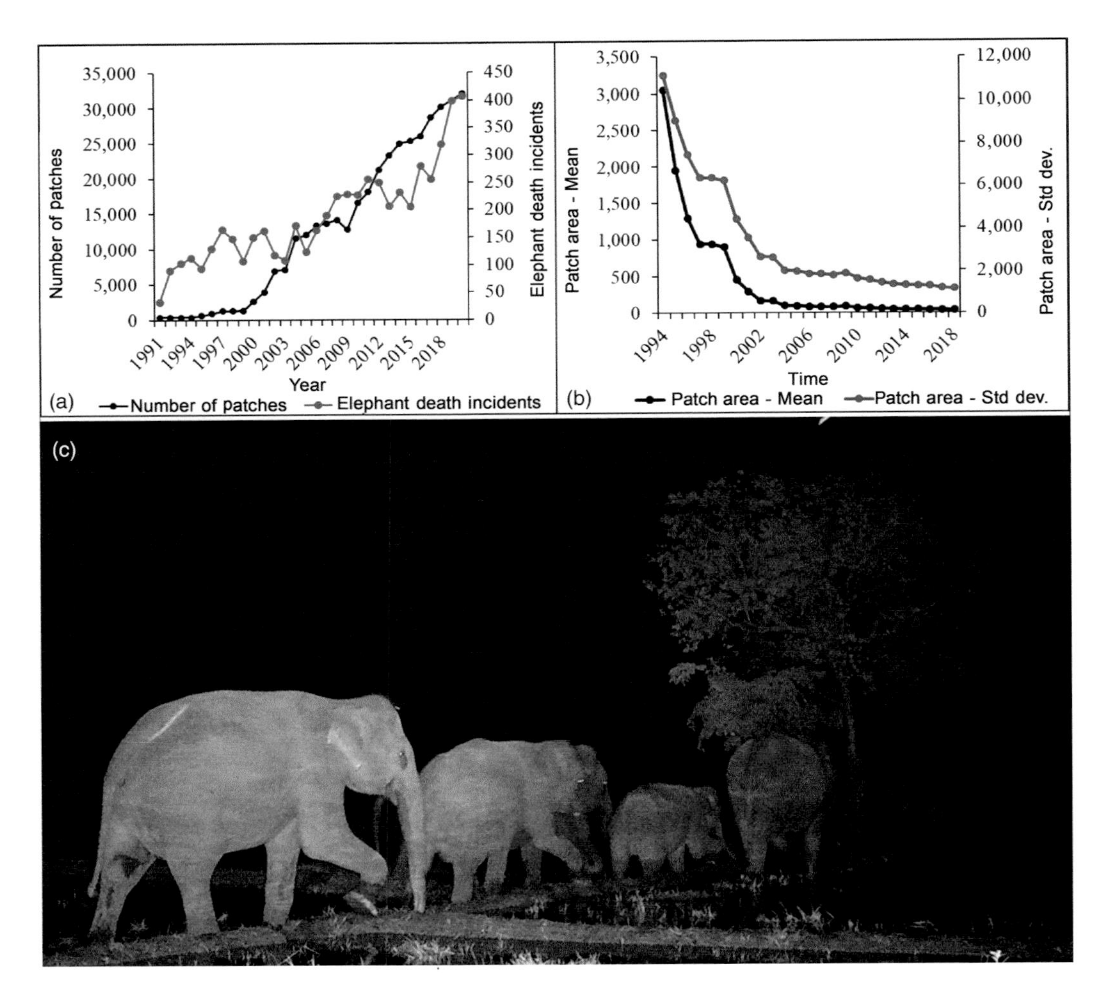

Figure 6.6 (a) Increasing forest fragmentation since 1991, quantified as a rising number of "patches," is significantly correlated with an increasing number of elephant deaths. (b) Increased fragmentation is associated with smaller average patch size ((a) and (b) reproduced from Rathnayake et al., 2022a, © 2022 The authors. Published by Elsevier BV, Open Access). (c) Elephant herd negotiating a paddy field outside Udawalawe National Park. Although males are more likely to be seen on the edges of elephant range than females (e.g. Figure 6.5a), camera traps and more extensive surveys show that herds most likely also do occur (photo by the Udawalawe Elephant Research Project). Elephant herds within the National Park are visibly active during the day, but outside the park they are active almost exclusively at night and are thus rarely seen by people.

adjustment to human landscape alterations sometime in the recent or distant past. A study outside Chitwan National Park in Nepal showed that tigers and people often use the same trails, but largely avoid encountering one another because tigers seem to avoid the paths during the daytime, when people are most active (Carter et al., 2012). As is the case with some African populations, elephants seem to be displaying a similar tendency (although as also noted earlier, negative encounters do occur and are increasing in frequency). Where the elephants actually are often may not be consistent with where people expect them to be (Kumar et al., 2018).

Fernando and colleagues propose a reasonable explanation for how elephants are persisting outside PAs, at least in Sri Lanka. When the landscape consists of a mosaic of seasonal and shifting agriculture, elephants can forage on the ample vegetative growth that takes place during fallow periods (Pastorini et al., 2013). Grasses and other fast-growing herbaceous species with potentially high nutritional value are maintained in abundance, while nonpalatable introduced species such as *lantana*, which are highly invasive inside PAs, are suppressed by a combination of fire and constant removal (personal observations). Shifting cultivation (c.f. "slash and burn", "swidden" or "*chena*" agriculture), which has for a long time been denigrated by land managers, creates a disturbance regime in an environment that would otherwise have a relatively poor understory and be dominated by trees. Such practices could in principle benefit herbivores, so long as they can have access to these habitats (see Chapter 8). Frequently, these areas also contain man-made water sources, some rather ancient (Pastorini et al., 2010). Together, such landscapes may support a larger number of elephants within a smaller area than one might expect based on rainfall or ecotype alone. Depriving access to these resources, as would be the case by attempting to restrict elephants inside PAs alone, is likely to be very detrimental (Fernando et al., 2008; Madsen et al., 2022). However, when elephants range outside PAs in the present-day context, on landscapes dominated by agriculture, the outcomes are also not favorable, as people report high levels of conflict with elephants (Figure 6.5b). The mere presence of elephants outside PAs, even in great numbers, does not in itself indicate anything about how well the populations are doing. Inside PAs, elephant herds are clearly active during the day, especially during wet seasons when they can forage almost continuously with few rest breaks and remain active at night (personal observations). If they have to forego daytime foraging outside PAs in the interest of safety, this would impose a substantial energetic cost. In Sri Lanka, where females by and large do not participate in crop foraging, they are certainly not offsetting this cost with a richer diet. They likely pay a heavy price for habitat loss that goes unrecognized.

In contrast to the African continent, where the patterns and causes of elephant mortality have received much attention owing to the role of poaching, there has been a dearth of research on the drivers of elephant mortalities in Asia. Indeed, even when mortality data are collected by governmental agencies, they are often unavailable to the wider community, with few exceptions. Since the early 2020s, Sri Lanka has been recording ~400 human-caused elephant deaths per year (Department of Wildlife Conservation, Sri Lanka), exceeding prior decades. A painstakingly detailed set of studies relating conflict incidents in Sri Lanka to land-use and land-cover (LULCC) change demonstrates unequivocally that land-use conversion lies at the heart of negative incidents (Figure 6.5c and d). The authors first characterized LULCC trends between 1993 and 2018 using satellite imagery, and then related the timing of conversions to conflicts incidents in official government records (Rathnayake et al., 2022b). The number of reported incidents, which included human and elephant deaths, human injuries, and property damage, was significantly positively correlated with increased rates of LULCC. Strikingly, 98% of conflicts occurred within 1 km of an area that had experienced a recent conversion. The number of documented elephant deaths also clearly increases with the degree of fragmentation. As injuries to elephants are generally undocumented, as well as indirect effects such as reduced body condition, lowered birth rates, and offspring survival, these additional negative impacts on elephant populations are impossible to quantify directly. The demographic consequences of living on human-dominated landscapes therefore remains a critical unknown.

One also must not forget, as it is so easy to do, that these developments are relatively recent. There is a common perception that "the majority of Asian elephant range has been densely populated by humans for millennia, and elephants frequently come into contact and conflict with people. As a result, most Asian elephants are behaviorally adapted to avoid humans" (Pastorini et al., 2010). This seems to exemplify the classic problem of *shifting baselines* (aka the predicament of the slowly boiled frog). A shifting baseline is the phenomenon by which changes to the environment or animal populations occur slowly enough that people fail to notice they are occurring. Observers today perceive several large Asian elephant populations to be existing in very fragmented, disturbed, and crowded landscapes. But the human population explosion is a relatively recent phenomenon relative to the evolutionary timescale (or even the generation time of elephants), as are industrial agriculture and forest cover loss. One merely has to witness the fact that the massive urban centers and associated infrastructures constructed during the heyday of Asian civilizations were almost entirely engulfed by forest cover until their excavation by archeological expeditions in the early twentieth century. Certainly, the footprint of these civilizations and their farming practices may have shaped the landscape, in the form of altered hydrology, soil, and plant communities. It is interesting to consider that since elephants appear to have a particular affinity for flood plains and riparian habitats of the sort which first led to the (multiple) domestications of rice (Clift & Guedes, 2021), humans may have effectively been creating ideal elephant habitat for centuries if not millennia through our agricultural practices. But it simply cannot be claimed that elephants confronted humans at similar densities as they do today because there were simply fewer people on earth and they were, by and large, more weakly linked economically (although these links may nevertheless have been far flung). We will dig into the evidence for this in greater detail in Chapter 8.

6.3 Moist Evergreen Forests, Rainforests, and Grasslands

The primary distinction between the ecosystems represented by this section and the preceding one is that plant productivity remains more or less constant throughout the year. Although most such environments might commonly be referred to as rainforests, at higher elevations they also give way to cloud forest and even grasslands. In Africa, they surround the equatorial midline of the continent, largely following a rainfall gradient. The remaining equatorial forests of Asia that contain any elephant populations are primarily found in Southeast Asia, much of the forests in and around the Indian subcontinent having been heavily reduced through human activity (see Chapter 8). Elephants in these environments are not as limited by water either in terms of surface water sources or moisture content in food. But although availability of forage may not represent a limiting condition, particularly dense vegetation can itself impede efficient movement. Moreover, this visual abundance can be somewhat deceptive, as individual plant species are very patchily distributed and preferred or high-nutrition foods such as fruit can be ephemeral. As a result, large animals maintain networks of well-worn trails, or pathways. Such game trails also feature in deciduous forests and even grasslands, very likely representing efficient routes between key resources. Their function becomes more obvious in more heavily forested areas, where they connect and highlight locations of special interest for elephants.

It is becoming difficult to determine what might have historically governed elephant space use under undisturbed conditions, in light of perennial and increasing human pressures. The distribution of elephants is reflective not only of where resources are but also of human activities. The development of linear infrastructure through forest habitats can be particularly devastating for the species reliant on these environments, even more so than for those that inhabit more open areas. Forests contain complex architecture with low light levels and microclimates to which interior species are adapted, thus linear openings can present a sensory barrier to their movement, not only a physical impediment (Laurance et al., 2009). Roads also tend to proliferate. At the same time, these openings facilitate hunting, logging, mining, and other forms of destructive resource extraction. Hunting is expected to have a more severe impact on forest species in the near term than even climate change, owing to the rapidity with which it can decimate animal populations (Abernethy et al., 2013). Many studies over the past decade have described "empty forest syndrome," the phenomenon by which forest stands are effectively emptied of their wildlife through hunting (Wilkie et al., 2011). One review noted that bird and mammal taxa larger than around 2 kg, barring few exceptions, have been either wiped out or severely reduced in abundance, even inside tropical nature reserves (Harrison, 2011). Indeed, the author observes, land that includes logging and oil concessions can harbor more wildlife than PAs if their managers actively curtail hunting.

6.3.1 West and Central Africa

The possibility that heavy reliance on seasonally available fruit might be responsible for the smaller group sizes observed in African forest elephants is a long-standing idea (White, 1994). But on the other hand, local elephant densities can temporarily increase when preferred species are fruiting. A study at Lopé Reserve (now Lopé National Park) in Gabon compared elephant dung densities in areas with fruiting *Sacoglottis gabonensis* trees to areas without this species, and inferred that elephants may be converging on areas with these fruit trees from a neighborhood of approximately 3,000 km^2 (White, 1994)! A separate study, also in Gabon, tracked a remarkable sample of 96 elephants from the vicinity of six different national parks over multiple years and found that the mean movement distance was 2,463 km annually, though the average home range size was just 195 km^2 (Beirne et al., 2021). Annual rainfall had the strongest positive effect on movement, whereas temperature seasonality had the strongest negative effects. Females had smaller home ranges than males. The study additionally looked at the amount of variation in three other aspects of movement: diurnality (day/night activity patterns), site fidelity (the amount of overlap in the annual home ranges of a given individual), and the proportion of exploratory behavior. Forest elephants tended to be slightly more diurnal, females more so than males. There was also an 85% overlap in home ranges across years, greater for females than males. The variation in exploratory behavior was similar to that of displacement, with rainfall having a strong positive effect. Interestingly, the authors highlight a lot of interindividual variation in characteristic movement patterns. Strikingly, out of the 98 animals that were originally collared, it seems that only two (one male, one female) got anywhere near villages over a period of more than five years; only one of these, a female with two calves, had a visitation pattern associated with crop availability (Mbamy et al., 2024). Beirne et al., moreover, found little support for NDVI (vegetation productivity) on movement. This is likely because the key food resource is not green vegetation, but fruit. If elephants are "following" fruit trees, in a sense, the typical observation of small group sizes (discussed in Chapter 3) seems to suggest that there is some limitation on the ability of families to move together as cohesive units, as savanna elephants do. It may be that for much of the year, forest elephants must forage in smaller groups over a large area with dispersed food sources, only periodically converging at particular sites when there is a high density of simultaneously fruiting trees or the stable mineral-rich clearings known as *bais*. Beirne and colleagues moreover point out that since forest elephants are limited by water, rainfall likely allows for longer and more directed movements away from reliable water sources.

Nouabalé-Ndoki National Park in Congo basin is a PA contiguous with PAs in adjoining Cameroon and the Central African Republic, and contains permanent trails several meters wide. Blake and Inkamba-Nkulu (2004) tested whether these trails were associated with fruit, browse, or mineral deposits. They found that the size and density of trails increased "dramatically" nearer to mineral deposits. The number of fruit trees favored by elephants was higher at trail intersections and positively correlated with the size of such intersections. On the other hand, there was no apparent relationship between trail characteristics and the presence of monocot browse species (palms). Noting that a similar strong association between elephants and mature fruit trees *Balanites wilsoniana* has also been observed in Kibale National Park in Uganda (Babweteera et al., 2007; Cochrane, 2003; Wing & Buss, 1970), researchers suggest that trails are a form of societal spatial memory that enables even less experienced individuals to locate important resources (Blake & Inkamba-Nkulu, 2004; Campos-Arceiz & Blake, 2011); more on this in Chapter 7.

Bais represent extremely rare but important point resources for African forest elephants, and their scale seems unique to the African context. Although mineral lick sites do occur in Asia as well, they are much smaller (discussed in Section 6.3.2). Unlike the distribution of water and food, which vary seasonally, *bais* are static

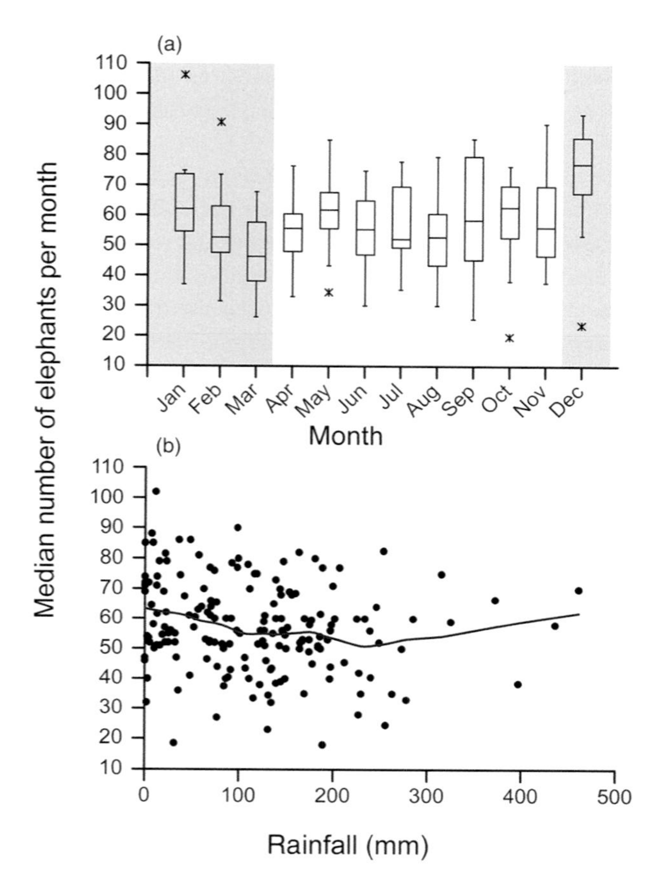

Figure 6.7 Visitation at Dzanga *bai* relative to rainfall. (a) The number of elephants observed in dry seasons (shaded backgrounds) were not appreciably different from wet seasons over a 20-year period. (b) Elephant numbers show a weak negative correlation with increasing rainfall (reproduced from Turkalo et al., 2013, *PLOS ONE*, Open Access).

locations. Nevertheless, it has not been studied whether the conditions within them, such as their moistness and mineral composition, change in time. A study analyzing acoustic recordings of elephant visitations at six different clearings in Gabon over a one-year time period found that elephant visitation rates at these clearings varied substantially on a daily basis across sites and different times of year (Wrege et al., 2012). Acoustic monitoring reveals that more than three-quarters of the visitations occurred at night, evidently reflecting the cryptic tendencies of many forest-dwelling species when they venture into the open (although one expects elephants historically would have had little to fear in these deep forest clearings were it not for the threat of people hunting them). Comparison of day versus night-time recordings show that there can be much larger fluctuations in the numbers of elephants visiting at night than during the day, although the two were correlated overall (Wrege et al., 2012). Yet the changes in visitation rates over time do not seem to be driven by rainfall (Figure 6.7), a finding that is in agreement with observations at M'beli Bai in the northern Republic of Congo (Fishlock et al., 2008) and the longer-term (20-year) observations reported from Dzanga bai in the Central African Republic based on daytime observations (Turkalo et al., 2013). The last study did find that visitation rates were marginally lower on days with more rain.

Perhaps the most intriguing insight from the Dzanga study, which is the longest individually based study of forest elephants, is that individuals and family groups seem to visit a given *bai* only sporadically. On average, seven to eight months might elapse between visits for any given individual or family, with some (especially males) being absent over multiple years. But as there are multiple clearings, individuals may simply be roaming between different clearings. As Turkalo and colleagues note, other studies have found that elephants of either sex seem to spend around a third of their time within 500 m of a *bai*, though not necessarily the same one (Blake, 2002; Momont, 2007). "Seasonality" need not just mean rainfall – other environmental variables may also be dynamic. It is not yet known whether the observed fluctuation in visitation is driven by attributes of the *bais* themselves or features of the surrounding forest, such as the timing of fruiting events for various species. A massive study of tree flowering and fruiting phenology encompassing 5,446 trees belonging to 196 species across 12 sites in the tropical forests of Africa found that although annual cycles were the most common strategy, there were also subannual and supra-annual patterns (Adamescu et al., 2018). Species exhibiting annual cycles were more common in West and West Central Africa, whereas those in East and East Central African forests showed cycles ranging from subannual to supra-annual. However, no regular cycle was observed for over 50% of individual trees. If biological changes such as fruiting events are asynchronous such that they are not directly tied to specific seasons as defined by rainfall, then it is possible that the observed fluctuations may at least be partially be due to such occurrences. Elephants are very likely actively exploring and seeking out fruiting trees, stopping by whichever *bai* happens to be nearest as they move among these sites. This form of scramble competition may preclude the maintenance of large social groups in close proximity (despite their ability to gather in the *bais* themselves), but the extent to which elephants maintain social contact within the forests, and over what magnitude of distances, is of course a mystery.[3]

As in other systems, elephants in these environments are increasingly being hemmed in by human activities. The current and anticipated proliferation of roads in equatorial Africa can have profound impacts on elephants and ecosystems (Laurance et al., 2017). Elephants may be described as having two possible strategies in response. The "siege" strategy entails restricting movements to avoid roads, whereas the "skirmish" strategy involves continuing to range widely while making risky road crossings

[3] One way to maintain predictability without active communication is to gather with regularity at an established place and time. Elephants at Loango converge on the beach for daily "tea parties" (P. Lee, personal communication). Perhaps they coordinate this behavior through acoustic and chemosensory cues (see Chapter 7), but may also simply do so out of habit.

(Blake et al., 2008).[4] Blake and colleagues tracked 28 elephants in six different wilderness areas within the Congo basin (including Congo, Central African Republic, and Gabon) via GPS collars to examine their responses to roads, some of which were in PAs and others which were not. One individual was subsequently excluded from analyses owing to poor data quality. Collectively, the remaining set represented 28.5 years of observations. The results are striking. Only a single individual crossed an unprotected road whereas 17 individuals crossed roads within PAs, although nearly all individuals had potential access to unprotected roads. That single individual, a female with a dependent calf, seemed to travel quite far out of her way to find a crossing point far from villages, then ran across at 14 times her usual travelling speed. Overall, home range sizes were significantly related to the amount of roadless area available to the subject, and showed no sign of a plateau in size even in areas that encompassed the largest intact habitat left on earth for these species.

One can draw at least two conclusions from these observations. Clearly, there must be a difference between roads of these two classes that is salient to forest elephants. And, at least for the time being, they are primarily employing a siege strategy. Elephants must perceive the risks of crossing unprotected roads as being greater than the potential access to additional forage that is gained by doing so. This would be especially true of roads that are along PA edges, where crossing would place animals in an entirely unprotected habitat – the question then is whether elephants are avoiding the roads themselves (owing to their vehicular traffic and accompanying human activities, etc.) or are they also avoiding whatever lies on the other side? This is not explicitly mentioned in this study, but others offer some additional food for thought.

The Sangha Trinational Conservation Area centers around land that belongs to the Republic of Congo to the east, Cameroon to the west, and Central African Republic to the north. The Sangha river, which is oriented north–south, acts as an international boundary between Cameroon and the other two countries respectively. A survey was conducted along a 60 km stretch of the river within a 50–100 m buffer of the riverbanks on both the eastern and western sides to assess the degree to which elephants occurred on either side, and related these to human activities on either side (Weinbaum et al., 2007). These included active or abandoned villages as well as a large logging town of around 8,000 people located 15 km to the north. The study found that overall, dung counts increased on both banks toward the south, which researchers attributed to the increasing distance away from the town (Figure 6.7). There was a weak but significant positive association with abandoned villages, and a tendency toward negative association with active villages. Interestingly, there was more dung on the western side than on the eastern (Figure 6.8), which was suggested to be because much of the Cameroonian side had been logged, promoting an abundance of the types of secondary vegetation elephants prefer. But surprisingly, this pattern actually flips toward the

[4] I cannot help but point out the very obvious and intentional use of war terminology here, conveying that roads are an assault on wildlife or wilderness to which wildlife are responding in a similarly warlike confrontational manner. This may be true, but such language already implies some form of conflict and derives from a heavily one-sided perspective. There is no analogue of war from the perspective of wildlife, who are merely trying to survive.

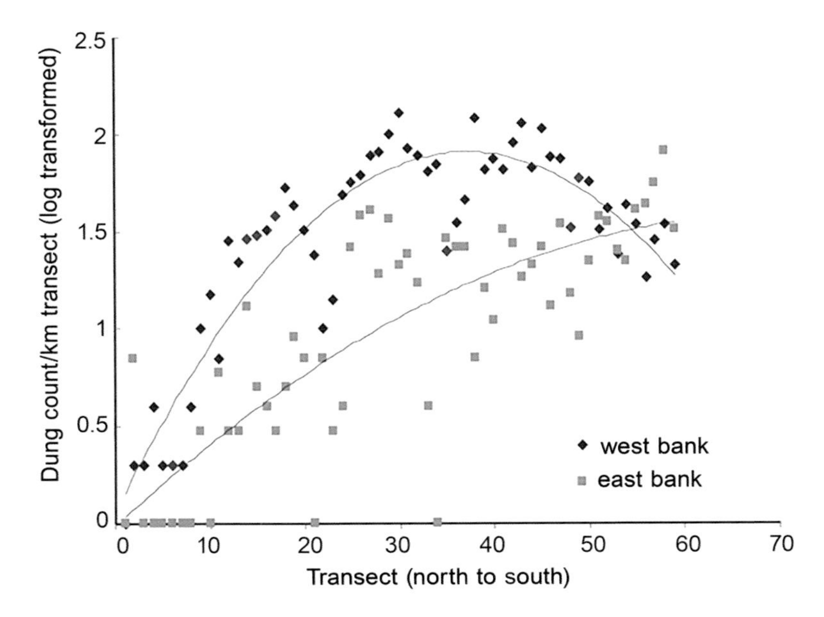

Figure 6.8 Forest elephant dung densities along adjacent banks of the Sangha river, in the Sangha Trinational Conservation Area. The north end of the transect is in close proximity to a large logging town of around 8,000 people. The two sides of the river belong to different national jurisdictions and contain different degrees of human activity, resulting in strikingly different patterns of elephant activity (reproduced from Weinbaum et al., 2007, used under Creative Commons license).

far south, despite being closer to another logging town on the eastern side, in Congo. Researchers report the local observation that anti-poaching laws have been heavily enforced by the presence of a nongovernmental organization on the Congolese side, which might "chase a number of poachers to the other side of the border where there is less on the ground enforcement." This particularly complicated, but perhaps not unique, example illustrates that a riverbank is not merely a riverbank to an elephant. Their choice of habitat, and crossing points between habitat boundaries, is dictated not only by the visible presence of forage but also the invisible landscape of fear determined by human laws, hunting pressures, and protective activities.

The concept of the "hunting shadow" refers to the zone surrounding an area undergoing hunting pressure that may become devoid of the target species. Importantly, the size of this surrounding zone will depend on the ranging behavior of the species itself – those that disperse only within short ranges are expected to experience narrower hunting shadows than those that disperse more widely (Figure 6.8). Despite the evidenced reluctance of forest elephants to cross unprotected roads, and their avoidance of areas with hunting pressure, they do occur outside PAs, notably in the managed forests that constitute logging concessions. Elephant occupancy of timber concessions was evaluated on the basis of interviews with concession workers in eastern Cameroon, over an area which included and extended beyond the site covered earlier by Weinbaum and colleagues (Brittain et al., 2020). In keeping with other studies, the probability that elephants occurred in a particular patch of habitat increased further from villages or roads and closer to rivers. Management units in different areas varied in their detection of elephants, which researchers attributed to variation in the actual presence of elephants rather than habitat type or respondents' abilities. The detection of elephants decreased by 30% during the six-year study window from 2008 to 2013, which was attributed to actual decreases in abundance, rather than changes in detectability.

6.3.2 South and Southeast Asia

Because very little remains of the rainforests of South Asia that would have originally been occupied by elephants, it is unfortunately now virtually impossible to know how elephants would have behaved in those environments when undisturbed. Genetic studies provide some insight into how populations may have dispersed and differentiated from one another, serving as an indirect indicator of space-use over various spatial and temporal scales. In India for example, where elephants are today found in distinct subpopulations, genetic structure suggests many of them may have been historically connected – and also hint at ongoing population fragmentation brought about by human activities (De et al., 2021; Vidya et al., 2005). However, the methodological details can vary between studies, making it difficult to draw definitive conclusions about either historical or recent barriers.

In landscapes that today contain a mix of natural and human-dominated areas, elephants choose to spend their time in different habitats during the day and at night, ostensibly in order to minimize their encounters with people. A study conducted in the Western Ghats of southern India illustrates how elephants behave in such fragmented landscapes. The study focused on the Valparai plateau, adjacent to the Anamalai Tiger Reserve, which consists primarily of plantations interspersed by remnant rainforest fragments, tracking herds over a five-year period and six different habitat types (Kumar et al., 2010). Although the rainforest patches constituted a much smaller share of the landscape relative to the area under various plantation types and settlements, elephants nevertheless appeared to use them to a disproportionately greater degree during the daytime. while venturing into tea plantations at night. They showed a preference for riparian areas overall, relative to their extent. There were seasonal differences in their use of different plantation types, being in coffee plantations more often during the dry season and Eucalyptus plantations more often in the wet season. These choices appear to reflect the availability of shelter, preferred forage, and avoidance of human activity. Coffee is harvested only for a few months out of the year, and Eucalyptus felled once in seven years, thus periods of human activity are low while both types of habitats contain cover and secondary vegetation that can be consumed. Tea plantations, which contain no forage as well as high levels of year-round human activity, nevertheless must be traversed as they are extensive and link other types of habitat; thus elephants opt to do so largely at night (though they are also visible during the day, to a lesser extent).

The forests of Southeast Asia were once home to such large numbers of elephants that the mind reels with trying to grapple with the probable extent of population declines. The iconography of elephants, present in the art and archeology of so many Southeast Asian cultures, offers partial testament to this likelihood (although they also exist alongside entirely mythological creatures and therefore this cannot be viewed as definitive proof of abundance any more than the affinity of British nobility for lions can be cited as evidence for the range of African lions extending to England). Laos, once known as "The Land of a Million Elephants," is now believed to contain a few hundred at best. Similar fates have befallen elephants in Cambodia, Myanmar, and Vietnam, not necessarily through forest loss but the emptying of forests through hunting and capture. Against this backdrop, the battle to protect the remaining populations will require us to take a hard look at the forest types that remain and whether they are actually serving the needs of elephants. If not, it might seem prudent to understand why.

Among the remaining mainland populations in Southeast Asia, peninsular Malaysia is one area in which there appears to be a significant amount of habitat, elephant populations, and where movements have been studied. Using GPS data for 48 elephants over a period of seven years, de la Torre et al. (2021) documented the movements of elephants in and around PAs. PAs largely consist of forest, whereas the wider landscape, labeled "managed elephant range," consists of both forests of various categories (including those that are logged) as well as plantations of rubber, oil palm, and other agriculture. Sixteen individuals (mostly females) were collared and then released at the same location, whereas 32 individuals (mostly males) were translocated from areas with a high number of negative incidents with people into PAs before being released. Using a statistical procedure known as a step selection function and various environmental covariates to model the characteristics of areas elephants used relative to the areas available to them, they created a map of the probability of habitat use. This procedure, which is indicative of animals' habitat selection (or preferences), is here and elsewhere sometimes confusingly labelled "habitat suitability." Doing so implicitly assumes that such selection is a primary if not sole determinant of whether or not a resource patch is suitable, but this can be misleading as we shall see. They also separately examined reports of hostile incidents with people. Here again, another questionable assumption was that solitary individuals are likely male whereas groups larger than six are likely females; the analyses excluded incidents involving groups between two and five individuals. In India and Sri Lanka, females were observed most often alone or with a calf (de Silva & Wittemyer, 2012; Nandini et al., 2018), and males are frequently seen in both mixed and single-sex groups (Keerthipriya et al., 2021), with all-male groups becoming more common in human-dominated landscapes (Srinivasaiah et al., 2019). With these warnings, let us plunge onward.

From the collar data, elephants of both sexes evidenced preference for forest gaps, secondary forest, and areas of regrowth as well as new plantations (Figure 6.9a and b). But they preferred to remain close to mature forest and avoided areas in which plantations dominated the landscape (Figure 6.9b). Males had a greater tendency to use more open areas and were more strongly influenced by the presence of water than females. Males especially seem to avoid forest interior habitat (Figure 6.9b), which might relate to their greater preference for lowlands compared with females, though both avoided very rugged terrain (Figure 6.9c). Over 60% of the crop and property damage incidents were attributable to large groups, whereas 53% of human injuries or death was attributed to solitary individuals. Not surprisingly, negative interactions were reported in areas that were more likely to be used by elephants. These observations suggest that females (or large groups) may be using crop or plantation areas and be involved in so-called instances of conflict at least to a similar extent as males, in contrast with observations in South Asia. The conundrum this raises is whether elephants actively select for plantations and other agricultural land-uses *even when* primary forest habitat is available. To understand why, studies from a single (albeit large) locality offer an incomplete perspective.

Borneo mirrors parts of peninsular Malaysia, having also undergone substantial land-use change in recent decades, though not as extensively as the mainland of Southeast Asia. Much of the elephant population is found in the lowlands of Sabah, away from the mountainous central highlands of the island. Like peninsular Malaysia, many areas have been converted to a patchwork of oil palm plantations and settlement. Studies of elephant movement paint an intriguing and perhaps somewhat inconsistent picture. One of the first studies to document home range sizes in Bornean elephants was based on movement data from five females tracked using satellite collars around the Lower Kinabatangan wildlife sanctuary and Ulu Segama-Malua Forest Reserves in Sabah (Alfred et al., 2012). They found ranges to be on average smaller in nonfragmented forest (at least 250–400 km^2), with a greater degree of monthly displacement, than in fragmented forest (at least 600 km^2). Researchers suggest this is influenced not only by the availability of forage but also of permanent water sources. Nonfragmented forest would therefore appear to have more resources than fragmented forest, although it must be cautioned that the sample size in this study was extremely small, both in terms of the number of individuals and the number of days over which data were obtained (ranging from 12 to 169 days among the five). However, they specifically noted that elephants in forests that were undergoing logging were moving further than those in forests where there was no such disturbance. The authors also observe that the latter were also associated with higher levels of negative interaction with people, as the increased ranging made it more likely for elephants to encounter people. The elephants also preferred to remain mostly below an elevation of 300 m, though they occasionally ventured above 750 m, and largely used dipterocarp forest (with the exception of one female, who was mostly using lowland swamp habitat).

A second study covered approximately the same areas plus Tabin Wildlife reserve, and was based on a larger sample of 29 collared individuals, 13 males and 16 females (Evans et al., 2020). Of these, 15 had been translocated from oil palm plantations to nearby protected forests 1–2 km away. Curiously, home ranges were found to be quite a bit smaller on average than other estimates, just around 150 km^2. Equally interesting, there did not appear to be any significant difference in home range size between the sexes (or between translocated and nontranslocated individuals), although females' ranges tended to be a bit larger than that of males. This is the opposite of what one

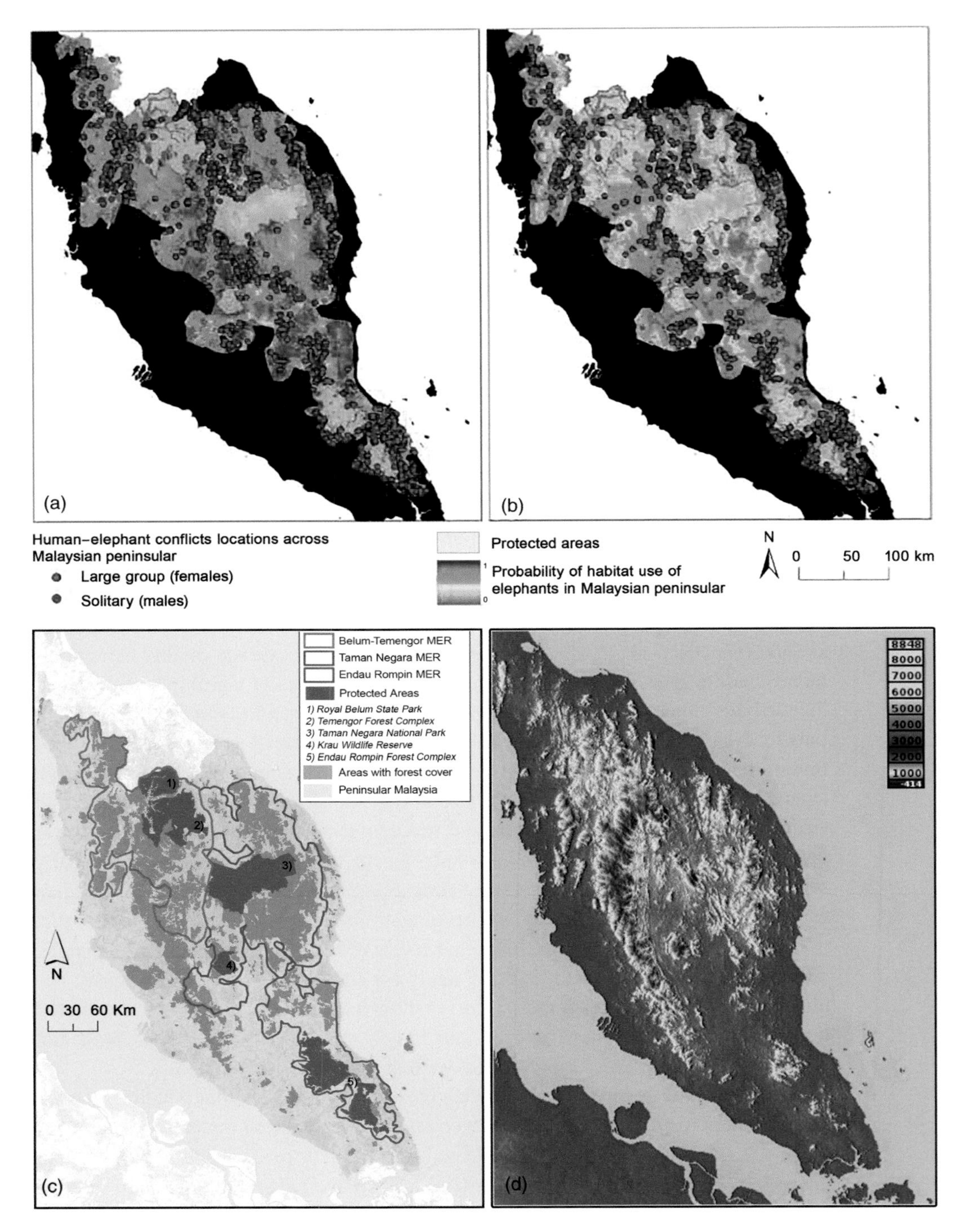

Figure 6.9 Elephant habitat preferences versus forest cover and elevation in the Malaysian peninsula. (a)–(b) The color spectrum represents varying probabilities of use as predicted on the basis of elephant movements in relation to environmental covariates for females (a) and males (b). The red and blue dots represent conflict incidents attributed either to groups of >6

expects, based on sexual differences in reproductive strategies, discussed in Chapter 4, as well as observations from South Asia. There was a tendency for the home ranges of elephants occurring in areas with a higher estimated population density to be larger than those occurring in areas with lower estimated densities. If so, this would suggest the need for additional space in areas of potentially higher intraspecific resource competition (though conversely, higher densities may be supported in more resource-rich areas so we must be cautious with interpretation). A novel aspect of the study was their characterization of so-called animal-defined corridors, areas in which individuals made relatively linear, high-speed, directed movements. Elephants showed these high-speed movements at a disproportionately higher rate in nonforest areas (mixed landscapes, agriculture, and plantations) compared with forest, although they used these areas more frequently than the latter. They also largely avoided roads and settlements and were further constrained by natural and anthropogenic habitat limits, with elevation being the most significant feature driving habitat selection among the factors considered – especially ridgelines. Contrary to what one might expect, canopy cover and slope were not found to be significant.

Still another study characterized the ranges of elephants in both peninsular Malaysia and Sabah over a concurrent timespan (de la Torre et al., 2022). In Sabah, this involved 55 elephants monitored over a ten-year period overlapping with the study by Evans and colleagues and from the same populations. Somewhat astonishingly, home ranges were far larger than previously reported from this region – on average over 1,500 km^2 for females and 2,200 km^2 for males. These were significantly larger than for the peninsular individuals in the same study, where females averaged 230 km^2 and males averaged 730 km^2. Similar to the earlier studies, most elephants ranged outside forest complexes that were under stricter categories of protection. In both regions elephants preferred more open lowland areas as long as they were still close to forest cover, avoiding steeper rugged terrain. The use of higher elevation ridgelines was also reiterated, especially among males. But the most significant observation in terms of developing informed conservation policy, is that elephants actively selected for forest types *other than* those in strict reserves, including commercial forest areas that were selectively logged.

There are at least two takeaways from all of these studies. First, home-range sizes vary quite a lot even in the same regions and the reasons for this variance are not well understood, therefore perhaps we should refrain from generalizing based on any particular study. But secondly, and perhaps more importantly, is what they tell us about elephants' use of so-called degraded habitat, which in the context of Southeast Asia usually means nonforest habitat. While elephants undoubtedly require "forests," not all forests are equal. From the movements of collared elephants (Evans et al., 2018),

Figure 6.9 caption (cont.)

individuals or solitary animals respectively. (c) Locations of forest (green), PAs (dark green) and managed elephant ranges (MERs, outlined). (d) Elevation gradient from low (green) to high (red) (panels (a)–(c) reproduced from Fig. 2 and Fig. 1 in de la Torre et al., 2021, with permission from John Wiley and Sons-Books; panel (c) from mapsfor-free.com).

it was seen that they preferred disturbed habitat, often bordering oil palm plantations, rather than the primary dipterocarp forests. As observed elsewhere, waterways were also especially important. Evans and colleagues point out that elephants appear to actively select for areas of "low" quality, as defined in terms of above-ground carbon stocks (i.e. trees). But with the exception of a few species, such as palms, Asian elephants don't eat the mature trees themselves – they primarily require grasses and other herbaceous vegetation. In Sabah, such vegetation is more prevalent in the commercially logged "disturbed" forests rather than the protected forests with their more intact canopies. The plantations are characterized by trees of low stature, 13 m or less. This can be at odds with conservation paradigms overwhelmingly emphasizing the protection of carbon- and biodiversity-rich primary forests and underscores the need to understand elephants' ecological requirements in PA designation and conservation planning.

That elephants seem to be drawn toward secondary regenerating or "disturbed" habitat found along the edges of primary habitat has been observed elsewhere as well and is usually explained in terms of the abundance of edible understory vegetation (Rood et al., 2010). In Aceh province on the northern tip of Indonesia, where elephant habitat preferences were assessed using sign surveys and direct observation by Rood et al. (2010), this again presents a predicament. This study modeled habitat "suitability" (once again, a term to be applied with caution) on the basis of a statistical procedure known as ecological niche factor analysis. This procedure relates elephant occurrences to environmental covariates, asking whether it deviates from the "average" habitat based on a random scattering of points. They did. Elephants' core ranges tended to occur disproportionately in highly productive forested valleys, though they also occurred, on occasion, at elevations exceeding 2,000 m above sea level. While slopes per se did not limit elephants' ability to use habitat, the overall ruggedness of the terrain did seem to put them off. The major problem? What the model finds as being the ideal ecological niche of elephants is also highly at risk of deforestation and land-use conversion – up to 80% of the suitable habitat is at risk. The authors therefore caution: "if forest encroachment continues at its current rate, elephants are deemed to prevail within suboptimal habitat conditions at the margins of their niche ... We believe that the occurrence of elephants within marginal habitats reflects the conversion of previously suitable habitat."

This assessment is corroborated by another study, also conducted in the same Sumatran province, by Wilson et al. (2021). The study areas were within the Leuser ecosystem, which contains 30,000 km^2 of dense forest, and involved collared individuals belonging to three (putatively different) family groups. Once again, there was a clear preference for lowland valleys and less rugged terrain, this time on the basis of animals' movements as opposed to indirect or direct observation. Unfortunately, lower elevations also contained more human settlements and other anthropogenic disturbances. Less disturbed forests could be found on higher, more rugged terrain. Elephants were in other words sandwiched between these areas of disturbance and steeper, more rugged terrain. The seeming preference for what seems to be forest "edge" in these situations therefore might be no more than a consequence of the loss of lowland rainforests, which would have once harbored much richer habitat for elephants. Given similarities in terrain and habitat characteristics, this may be contributing to the habitat selection preferences of the Sundaic elephants as well. Moreover, elephants are now engaged is a phenomenon known as *niche construction*, whereby they actively engage in carving out living space from among the options available to them. Modern-day agricultural plantation landscapes, with their monocultures of high-nutrition crops, are a relatively new development in elephants' evolutionary history. Mixed-use landscapes can be an extremely important type of landscape for Sumatran elephants, as other studies have also emphasized (Imron et al., 2023). But although some elephant populations may set about busily exploiting these new resources, human judgment will determine whether these new niches are ultimately ones where elephants can persist.

6.4 Concluding Remarks on Habitat Use in Space and Time

Habitat use is a behavior, therefore its study and relevance to conservation issues should be informed by concepts from the field of behavior. One of the issues with studies of habitat use often is a lack of reflection on what the habitat choices of an animal truly reflect, along with a somewhat cavalier use of the term "suitability." It is important to recognize that the area an animal uses does not depend solely on its predisposition to forage in certain areas or move in certain ways in a manner reflective of either some ideal distribution or ideal preferences. The animal is subject to numerous *constraints*. These constraints arise in the most immediate way when there is competition with another individual, which may be mediated through social behaviors or conventions, and these may be the most observable, such as the spatial exclusion of individuals from prime habitat through dominance interactions. However, there are far more subtle constraints, such as whether an individual or even a whole population is able to preserve the knowledge of where and when resources occur – animals are not born knowing this; the knowledge must be learned and transmitted both horizontally (to members of the same or other social groups) and vertically (to progeny). External cues, such as pathways built up over time that connect critical resources, can be a crucial means of preserving this knowledge (Blake & Inkamba-Nkulu, 2004; Von Gerhardt et al., 2014). One of the greatest threats to the persistence of elephant populations in areas where land-use changes and habitat fragmentation proceed rapidly and are coupled with the added threat of hunting, is that these knowledge systems may be irreplaceably lost.

Second, the mere use of a habitat by an animal does not make it automatically "suitable." I would propose that this term be applied far more selectively, with recognition of the fact that an animal may either voluntarily choose or be forced to use habitats that are suboptimal in various ways. They may be suboptimal because they lack adequate resources but represent the best of available options. They may be suboptimal because the resources themselves are toxic or harmful in some way. They may be suboptimal because they attract animals into dangerous circumstances, putting them in harm's way. These latter instances represent ecological traps that may ultimately act as population sinks, that is, the opposite of suitable habitat. From the perspective of both conservation and management, the "suitability" of an area of potential habitat must be evaluated ultimately with respect to its effect on the overall population, and this may entail the inclusion of variables such as human acceptance, not merely ecological features.

Lastly, by now there is ample evidence that PAs alone do not necessarily suffice to protect elephants, or for that matter other wildlife. They are certainly necessary as refugia from harm and especially important when centered around vital resources. But in the case of elephants specifically, PAs are typically too small and may encompass habitat types that are inadequate for them, especially when managed with some artificial notion of what constitutes the "natural" condition. Instead, we must ask *what are the dynamics of ecosystems* that appear to able to support healthy elephant populations? Can we distinguish the conditions that gave rise to these ecosystems? Can we maintain them? If not, what alternatives are there to mimic or replicate such dynamics? For instance, it is both possible and likely that systems of traditional pastoralism, silviculture, and shifting cultivation were responsible for enabling elephants to have thrived historically in their many diverse ecological contexts, and that halting these practices – whether through "protection," privatization, nationalization, or land-use change – has contributed to reducing the capacity of these landscapes to accommodate elephants in the present day. Yet these practices may never be fully recovered given current legal systems and human population densities. The first question is whether they can be revived to any degree at all in a just and equitable manner. If not, it is imperative that we identify alternative mechanisms to promote ecological regimes that can once again accommodate elephant populations while also minimizing their potential impacts on human communities and livelihoods. It goes without saying that this is a tall order, better achieved if different stakeholders (including scientists, conservationists, government agencies, private entities, and local citizens) can be brought together to achieve mutual objectives rather than working at cross-purposes.

7 Communication and Cognition

7.1 *Umwelt* and *Innenwelt*

There can be no discussion of animal minds without first acknowledging that some experience the world in quite a different way than humans do, and that the things that matter to them may be difficult if not impossible for us to directly appreciate. Jakob von Uexküll's notion of the *umwelt* remains a valuable term that, in many ways, prompts us to grasp the ungraspable. In a now-famous example, he circumscribes the rather impoverished universe of the tick, which is concerned only with the perception of odor, temperature, and touch, and each to only a very narrow degree. A tick neither senses nor cares about much else, including other ticks, save for the brief interval when it is seeking a mate. The *umwelt* of the tick consists of objects that have attributes falling within this restricted domain. The actual experience of the tick, its *innenwelt*, is known only to the tick, try as we might to imagine it.

The more similar an organism is to us, however, the easier the task becomes. Hence we have little trouble imagining the worlds of other primates, which, like ourselves, go through life paying close attention to sights and sounds. We may be especially aware of those associated with other conspecifics and external threats, which for much of human history also included predators. Still other species have augmented senses that once again challenge our creativity. As the philosopher Thomas Nagel wondered, what is it is it like to be a bat, experiencing the world through sound (Nagel, 1974)? Or to be a bird, navigating by virtue of magnetoreception and experiencing time itself distinctly from our own ponderous pace (Tudge, 2009)? Or, for that matter, the family dog, which smells the identities of all its neighborhood friends and foes and can tell who passed by yesterday? These are all alien minds, despite arising through the same evolutionary process that shaped ours. By virtue of our clever technologies and experiments, we gain a small inkling of the parallel universe of environmental stimuli that other species experience and react to, and try to interrogate it in the study of cognition.

Elephants also inhabit an *umwelt* different from ours. It is dominated by scents, touch, sounds, and other vibrations. Sight is also present but may be far less important than it is to us. No doubt, an elephant must additionally have quite a unique sense of taste – but this is something we seldom think about, except when it is a taste for human crops or commodities. In each of these domains, an elephant's sensitivity is distinct from our own. Yet, like ourselves, elephants also care about other elephants.

However, the pace of an elephant's life is slow: From many perplexing hours of observing elephants in the wild with no clue as to what triggered some behavior, I would venture that its reaction time is decidedly not the same as ours, thus cause and effect may not be immediately obvious to the human observer. For all these reasons, to try and tap into the elephants' *innenwelt* is a tricky business. Nevertheless, for better or worse, the degree to which we humans tend to sympathize with or demonize other species is colored to some degree by our perception of their cognitive sophistication (alongside other influences, such as our own cultural influences). Unfortunately, our views may in turn be biased by our own preferences; vegetarians, for instance, appear prone to view species with more anthropomorphism than nonvegetarians (Leach et al., 2023). Despite all this, to conserve a species, and deal appropriately with the realities of living near it, we must first understand it and try to see from its perspective as best we can.

Cognition is tightly linked with communication, making the latter a useful means by which to interrogate the former. Some primatologists have been able to take advantage of this to devise creative experiments to shed light on the minds of nonhuman primates (Cheney & Seyfarth, 1990, 2008). If the cognitive apparatus is the means by which an animal comprehends the world, the communicative apparatus is the means by which it makes itself known to that world. Modes of communication therefore often mirror dominant modes of perception. The more important a particular sensory domain, the more important it is also likely to be as a channel of communication. Thus, we humans communicate primarily through visual and acoustic means. Cephalopods (squid, cuttlefish, octopi), on the other hand, limit themselves to the first of these two (and, I might add, elevate it to an art form unsurpassed by any other nonhuman animal). It stands to reason therefore that elephants would also communicate primarily through acoustic and chemical channels. The *social complexity hypothesis* posits that social complexity drives the evolution of complex communication (Freeberg et al., 2012). If so, it would seem that elephants should be prime candidates for the latter, as some researchers have proposed (Hedwig et al., 2021). In this chapter I discuss what little we have been able to glean of how elephants perceive the world, how they express themselves, and the social components of their *umwelt*.

Before delving in, it must be stated that tests of cognitive abilities in any species rely very heavily on our own ability to adequately test them (for a more extended discussion of this with respect to elephants, see Jacobson & Plotnik, 2020). One has to be confident that the subject is actually attending and reacting to the features that experimenters intend them to, for the reasons the experimenters intended rather than something else, and that it is sufficiently motivated to participate at all. One is typically also dealing with a very small number of individuals and it is difficult to know whether the results represent a species-typical pattern or something more narrowly limited to the population being tested, shaped by an individual's own unique experiences and history. The latter is especially true of captive animals, which are subject to conditions distinct from that of conspecifics in the wild, simply by virtue of being taken care of by humans. For all these reasons, I prefer to refrain from overgeneralizing, and especially encourage keeping in mind that absence of evidence is not evidence of absence.

Lastly, I must qualify that it is not my aim to provide an exhaustive catalog of behaviors and communication signals in any species, but to highlight features important for understanding the world of elephants to facilitate ongoing efforts to live with them (see also Ball et al., 2022). In particular, humans employ various deterrent methods aimed at driving elephants away from places where we do not want them. How effective are these likely to be in the long term from a sensory and psychological perspective? What consequences might there be? Likewise, when we attempt to rehabilitate or reintroduce elephants into the wild, what considerations might we need to make in order to ensure that elephant populations are well-functioning and likely to survive? We are encouraged to think about the work discussed in this section not merely as an exercise in trying to understand elephants, which requires some effort, but also trying to understand how they are likely react to our actions, for better or worse.

7.2 Sensory Channels

7.2.1 Scent and Taste: Chemosensory Cues and Signals

As anyone with a dog knows, scent trails left by conspecifics can convey a fascinatingly rich trove of information. It is perhaps fitting to start with chemosensory modes of perception and communication, because they involve the most distinctive feature of the proboscidean lineage, the trunk. Elephants use their prodigious and highly versatile appendage in a variety of ways: mate-searching, foraging, movement coordination, and as yet poorly understood social situations. Elephants also have a very well-developed vomeronasal olfactory system, commonly investigating various objects of interest by touching them and inserting the trunk tip into the mouth. Both sexes make extensive use of their trunks to investigate their surroundings, their food, and their potential friends or competitors; the focus of interest and elicited responses may vary by age and sex (LaDue et al., 2018; Merte et al., 2009; Rasmussen et al., 2007; Rasmussen & Greenwood, 2003). Asian elephants are among the few mammals in which the use of pheromones have been identified: frontalin, which is associated with musth (discussed further later in this section), and dodecan-1-yl, a preovulatory pheromone (Schulte & Ladue, 2021, and references therein). Pheromones are chemical compounds that are used for specific intraspecies signaling purposes. Given the behavioral similarities among elephants, I would wager along with Schulte and LaDue that these pheromones are also employed by African elephants.

In a rather cute (albeit carefully designed) experiment involving the Amboseli population, researchers observed how female African savanna elephants react to the urine of either kin or nonkin when encountered in either unexpected or highly predictable locations, using an expectancy-violation paradigm (Bates et al., 2008). They initially predicted that elephants would react more strongly to samples from kin who were absent and known to be at least 1 km away or to group members that were trailing the subjects, as these represent the more surprising situations. On the other hand, elephants should not have had any expectations about individuals that were unrelated to them or were known to be ahead of the subjects, therefore they predicted these samples to be of less interest. To test this, they collected urine samples mixed with earth (or else a control sample of water + earth of similar consistency), sneakily placed them in the path of oncoming elephant groups and measured the responses of the first female to encounter the sample. Given the opportunistic and time-sensitive nature of these experiments, one can only imagine the experimenters' delight upon completing each successful trial!

While they did not see a statistically significant difference in the amount of time in that the initial females spent investing samples under each of the five conditions, there was a tendency for the group as a whole to spend more time on samples belonging to a group-member that was trailing (as predicted) as well as to those that were ahead (not as predicted). There was less time spent on those that were absent, whether or not there were kin, similar to the control condition. However, individuals did reach their trunks significantly more frequently toward the samples of absent kin and those that were supposed to be behind them. It's somewhat irresistible to interpret these reactions as the proboscidean equivalent of an incredulous double-take. Elephants do therefore appear to be able to discriminate individuals on the basis of their odor and have some expectations about where they ought to be, at least in the immediate vicinity.

Elephants can discriminate among individuals other than their own species – zoo-housed Asian elephants discriminate familiar versus unfamiliar humans on the basis of sight, smell, and sound (Polla et al., 2018). What is perhaps more surprising is that elephants might be able to extend this power of discrimination to classify human groups. In a separate set of experiments on savanna elephants, Bates and colleagues tested the ability of elephants to distinguish among human ethnic groups (Bates et al., 2007). This is salient because humans can represent predators. In this case the two human subgroups were Maasai, who might occasionally spear elephants in response to the death of livestock, and Kamba agriculturalists, who typically did not pose such a threat. However, the elicited reaction did not seem to depend on whether the specific family groups had experienced spearing incidents in the past, though those with such experience did show greater variance in how fast and far they moved in response. Elephants and Maasai also generally inhabit the same landscapes and therefore have opportunities to interact, so the effect might be more to do with other types of encounters, such as competition over water sources. Elephants retreated in fear when encountering garments that had been worn by Maasai men, traveling faster, moving further away, and taking longer to relax compared with conditions involving garments worn by either Kamba men or unworn controls.

One of the most striking ways in which elephants employ chemical signals is the condition of musth (Rasmussen, 1988), in which mature males signal their reproductive motivational state, discussed extensively in Chapter 4. In all three species, the males themselves dribble a constant stream of urine and secrete a viscous fluid from their temporal glands, leaving highly visible and pungent stains around the genital area and temples respectively. The combined odor can drift on the wind for many kilometers, while leaving a more precise scent trail on the ground. Bulls can follow one another around using scent trails (Allen et al., 2021a), possibly choosing whether or not to confront one another. Frontalin, which appears to qualify as a sex pheromone, is the primary compound associated with musth (Rasmussen & Greenwood, 2003). The composition of musth-related secretions changes with age, with younger males producing frontalin-free, honey-like odors that don't appear to elicit as much attention from older males as the more pungent odors of more mature bulls (Rasmussen & Greenwood, 2003). Why this change takes place is not very clear. One might speculate that it is in some sense an experimental or training period, in which individuals haven't quite settled into a clear annual pattern and the accompanying signals let others know that they are not quite yet to be considered a competitive threat. As discussed in Chapter 4, even mature bulls might require decades before they settle on a regular ranging pattern during musth periods. It is possible that there are corresponding subtle shifts in the composition of musth secretions with age that have not yet been identified.

Whether in musth or not, male elephants routinely inspect females to assess reproductive availability, first touching their genitalia with the tips of their trunks, then dipping their trunks back into their mouths for contact with the vomeronasal organ (Rasmussen et al., 1982). It is now known from studies of Asian elephants that females release a preovulatory pheromone, (Z)-7-dodecenyl acetate (Rasmussen, 2001). The chemosensory apparatus appears to be important very early in life for elephants, who have highly developed vomeronasal organs at birth and sensitivity when as young as six weeks old (Göbbel et al., 2004; Johnson et al., 2002). One reason chemical signals or cues may be so useful in inter- and intraspecific communication is that they are difficult to fake, being byproducts of the physiology of the individual (Schulte et al., 2007). The function of musth has therefore usually been explained in game-theoretic terms, as a mechanism for honestly signaling condition (Hollister-Smith et al., 2008; Poole, 1989a; Schulte & Rasmussen, 1999; Sukumar, 2003; Wyse et al., 2017). But this explanation somewhat glosses over two distinct features of the state. First and foremost, it is a period during which bulls focus their efforts on reproduction rather than foraging; that is, it is a period in which bulls severely restrict their feeding. This is *accompanied by* the various chemical (and sometimes acoustic) signals of physical state. We have previously discussed the alternation of breeding and foraging motivational states in males, necessitated by elephants' reproductive physiology and life history. Once individuals start to exhibit discrete and desynchronized foraging and reproductive periods, it becomes advantageous to broadcast as well as heed such signals in order to avoid costly confrontations. If musth, that is, individualized periods of heightened sexual activity, evolved primarily as a means to resolve the foraging–reproduction tradeoff inherent in the male scramble-competition for breeding females, it is possible that the *signaling components* of musth emerged subsequently in response to the asynchrony of breeding periods in males, inverting traditional explanations concerning the evolution of musth. Given that musth signaling is evident in both the Asian and African clades, it must have evolved prior to their divergence. Ancestral proboscideans, subject to the same reproductive and physiological constraints as modern-day elephants, likely evolved the alternation of states very early in their history and subsequently acquired the means to communicate it. Both sexes could then take advantage of the potential for information transmission about male condition and quality, for competitor assessment as well as mate selection.

The use of olfaction in the foraging domain has in general been far less studied but provides another excellent reminder that nonhuman animals may apply sensory abilities that are familiar to us in ways that are markedly distinct. For instance, we might be capable of appreciating the difference between the pungent aroma of, say, one cow from that of an entire herd of cattle (as anyone who has driven along certain stretches of California's central valley highways can surely attest). But we would have a difficult time smelling the difference between one and three. A study of captive Asian elephants, on the other hand, reveals their capacity to discriminate among different ratios of food items as small as sunflower seeds (Plotnik et al., 2019). In this experiment, elephants had to choose between two otherwise identical buckets containing differing proportions of sunflower seeds, discernible only by virtue of holes punched into the lids of the container. Elephants were of course motivated to select the bucket containing the greater quantity, having a 50% chance of selecting the correct bucket by chance. As one might expect, performance improved as the disparity grew larger, from an average probability of success being just above 50% when the ratio of seeds between the two buckets was closer to 1, to an average above 70% as it exceeded one-third (i.e. one bucket had three times more seeds).

It is easy to imagine the usefulness of being able to judge quantities through olfaction in natural contexts. For instance, elephants may use this ability to forage more efficiently by determining where fruits or other seasonal food sources are more abundant. Unfortunately, it also provides a potential mechanism by which elephants can discern when rice paddies are ready for harvest, to the frustration of farmers. Elephants in some localities are observed to time their visits to agricultural areas with crop cycles (Fernando et al., 2022; Webber et al., 2011), but not all fields are planted at exactly the same time. Whether they make fine-scale adjustments to target specific fields is less clear, though their timing seems to suggest so (Gross et al., 2018). It has been proposed that disguising the smell of ripening paddy might be one means of avoiding crop loss (Santiapillai & Read, 2010). Extending this idea, the use of olfactory deterrents has been an ongoing area of research. Researchers in Kenya and Uganda used locally sourced ingredients to create foul-smelling liquid and reported significant reductions in crop foraging compared with control sites (Tiller et al., 2022). The burning of chili briquettes (consisting of chili, elephant dung, and water) was tried in Botswana, but only repelled elephants while smoldering, and therefore could only be considered a short-term tactic (Pozo et al., 2019). Incense sticks have likewise been tested more informally in both Asia and Africa and show some promise, but have not been evaluated systematically. Before these approaches can be used at scale, however, there are important factors to consider such as the marketability of the methods and the duration of time for which they would be effective, given the propensity of elephants (and most other wildlife) to habituate to deterrent methods once they become more common.

A relatively new area of research takes a different tack by trying to evaluate what secondary compounds present in plants may naturally deter elephants (and other herbivores). We have already had a brief look at this in Chapter 5 with respect to the possible role of geophagy in neutralizing tannins and other leaf compounds. Often, herbivore defenses are associated with plant signals which can either act directly on the herbivore or indirectly by provoking a reaction from another species (Mithöfer & Boland, 2012). One well-studied and example of the latter is that of the ant-acacia mutualism, where herbivore damage to the plant can elicit chemical distress signals that rally the resident ants to its defense (Agrawal & Rutter, 1998). There is now accumulating evidence based on studies of diet choice by savanna elephants that they are sensitive to potentially toxic plant secondary metabolites, notably monoterpenes (Bester et al., 2023a; Schmitt et al., 2020). Choice experiments with eight monoterpene compounds commonly found in Southern African woody plants showed that elephants avoided all but one of these at high concentrations and five at all tested concentrations (ranging from 5% to 20%). The efficacy of deterrence can also depend on the combination of compounds present, in addition to their concentration (Bester et al., 2023b). This neatly complements an independent study conducted in Nepal which found that medicinal and aromatic plants incurred less damage from Asian elephants (Gross et al., 2017). Encouraging the cultivation of these naturally resistant plant species in areas with high potential for elephant damage could therefore be one mechanism for reducing economic losses and negative human–elephant interactions. The practical challenges of doing so, with which I am becoming all too familiar in our own research, is more to do with economics than ecology.

7.2.2 Vibration: Acoustic and Seismic Domains

Vibrations can travel through air and ground, respectively perceived as acoustic and/or seismic cues and signals. The use of sound as a means of communication has, of course, received a lot of research attention, given that elephants have a varied and lively vocal repertoire (Berg, 1983; de Silva, 2010b; McKay, 1973; Nair et al., 2009; Poole et al., 1988; Stoeger & de Silva, 2013). Much of this effort has focused on the perception and function of vocal signals specifically, the greatest proportion of this in turn being on African elephants (McComb et al., 2000; Soltis, 2010). The reason for this is largely pragmatic – acoustic signals lend themselves well to playback experiments, which are easier to carry out in environments where subjects are relatively easy to approach and observe. However, the perception of other types of sound, as well as the actual hearing abilities of elephants, has received far less attention. Likewise, far less is known about elephants' use of seismic cues. As in other sections, I will not attempt here to exhaustively discuss literature that has already been reviewed elsewhere. Instead, I will provide an overview of work that provides background for situating the cognitive abilities of elephants in relation to other species, as well as context for interesting questions in need of further exploration.

African elephants produce vocalizations falling into eight acoustically distinct categories and Asian elephants produce roughly ten, with additional variation within these categories across species (Table 3.1, Stoeger & de Silva, 2013, and references therein). The finding that some elephant vocalizations contain infrasonic components (i.e. frequencies below the 20 Hz hearing threshold of humans) has been a topic of special fascination for researchers and the general public alike (Herbst et al., 2012; Payne et al., 1986; Poole et al., 1988). These low frequency vocalizations (termed "rumbles" or "growls") are produced through myoelastic-aerodynamic vocal fold vibration, a passive process in which the vocal folds vibrate as air moves through them – the same mechanism as in human voicing and distinct from the neurally controlled process that produces purring in cats (Herbst et al., 2012). As with many species that vocalize in this way, the frequency of oscillation scales with body size, thus elephants are able to produce these low frequencies by virtue of having very long vocal cords. Elephants add further variation by opening the mouth (oral rumbles) or keeping it shut (nasal rumbles/growls; Stoeger et al., 2012a). These vocalizations are used in contexts ranging from a variety of social situations from infancy to adulthood, from contexts ranging from distress calls by calves, to mate-search and competition (Soltis, 2010; Stoeger & de Silva, 2013). In both Asian and African elephants, vocalizations can vary systematically with respect to the degree of arousal (or disturbance) experienced, with more irregular features in situations of greater arousal/disturbance (Sharma et al., 2020; Soltis et al., 2009, 2011).

However, although the fundamental frequencies of such calls may be as low as 10 Hz, they often contain upper harmonics that are audible to humans (at least at close range). The functional utility of lower versus upper harmonics (and vocalizations that correspondingly emphasize different parts of the acoustic spectrum) is not very well understood. The lower harmonics are determined by the sound source (i.e. vocal cord length) whereas upper harmonics can be changed by the rest of the vocal tract and may be subject to active control (i.e. the oral and nasal cavities which act as filters, as well as placement of the tongue). For instance, variation in fundamental frequency owing to scaling with body size can convey cues relating to age, sex, and competitive ability (Stoeger et al., 2014; Stoeger & Baotic, 2016), whereas qualitative features of upper harmonics may provide cues on individual identity (McComb et al., 2003). Although variations in frequency (or pitch) can be controlled to a certain degree, these other attributes have constraints that are largely beyond conscious control. In humans, this is analogous to the distinctions between a tenor and a soprano – both may sing melodiously, but each has an individual range and vocal quality that is distinct from the other. The additional layer of sophistication that humans introduce through the intentional and consciously controlled creation of phonemes, which form the basis of words, depends on manipulation of the filter component rather than the source (patterns that phoneticians refer to as *formants*). These distinctions raise a number of interesting and as yet unresolved questions with respect to how and why elephants use different parts of the acoustic spectrum.

One reason infrasonic cues and signals draw attention is that they suffer less attenuation, which allows them to travel further in more open environments (Garstang, 2004). This seems to provide one possible explanation for how elephants might make coordinated movements over distances beyond visual and olfactory range (though we have no way of knowing the latter as yet). Because low-frequency vocalizations are often used in such situations (de Silva, 2010b; Leighty et al., 2008a), are known to contain individually variable attributes (de Silva, 2010b; Leighty et al., 2008b; Soltis et al., 2005), and these attributes are likely used by elephants in discriminating among conspecifics (McComb et al., 2000; Soltis et al., 2005), they are good candidates for a long-range communication channel. The evidence, however, remains circumstantial as most if not all observational and experimental studies have been conducted at relatively close range. Infrasonic signals cannot simply be thought of as an elephantine telegraph system for a number of reasons. This is because the distance a sound can potentially travel *and over which it can be detected* depends not only on its frequency but also its power level, the environmental conditions, and the sensitivity of the detector. Playback studies by McComb and colleagues found that while savanna elephants can recognize social affiliates over distances of 2.5 km, they usually succeeded at distances between 1 and 1.5 km (McComb et al., 2003). As they observe, this is because the individually distinct acoustic features that convey the caller's identify are likely to be "well-above the infrasonic range," since they found that components above 115 Hz lose much of their identifying features at around the same distance.

It is exceedingly difficult to assess the power level of vocalizations produced by wild elephants. I know this because I've tried. Not only must the measuring instrument be close to the vocalizer to get an accurate reading of the source, but there must also be no competing background noise (such as birds, wind, or vehicles) that distort the measurement. Nevertheless, a study of forest elephant rumbles from recordings obtained in Gabon suggests that vocalizations under typical ambient conditions would not be detected beyond a kilometer, with the important harmonic structure degrading possibly within just 100 m (Hedwig et al., 2018). Personal observations also indicate that the low-frequency vocalizations of Asian elephants labeled "growls" ("nasal rumbles") which contain no energy above 500 Hz are relatively soft vocalizations that also function within a similar range. Additionally, the single study of elephant hearing, which was conducted on a seven-year-old Asian elephant, seems to show that this species at least does not actually hear all that well below 16 Hz, the lowest tested frequency (Rickye Heffner & Heffner, 1980). The subject needed a volume in excess of 60 dB to detect frequencies under 16 Hz and over 8000 Hz, whereas there seems to be much more sensitivity between 250 and 4,000 Hz and especially around 1,000 Hz. A similar characterization is not available for savanna elephants, and it is possible that they have evolved higher sensitivity to lower frequencies.

A final consideration, which has not received much research attention, is that sound localization in elephants, as in humans, depends on the distance between the ears and the time and intensity difference registered by the brain as sound passes between them (Heffner & Heffner, 1982). Lower frequencies, with a period that is longer than this time difference relative to head size, are more difficult to localize. While an elephant may therefore perceive and perhaps even identify low-frequency vocalizations by conspecifics, it is not at all clear over what range it would be able to effectively spontaneously *locate* them (without the use of other cues, or additional information such as prior knowledge and expectations). Therefore, the potential for low-frequency vocalizations to be used in movement coordination may be much more limited and dependent on environmental context than is usually appreciated.

In the case of forest elephants and Asian elephants, which typically inhabit denser habitats than savanna elephants, it is possible that low frequency vocalizations are more likely to serve as a cryptic communication channel than a long-distance one. As discussed in Chapter 2, these species appear to have more fluid social affiliations than those of savanna elephants. I hypothesized that this might be driven by the resource distribution (food as well as water) and reduced chance of predation in such environments, compared with the structure of resources and abundance of large predators on savannas. It is also possible, however, that the ecological conditions of forests, namely their high rate of interference with sound transmission, simply do not allow elephants to coordinate with social affiliates over long distances. All of these explanations are mutually reinforcing rather than competing, therefore they are difficult to test for all practical purposes. However, elephants produce a variety of other types of vocalizations that are not limited to low frequencies. These prove to be quite interesting in their own right.

Another class of vocalizations that may function over long range (especially in forested environments) are those variably known as bellows, screams, roars, or long roars, which are much higher-frequency broadband vocalizations that may or not contain tonal regions (de Silva, 2010b; Pardo et al., 2019). In addition to being extremely loud (though their power levels have not been formally measured for the aforementioned reasons), in Asian elephants these vocalizations are made precisely in situations where groups are fragmented and dispersed, and frequently when a "lost" individual, such as a calf, is attempting to locate its social group. Interestingly, these calls can also be used in combination with rumble- or growl-type vocalizations (Pardo et al., 2019). Comparing these "combination calls" by the three elephant species, one expects structural features to be more similar between the two African sister species compared with the Asian. But this was not the case. The proportions of various combinations varied both across species and among different populations of the same species, along with the behavioral contexts they were associated with (Pardo et al., 2019). For example, two-component calls were more common than three-component calls among Asian elephants and savanna elephants, whereas forest elephants also had a similar proportion of three-component calls. Certain sequences, such as the rumble–roar–rumble occurred more frequently in savanna elephants in the context of group separation, whereas they occurred more frequently among Asian elephants in the context of disturbance. This suggests that elephant vocalizations are both structurally and functionally quite flexible, to a degree that may be less constrained by phylogenetic history and more related to the local social and ecological context.

Several captive savanna elephants have demonstrated the ability to reliably produce specific, different types of vocalizations on cue from their trainers, demonstrating volitional control of vocal production (Stoeger & Baotic, 2021). Infant savanna elephants moreover produce a set of vocalizations that are less differentiated by call type and sex than those of adults (Stoeger-Horwath et al., 2007). The fact that elephants have an exceptionally complex and malleable filter apparatus (i.e. the mouth, tongue, nasal cavity, trunk), together with apparent flexibility in vocal production gives rise to the intriguing possibility that vocal communication signals are *learned*. An important additional prerequisite for doing so would be the capacity for vocal learning and imitation.

Excitingly, elephants are among the small handful of clades that have demonstrated this imitative capacity, at least in captivity. Two individual savanna elephants were observed making species atypical vocalizations (Figure 7.1a): one producing those resembling the "chirps" of Asian elephants and the other producing sounds reminiscent of trucks passing along the nearby highway (Poole et al., 2005). Even more surprising, a male Asian elephant by the name of Koshik in a South Korean zoo became a minor celebrity for his apparently spontaneous and self-taught ability to imitate several words of Korean (Stoeger et al., 2012b), much to the astonishment (and perhaps initial disbelief) of keepers and researchers alike (Figure 7.1b). These examples demonstrate vocal *production* learning, not merely flexibility in the contextual usage of signals that naturally exist in the animals' repertoire. Piecing these observations together, there are intriguing signs that vocal learning may be very important for elephants. The low-frequency portion of the sound spectrum may, however, be less useful in conveying social, emotional, and contextual information than higher-frequency components and combinatorial sequences. The development of the full vocal repertoire, and the potential for contextual learning of sound production remain areas in need of future research, especially under natural or semiwild conditions. Tantalizingly, playback experiments on savanna elephants show that they use distinct name-like calls for one another (Pardo et al., 2024). Such names would need to be learned and replicated among individuals, offering one potential function for vocal learning that is suggestive of additional hidden cognitive capacities.

Research on other vocally adept mammals, notably whales, suggests additional possibilities to consider. Among sperm whales, which can be assigned to vocal clans on the basis of stereotyped patterns of clicks known as *codas* (Amorim et al., 2020), the signal differences between matrilines are much greater than geographic distinctions (Rendell et al., 2012). This vertical intergenerational transmission of signals suggests a possible genetic component in the acquisition of these vocal signals, though a role for learning cannot be ruled out. Among orcas, there is evidence from multiple studies of a capacity to copy vocal signals both from conspecifics and even other species, such as dolphins and sea lions (Janik & Knörnschild, 2021). Longitudinal studies of wild orcas and, most famously, humpback whales, have shown the capacity for horizontal/oblique transmission of changes to vocal signals, that is, among unrelated individuals of varying age classes, rather than from parent to offspring (Deecke et al., 2000; Garland et al., 2011). The latter, a clear form of social learning, constitutes a form of culture. I will return to this concept later in the chapter.

Given the potential (and understudied) importance of vocal learning for elephants, there are a world of questions that beg for exploration. What developmental stages might there be? Do elephants, like humans and birds, have sensitive periods for acquiring species or even population-typical signals (i.e. dialects)? What happens if they experience social disruption at such times? These questions are especially salient

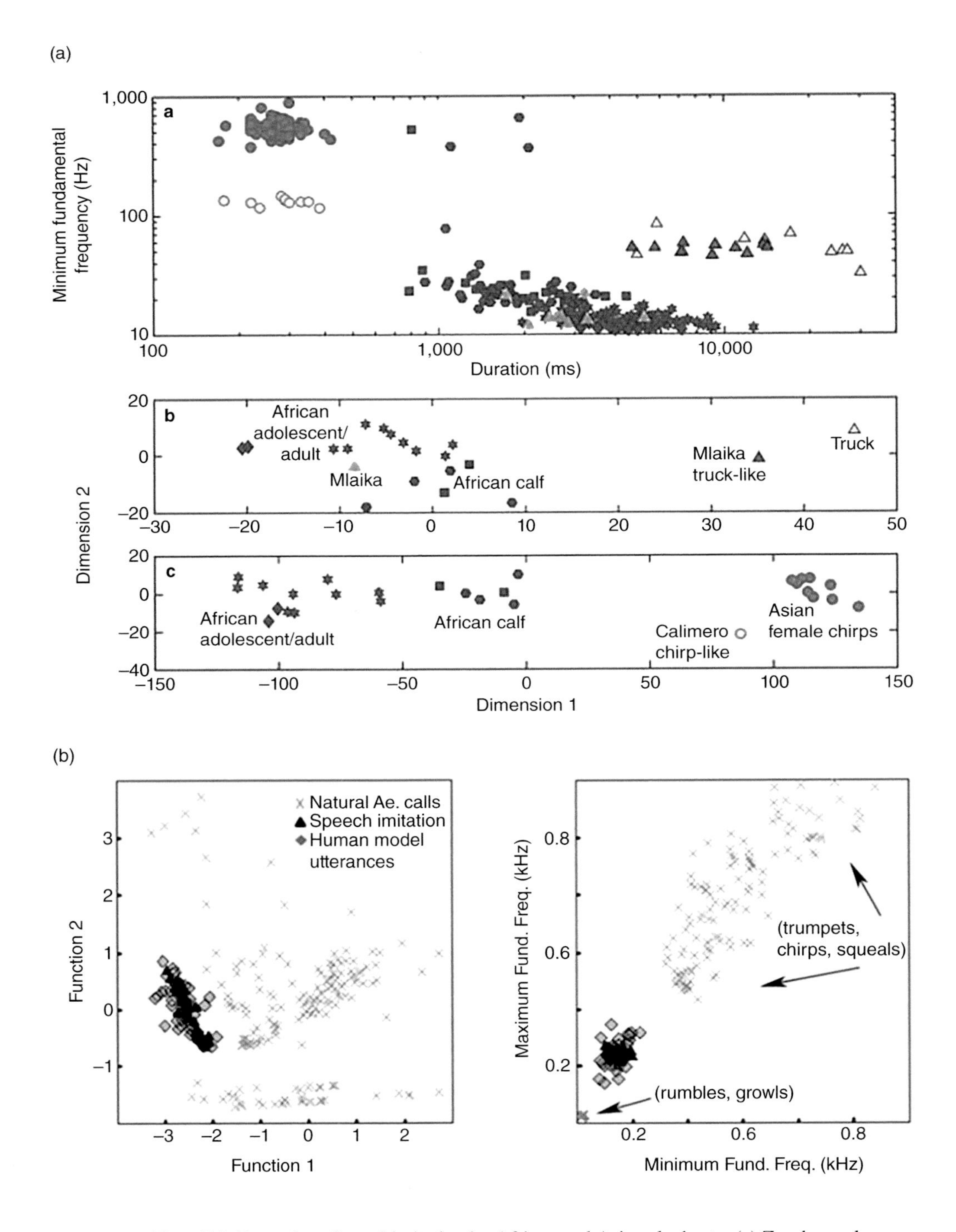

Figure 7.1 Examples of vocal imitation by African and Asian elephants. (a) Zoo-housed savanna elephant Mlaika produced truck-like vocalizations (dark gray triangles) that cluster more closely with the sound of trucks (outlined triangles) than with other savanna elephant vocalizations

for cases in which elephants have been brought into human care from the wild at a very young age. There are numerous rescue efforts in both Asia and Africa that attempt to rehabilitate calves that may have experienced injury or become separated from their social companions with the eventual aim of returning them to the wild (i.e. the aims of "rescue–rehab–release": Goldenberg et al., 2022). The ability to communicate with conspecifics is an essential behavioral competency; but we have as yet virtually no understanding of whether individuals adequately develop these skills. Indeed, our efforts may inadvertently complicate matters further by moving individuals between populations.

I now turn to a related but much more mysterious mode of perception, seismic sensation. Though physical vibrations, known as Rayleigh waves, are often produced through direct impact on the ground, they can also accompany powerful low-frequency sounds (Günther et al., 2004). While humans may perceive (for better or worse) the power of vibrations in the solitary vicinity of especially modified cars or the frenzied atmosphere of collective musical gatherings, we evidently do not use it sufficiently often for it to merit its own vocabulary. Humans have no natural seismic sense or detector, other than our bodies themselves, which consist mostly of water, and our bones, which may rattle. And yet who does not appreciate the visceral and almost primal quality of a well-tuned bass or beating drum? We can therefore, perhaps faintly, imagine what it must be like for a creature that walks the earth sensing vibrations through its ponderous yet delicate feet. An elephant is an animal that spends most of its life standing, thus it is potentially subject to a constant stream of seismic sensations from near and far. The challenge, of course, is that so many occurrences in the environment may produce seismic vibrations, and with our own impoverished modes of perception, we may not imagine how it could be possible to discern one source from another. Might it be possible that elephants perceive the seismic domain with as much nuance and clarity as they (and we) do the realm of sound? This is a fascinating question to ponder, with few studies scratching the surface of possibilities.

The fatty footpads of African elephants contain specialized adaptive mechanoreceptors known as Vater-Pacinian corpuscles, which are absent in the trunk (Weissengruber et al., 2006). These receptors are sensitive to low-frequency vibrations. Because seismic waves can potentially propagate even further than airborne waves (depending on the substrate), it is a good channel for receiving information over long distances. Playback experiments conducted at Etosha National Park in Namibia have demonstrated that savanna elephants are also capable of detecting these seismic signals (O'Connell-Rodwell et al., 2004, 2006), by way of the

---

Figure 7.1 caption (cont.)

(other dark symbols); savanna elephant Calimero produced chirp-like vocalizations (light circles) that are more similar to those of Asian elephants he was housed with (dark circles) than to the other sound categories. Reproduced from Poole et al. (2005) with permission from Springer Nature BV. (b) The male Asian elephant Koshik's vocalizations more closely cluster with human utterances than with natural elephant vocalizations (figures modified from Stoeger et al., 2012b, with permission from Elsevier, Copyright © 2012 Elsevier Ltd.).

specialized sensors of the feet, which are transmitted to the body via bone conduction (O'Connell-Rodwell, 2007). The experiments used putative alarm calls, rumble vocalizations recorded during the presence of lions. Importantly, elephants appear to be able to discriminate between familiar and unfamiliar individuals using this mode of perception (O'Connell-Rodwell et al., 2007). In response to alarm calls from familiar individuals, they significantly increased vigilance behavior and reduced spacing among individuals in the group. They did not do so in response to playbacks of unfamiliar individuals or various types of controls. The latter finding indicates that savanna elephants are capable of astonishingly refined distinctions in the information transmitted through seismic channels, perhaps on a par with their capacities in the acoustic domain. What is more, it suggests they are making judgments as to the reliability of information based on familiarity – a cognitive feat that is certainly transferrable across perceptual modes. O'Connell-Rodwell and colleagues point out that if elephants are capable of making such refined classifications of seismic stimuli, they are likely capable of categorically distinguishing many other types of signals and cues, including both natural (e.g. thunderstorms) and anthropogenic (e.g. vehicles) sources (O'Connell-Rodwell, 2007). A separate study supports this. Elephants were shown to move further away in response to stimuli that were a *combination* of human and elephant-derived seismic stimuli than either one separately (Mortimer et al., 2021). Savanna elephants may also be able to likewise detect and classify the movements of animal groups, including disturbances that might result when they are fleeing predators.

Alas, there is much less known about the potential for seismic communication among African forest elephants and Asian elephants. Asian elephant footpads also harbor the same types of mechanoreceptors as found in African elephants, hinting that this is a possibility (Bouley et al., 2007). But this in itself isn't sufficient demonstration of function as human fetuses also have Pacinian corpuscles in the feet (Jin et al., 2020). Although an early study suggested that vocalizations and "foot stomps" of Asian elephants could be detectable with seismic instrumentation (O'Connell-Rodwell et al., 2000), the substrates of forest environments will differ in their transmission capabilities from the conditions just discussed and behavioral experiments in the wild have thus far not been conducted. Even if they are equally capable of perceiving seismic signals or cues, because these two species appear to be on the whole more cryptic, it remains an open question how well they are perceived and whether signals from conspecifics would adaptively serve them to the same degree.

Elephants most certainly use acoustic (and perhaps seismic) information to make judgments about threat levels. As discussed earlier with respect to the role of older individuals in savanna elephant society, recordings of lions were played back to experimentally assess how individuals responded to potential threats. They differentiated between one and three lions roaring, with older individuals showing more sensitivity to the sex of the roaring lions (McComb et al., 2011). Both savanna elephants and Asian elephants appear averse to the sound of bees (King et al., 2007, 2018), though the response is far more pronounced in the former. Savanna elephants even produce alarm calls in response to bees, which appropriately elicits a headshaking and dusting behavioral response (King et al., 2010). In India, Asian elephants have been shown to respond differently to the sound of leopard growls as opposed to those of tigers (Thuppil & Coss, 2013). In response to leopards, which do not pose much of a danger, they tended to vocalize and seek to confront the source, whereas in response to tigers, they preferred to sneak away. This is reminiscent of the differential responses of Diana monkeys (*Cercopithecus diana*) to leopards and eagles, which are ambush predators, as opposed to humans and chimpanzees, which are pursuit predators (Zuberbühler et al., 1997). The former elicited loud calls from male Diana monkeys, whereas they quietly retreated in response to the latter. These observations highlight the varying and quite sensitive behavioral responses that elephants and other species can make in response to subtle and not-so-subtle differences in the types of threats they encounter.

Even so, there remains an outstanding and somewhat tragic puzzle as to why, with all of this apparent sensory capability, elephants still continue to be susceptible to collision with motor vehicles and trains. Both produce very loud low-frequency cues that are audible from a considerable distance even to a human, let alone an elephant. No doubt heavier vehicles and trains also produce seismic vibrations by their movement alone, again detectable by such weakly endowed creatures as ourselves. There has been a lot of interest among engineers and conservationists as to whether elephant vocalizations can be used to detect the animals themselves and warn them away from harm, yet there is no reason to suppose that elephants would be actively vocalizing when they are approaching roads or tracks, and whether there is any human-generated signal that could effectively function as a deterrent from imminent collision (more on this later). But more importantly, why don't the animals themselves automatically retreat from cues generated by these vehicles and trains?

As any adult responsible for raising young humans knows, it is likely to do with learning and experience. In order to learn that a given sound is indicative of a dangerous fast-moving object, one has to learn or be taught this association – and for that, one has to survive the encounter. Elephants, which are accustomed to using and perceiving low-frequency signals of many kinds in the environment that are not necessarily threatening and perhaps even largely irrelevant, may not have evolved an immediate flight response in reaction to such cues – on the contrary, it's possible they even disregard such cues most of the time. I speculate that novelty might actually stimulate curiosity as opposed to fear. Young or inexperienced individuals would be especially curious and especially vulnerable, a bad combination. Safely crossing linear infrastructure is one situation in which the experienced judgment of older and wiser individuals would prove extremely valuable, on either continent. Of course, from the perspective of conservation, it would be better if they didn't have to undertake such risks at all, with the incorporation of intentional wildlife crossing points (overpasses, underpasses, viaducts) in hotspots of potential collision (Koskei et al., 2022; Okita-Ouma et al., 2021). Though such crossings might be costlier to build initially, but if designed appropriately, in the long term they are likely to be far more sustainable, easier to maintain, and be functionally useful not only for elephants but also other species, in contrast to technically sophisticated warning systems targeting only elephants, which may be challenging to maintain and entail a much steeper learning curve on the part of the elephants.

Loud noises of some sort are among the most frequently used deterrents humans employ against elephants when the latter show up in undesirable places. They can range from simply shouting to banging pots and pans to extremely unpleasant explosive sounds from firecracker-like devices. Most directly, these signals convey to the animal that it has been seen and may force it to reconsider the wisdom of approaching whatever is making the noise. Although the sound itself may evoke a fearful flight response initially owing to its novelty, unless it is followed by an actual negative interaction it may in time be ignored. Most species are quickly able to learn to disregard false threats (one pictures crows perched on scarecrows). This might provoke people to subsequently chase animals and escalate to other methods (discussed in Section 7.2.3). Signals representing actual threats, such as predators or other dangers and nuisances, may be more effective since they are more difficult to habituate to.

This is the reasoning behind experiments in various types of sounds that have been used intentionally on Asian elephants to assess their efficacy in deterrence. In another study in the Nilgiri Biosphere Reserve, southern India, automated playbacks of lion, tiger, leopard, and human vocalizations were conducted as elephants entered crop fields. It is worth noting that these experiments were conducted in a landscape in which tigers and leopards would have some presence, since the study villages were on the edges of the Bandipur tiger reserve; thus elephants would have had experiences with these predators. Lions would have been long absent, however, as Asiatic lions are now limited to a small (but recovering) population around Gir National Park, which is in a distant state. Nevertheless, lion growls had an efficacy of over 83%; tiger growls had an efficacy upwards of 90%, whereas leopards worked a little over 72% of the time. Recordings of people, on the other hand, worked just over 57% of the time. In Sri Lanka, where there are neither lions nor tigers (but where there are leopards), experimenters played back recordings of a female group, disturbed hornets' nest, lone female, or chainsaw to lone adult males being food provisioned along the fenced boundary of Udawalawe National Park. Oddly, the recording of the female group elicited the only statistically significant flight response (Wijayagunawardane et al., 2016). This study was performed with a highly artificial and ethically unjustifiable set-up, however, with playbacks being administered to 22 wild male elephants who were fed by the researchers at the park's boundary electric fence, where authorities have explicitly forbidden elephant feeding owing to its potential to result in negative human–elephant interactions (discussed further in Section 7.4). With no description or contextual details on the elephant vocalization recordings, it is impossible to interpret these results.

The automated playback paradigm described earlier has been extended to other contexts. A study in Europe also found that playbacks of predator vocalizations (human, wolf, dog) were effective at reducing crop damage by large herbivores, including various species of deer, moose, and wild boar (Widén et al., 2022). In this case, recordings of humans were found to be the most effective. More broadly, it has been proposed that the manipulation of fear could be one of several ways to creatively apply research findings toward applied conservation aims (an "applied ecology of fear"), such as reducing conflicts with wildlife (Gaynor et al., 2021). However, it is crucial not to oversell the potential of these interventions by ignoring the adaptive potential of animal populations. To date, no study has assessed the degree to which experience enhances or tempers the effectiveness of these stimuli, and over what timeframe. We have seen indications that elephants can be fearful of a predator they have never encountered (lions, in India), possibly because they are capable of extrapolating from their actual experience (tigers, in India). If the population elephants that would be exposed to these novel or artificial stimuli (including recordings of actual predators) is sufficiently large such that the same individual is unlikely to encounter the recording again, the same stimulus may remain effective for some time. However, if it is only a few individuals that are repeatedly being exposed, then they too will likely habituate regardless of their initial fear. It is highly unlikely that recordings alone will do the job. The stimuli would need to be continually varied and updated, and some form of physical reinforcement would be required to maintain their biological salience and potency. Rather than being the relatively cheap solution as advertised, they may simply be a means of buying time – a more high-tech equivalent of banging pots and pans. Although we might develop similarly adaptive responses to keep elephants on their toes, it amounts to just another type of arms race.

7.2.3 Visual and Tactile Perception

This section might initially appear to be an odd grouping. Elephants are not renowned for their eyesight, nor do the proportions of their eyes relative to the rest of their sensory apparatus suggest them to be spectacularly well-endowed in this regard. Although we suspect, on the basis of simple observations of both anatomy and behavior, that elephants may not see very far, this has not been evaluated. I've combined the visual and tactile here because elephants, rather like primates, may have the somewhat unique requirement of hand–eye coordination – except that for elephants the appendage in question is the trunk rather than the hand. The Asian elephant trunk tip contains both the deeply embedded Pacinian corpuscles as well as Meissners's corpuscles, which are closer to the skin's surface (Rasmussen & Munger, 1996; Stoeger, 2021), and these likely allow for a very fine sense of touch. Based on early studies of the retina of savanna elephants and the placement of elephants' eyes, which are visible from both the sides and the front, it has been suggested that elephants make use of both monocular and binocular vision (Stone & Halasz, 1989). It is thus speculated that it would be highly adaptive for elephants to be able to coordinate the use and placement of the trunk using sight (Jacobson & Plotnik, 2020). This is supported by the observation that savanna elephants appear to have fairly good visual acuity at least at close range (Shyan-Norwalt et al., 2010).

Although the trunk is the most obviously used appendage in tactile investigations, signals can involve all parts of the body including head, limbs, and tail

Figure 7.2 A comparison of similar gestures across species. The "trunk-over" gesture is performed by individuals of all age and sex classes among Asian elephants: (a) one young male to another young male, (b) a male in musth to a female, and (c) one adult female to another adult female. It is interpreted as a signal of dominance or control, though among calves it is expressed during play. Similar gestures are expressed by African savanna elephants but are classified as "rest-head" (d) and "reach-over" (e)–(f). These usually occur during social play and affiliative contexts, or else may be used by males in the context of courtship (photos (a)–(c) by the Udawalawe Elephant Research Project and photos (d)–(f) courtesy of Joyce Poole and Petter Granli of ElephantVoices).

(Makecha et al., 2012). The feet are also used very noticeably. Aside from detecting vibrations, the feet are often gently placed on objects during exploration, and thus also appear to be highly sensitive. Apart from sensing the world, physical contact is an important means of intraspecific signaling. Affiliative signals include gentle trunk touches, head or body rubbing (which can be accompanied by vocalizations, so these may be multimodal signals), brushing the tail against another, and leaning against one another while resting. Among Asian elephants, individuals of both sexes signal dominance by placing their trunk over the subordinate (Figure 7.2), in some instances making a point of placing pressure on the neck (almost biting) using their tushes. This is convenient, as dominants are usually taller. African savanna elephants also employ similar gestures, termed "reach-over" and "rest-head," but these are usually affiliative among most age/sex classes (J. Poole, personal communication). When these behaviors are performed by males in the reproductive context, they may be accompanied by attempts to mount, and are viewed more as a means of controlling females during mating attempts (Poole & Granli, 2021). Likewise, a neck-pressuring behavior also exists, which does appear associated with dominance display (J. Poole, personal communication). This indicates that these particular signals themselves are likely conserved from the last common ancestor, but that the *function* of signals is malleable and perhaps different across species.[1]

Elephants use a variety of postures and gestures that are visible at a distance, but we should allow for the possibility that some of these are not actually intended as visual displays. The same signals could be perceived in multiple modalities, or serve functions other than visual signaling. For instance, the high frequency of ear-flapping (LaDue et al., 2022; Poole, 1989a) and high-headed posture of musth bulls may be not so much visual displays as the byproduct of signals that help spread the odor of the temporal secretions and enable the male himself to better perceive his surroundings and potential competitors (Jacobson & Plotnik, 2020). Other signals, such as the trunk-over display, are also very obviously perceived by the subordinate individual as tactile sensations. There are few behaviors that can be interpreted clearly as intentionally visual displays.

Though the role of visual displays in many cases remains unclear, elephants may prove themselves quite capable on the receiving end. Savanna elephants do seem to be able to classify potential threats to a fairly refined degree by using relatively subtle visual cues. In the previously mentioned study by Bates et al. (2007), elephants discriminated among human garments by both odor and color, red being associated with the Maasai, who have a complicated relationship with elephants and sometimes express antagonism against them. As Bates and colleagues themselves note, however, elephants have dichromatic vision (i.e. similar to "color-blind" people; Yokoyama et al., 2005), and therefore only experience red as a "relatively drab hue". Elephants therefore appear capable of learning associations between visual stimuli and potentially dangerous (or even mildly distressing) objects other than nonhuman predators, but exactly how they make this determination remains to be seen. This is relevant, because people often intentionally use visual stimuli to attempt to scare or deter elephants from undesirable behavior, such as foraging or moving through crops and settlements. This includes hand-held flashlights, fire, and more sophisticated alarm systems that trigger some sort of light display (Adams et al., 2021). With the exception of fire, these methods largely rely on an animal's fear of novelty rather than fear of the light source itself. For all the same reasons discussed in Section 7.2.2, all such purely visual methods are likely to fail eventually once the novelty wears off. Fire is the only exception (perceived through sight, odor, and heat), which for obvious reasons one cannot advocate the use of as a deterrent. Moon phases may also contribute to the amount of cover available for crop foraging, although studies in Africa and Asia show mixed results. Studies conducted in Ghana, Tanzania, and Kenya all found that

[1] In mammals, gestural sophistication is more widespread than vocal complexity. It is an intriguing possibility that gestural plasticity might be a necessary but not sufficient antecedent to vocal plasticity. We earlier considered the possibility of vocal dialects, but there is now a suggestion by primatologists that there may also be *gestural* dialects, as exemplified by leaf-modifying behaviors shown by neighboring communities of chimpanzees. Given the dispersal potential and relatively fluid social organization of elephants (even of *L. africana*), I would not expect to see such great variation within species, which immediately argues against the need for gestural plasticity within species at least. But this is a little-explored topic.

there tended to be fewer incursions during the brighter phases of lunar cycle (Barnes et al., 2006; Corde et al., 2024; Gunn et al., 2014). In Sri Lanka, the frequency of incursions into crop lands by males had no relationship to moon phase, but there was a weak effect for family groups (Fernando et al., 2022). However, lunar effects are difficult to dissociate from human behavior, which includes night guarding.

7.2.4 Getting Our Signals Crossed

Those of us who feel so inclined might take to the woods to escape the bustle and hum of the city – only to contend with garrulous visitors who speak over the sound of quiet waters, winds, and birds. Sadly, unique soundscapes around the world are threatened no less than the physical environments that host them – and perhaps even more so. Noise pollution respects no protected area boundary. Only in the deepest caves may one today find refuge from anthropogenic noise. For species that rely on sound for the essential activities hunting (or avoiding being hunted), mating, socializing, finding their way around, these sources of pollution represent not merely a nuisance but a constant source of stress and risk to survival. There is increasing recognition that the acoustic environment is being polluted by anthropogenic sources of noise (Kunc & Schmidt, 2019; Sordello et al., 2020), which can variously affect where animals choose to forage, court, breed, or make their homes, among other things. Indeed, the temporary respite from certain types of noise such as flights and shipping brought about during the COVID-19 pandemic offered some rare opportunities to study these effects, about which I am sure we will see more publications in years to come. But many studies specifically concerning the effects of noise pollution on the communication systems of terrestrial species focus on higher frequencies, because the bulk of them focus on birds (Duquette et al., 2021). Given that low-frequency communication channels are used not only by elephants but also many other terrestrial species, and these frequencies are subject to near constant anthropogenic background noise, the potential impacts are vastly underappreciated and in dire need of further study. In Gabon, recordings from passive acoustic monitoring stations have been used to train artificial intelligence algorithms that allow researchers to not only ascertain where forest elephants are, but also to detect sounds of anthropogenic threats, such as gunshots associated with poaching (Wrege et al., 2010, 2017). In this way acoustics can itself become a tool for conservation.

Anthropogenic light sources are another well-recognized pollutant. This again is because we are ourselves highly acoustic and visual creatures. Not only moths but also astronomers appreciate the night sky. Given the tendency of Asian elephants to be crepuscular or nocturnal around human habitation, light sources might potentially impede nighttime movement at fine scales beyond the presence of human activities themselves. But what if we also consider the other domains discussed in preceding sections? Might we discover similar concerns? Whatever the domain of sensation and perception, there is a potential for contamination and pollution; therefore we must then also consider seismic noise. As there is so little research on this subject, I take the liberty here of making a leap of imagination – grounded in what we know empirically.

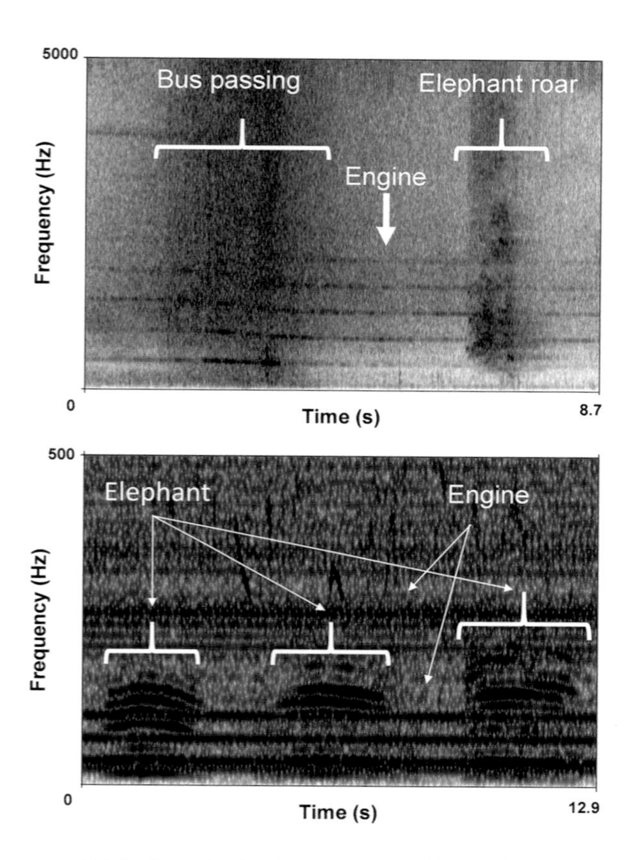

Figure 7.3 Noise pollution in the upper and lower frequency ranges. Upper panel: A bus passing by at high speed, followed by an elephant roar, recorded in Johor, peninsular Malaysia (the spectrogram was generated by the author from a recording by Lim Jia Cherng and Ee Phin Wong, provided by Yen Yi Loo of the Malaysian Elephant Management and Ecology (MEME) group). Lower panel: Engine noise from slow-moving and stationary jeeps carrying tourists at Addo National Park, South Africa (the spectrogram was generated by the author from a recording provided by Angela Stoeger and Anton Baotic).

If we imagine ourselves, for a moment, in the form of an elephant, we might appreciate that constant sources of low-frequency noise, such as the engines of tourist vehicles, highway traffic, or airplanes, may present an irritant that interferes with the basic capacity to communicate with one another at close range and find one another at longer ranges. When recording elephants inside Udawalawe National Park during daylight hours, I was myself often dismayed to discover constant low-frequency bands of noise overlapping exactly with the regions of elephant vocalizations I was trying to capture (Figure 7.3). As alluded to earlier, the omnipresence of seismic cues might potentially cause individuals to disregard relevant information concerning danger, such as those posed by heavy machinery, traffic, or trains. There is some hope that we might develop quieter technologies, such as electric motor vehicles, but this comes with the need for additional safety precautions in itself (quiet engines may increase the chance of wildlife collision). On the whole, the effect of sensory pollutants (be they visual, olfactory, acoustic, or seismic) is an iceberg whose tip we are only beginning to notice and about which there is still much more to be learned.

7.3 Elephant Minds

Now that we have some basis for understanding how elephants perceive and interact with their world, I turn to the internal cognitive processes that lie "under the hood" so to speak. Elephant brains are the largest among terrestrial animals in terms of absolute size (5.5–6.5 kg for Asian and African elephants respectively), but not so unusual relative to their body size (of between 5,000 and 6,000 kg), scaling closely to the curve of brain to body mass ratio for mammals. The "encephalization quotient" (EQ) resulting from this relationship is around 1.3–2.3 for elephants (Byrne et al., 2009), placing them among the likes of squirrels, camels, walruses, and macaques. By comparison, humans have an EQ of 7.4–7.8 while dogs have (on average) 1.2. Clearly, the EQ disparities seem to miss some dimensions along which humans perceive elephants and dogs as being different from squirrels, camels, and walruses (although we may certainly be mistaken). Used for a long time as a proxy for so-called intelligence, the EQ is however criticized as a rather crude metric by which to measure anything at all. Why should the amount of grey matter, relative to other matter, matter? There have been few controlled tests of how brain size actually corresponds to cognition, but evidence suggests that brain size does correlate with problem-solving ability across a diverse range of carnivores (Benson-Amram et al., 2016), even more so than other attributes such as social structure or dexterity. But equally relevant is the structure of the brain itself, how it is organized, and the functions of particular sets of neurons. While elephant brains have three times more neurons than human brains, 97.5% of the neurons in the elephant brain are in the cerebellum (Herculano-Houzel et al., 2014). Some parts of the brain are dedicated to tasks other than cognition altogether, such as motor control. African savanna elephants have ~64,000 facial nucleus neurons, compared with ~54,000 in Asian elephants, a discrepancy that has to do with the more extensive control of the ears in savanna elephants; both species easily beat other land mammals, thanks to their possession of a trunk (Kaufmann et al., 2022). It is therefore inadvisable to use any gross anatomical feature as a proxy for so-called intelligence.

Intelligence is not a generalized phenomenon, but consists of varying capacities. A concept that is especially interesting to consider is that of *cognitive buffering*. The fairly straightforward idea is that species that can use cognitive skills to innovate in the face of environment variability can increase their chances of withstanding change and disturbance (which is hypothesized to spur the evolution of larger brains). It implies a certain degree of behavioral flexibility, but this is a term that has been used in various different ways in comparative cognition (for a thorough review see Lea et al., 2020). The most relevant here is the *behavioral flexibility hypothesis* (see also *adaptive flexibility hypothesis*, Wright et al., 2010), which contends that "large brains confer an advantage when responding to variable, unpredictable, and novel ecological demands through enhanced behavioral flexibility, learning, and innovation" (Marino, 2005: 5306). In energetic terms, animals that can change their foraging tactics to fit changing circumstances can potentially reduce the amount of variation in energy intake, even if the resources around them vary seasonally (van Woerden et al., 2012). Humans, of course, are a good example of a species that can do this, but so can many primates and birds (Baldwin et al., 2022; van Woerden et al., 2012). It is natural to wonder if this is true for many species with generalist diets or habitat preferences, since both allow for a wider set of opportunities to explore and exploit. A review of 193 North American bird species found that innovativeness (in terms of food choices or foraging technique) was significantly correlated with habitat generalism, but not diet breadth (Overington et al., 2011). However, another study including 765 bird species worldwide found a positive relationship between habitat generalism and innovation in food type (i.e. they consumed more novel food types), but not technical innovation rate (i.e. novel means of searching for and handling food) or brain size. On the other hand, diet generalists had all three attributes: higher food innovation rates, higher technical innovation rates, as well as larger brain sizes (Ducatez et al., 2015). The strongest evidence that enhanced cognition helps to cope with environmental variability comes from a study that reviewed the survival success of bird species introduced to new environments, relating these outcomes to observed feeding innovations and brain size (Sol et al., 2005). They found that larger-brained species innovated more and fared better when encountering novel environments.

Although the living elephants are no longer speciose enough to serve as test subjects for these hypotheses, they would appear to exemplify the triumph of cognition as a survival tool – but modern-day challenges risk converting these cognitive assets into liabilities. Attributes that could be adaptive in some contexts (such as the disposition to explore new objects and environments) might also make certain species and individuals of particular species more prone to having negative interactions with people (Barrett et al., 2019). In the following sections, I take a look at experimental and observational evidence for the mental capacities of elephants. Often, cognitive abilities have to be studied in captivity for the degree of control it affords in experimental testing (though paradigms such as playback experiments can be used to interrogate subjects in the wild, as already discussed). Here, more so than in other chapters, I therefore turn to studies both in captivity as well as in the wild for the insights they provide.

7.3.1 Awareness of Self and Others

For decades, the test of mirror self-recognition has been used as a high bar indicative of self-awareness, treated as an all-or-nothing phenomenon (de Waal, 2019). Povinelli (1989) initially tested whether two zoo-housed female Asian elephants were capable of self-recognition using a mirror but failed to find evidence of it. But the mirrors used in the study were small relative to elephant body size and mounted out of reach. Reasoning that large mirrors that the animals could interact with more freely might work better, another set of researchers tested three other adult female zoo-housed

Figure 7.4 A young female Asian elephant investigates her mirror reflection at the Golden Triangle Asian Elephant Foundation, Chiang Rai, Thailand (photo by Joshua Plotnik).

elephants again many years later. All three showed investigative self-directed behaviors in front of the mirror (Figure 7.4), such as exploring inside their own mouths and tugging at an ear, indicative that they were using the mirror for self-examination. However only one out of the three passed the quantitative "mark test," in which she repeatedly touched a visible marking applied on one side of the face significantly more frequently than an invisible sham mark placed on the other (Plotnik et al., 2006). The success of a single individual is sufficient to demonstrate the potential of a species. In this case, the two that did not pass the "mark test" nevertheless behaved in ways that suggest they are aware of seeing their own reflections, even if they have no interest in the mark itself. African elephants have not been tested yet, but it is only a matter of time.

This example, among many others, demonstrates the need for caution in interpreting the outcomes of experimental tests of cognitive capacities one way or the other, as alluded to earlier. And this caution is extremely general. Though the list of species that show various self-directed behavior in front of mirrors remains limited, it is growing, and includes species ranging from house crows to horses (Baragli et al., 2021; Buniyaadi et al., 2020). For instance, it has now been shown that cleaner fish (*Labroides dimidiatus*) are not only able to learn to recognize themselves in a mirror, but they also go a step further (Kohda et al., 2023). First, ten fish were given the mark test, in which a marking resembling an ectoparasite was placed on the throat, and all passed the test by proceeding to scrape at their throats after observing their reflections. Subsequently, they were shown images of their own faces or that of unfamiliar individuals. Naïve fish were initially prone to attacking both their reflections and images of themselves, just as they did with those of unfamiliar conspecifics. But those with more experience decreased their aggression both to the reflection as well as images of themselves, but not images of unfamiliar fish! The authors therefore suggest that the fish remember their own faces, as we do.

If this ability exists in basal vertebrates, such as fish, the potential for self-recognition in the animal kingdom is indeed vast and perhaps largely limited by our ability to adequately put other species to the test. Mixed success rates even among sister taxa that belong to the same clade, and methodological differences across studies make it difficult to draw general conclusions, but for the moment it seems that some measure of self-awareness has been independently evolved across multiple disparate clades (Vanhooland et al., 2022). Mirrors also rarely occur in nature, with self-recognition depending on making an association specifically with a visual stimulus. Prior to the invention of mirrors and sound recordings, we humans were nevertheless aware of ourselves and our voices (indeed, seeing or hearing a recording of oneself takes some getting used to!). In what other sensory domains of which we are both unaware and incapable of testing might other species be capable of recognizing themselves? The desire to protect species such as elephants is certainly strengthened by the perception and appreciation of cognitive ability; but it is worth keeping in mind that what we consider special or unique may be more widespread than we currently appreciate. Moreover, such abilities need not be all or nothing, but rather a matter of degree (de Waal, 2019).

Elephants are also known to exhibit comforting and reassurance behavior when companions are distressed, through vocalizations and touching (Plotnik & de Waal, 2014). It is difficult to say in many situations whether they are actually trying to reassure themselves or the other individual, since agitation and excitement can be contagious, but it is reasonable to think it's some combination of both. Moreover, elephants try to help others that are injured or incapacitated, as countless amateur and professional videos attest. More rarely observed but even more striking, is their response to dead and dying conspecifics. The field of comparative thanatology attempts to characterize and identify the function of diverse behaviors exhibited by many taxa in response to encountering the death of another (Anderson, 2016). These range from "mechanistic, hard-wired functional responses performed in the absence of any conscious emotional components" as exhibited by insects, to "complex, socially malleable patterns that are likely to incorporate emotional states including sadness and grief," (Anderson, 2016, p. R555) attributed to socially complex (often mammalian) species including apes, cetaceans, and elephants.

While a single incident is an anecdote, a collection of them constitutes a dataset. Brilliantly, researchers have been able to compile observations of such thanatological behavior in Asian elephants from videos publicly posted on the internet (Pokharel et al., 2022), as well as through direct observation (Sharma et al., 2019). This includes vocalizations directed at the fallen animal or at bystanders and touching the conspecific variously with the trunk or feet. Larger animals may attempt to lift or carry smaller ones, whereas smaller ones may try to climb on or mount larger ones; occasionally, individuals temporarily guard the bodies of those that are dead, showing defensive responses to the approach of people – even if they do not have any known relationship with the deceased. Even more infrequently, we have observed Asian elephants inspect and handle the bones of a dead conspecific, long after the body had

Figure 7.5 Responses to death among Asian elephants. (a)–(c) An adult female, subadult, and juvenile linger for at least one hour near the body of a dead infant with no known relationship to the group. (d) A female stands over her dead newborn. (e) A group investigates the body of a subadult. (f) A different group approaches the body of an adult female. (g)–(h) The bones of a dead elephant were a source of interest for another elephant group, in which an adult female mouthed one of the fragments. The relationship between the social groups and dead individuals were not known for (e)–(h) (all photographs were taken at Udawalawe National Park, courtesy of the Udawalawe Elephant Research Project).

decomposed. The rarity of such observations isn't necessarily indicative of the rarity of their occurrence (though it might be), but rather of the low probability of human observers being present in order to witness such behavior among free ranging animals, especially when the bones and carcasses themselves are not often visible. The individuals participating in these inspections belong to both sexes and a wide range of age classes, as do the targets of their attention (Figure 7.5).

Forest elephants also show an interest in the remains of conspecifics, though this behavior has been difficult to witness owing to their habitat characteristics and cryptic nature. A group working in Gabon opportunistically documented this behavior with the aid of trail cameras placed in the vicinity of a carcass and recorded five visits by small groups over a period of 65 days. Most of the interest was shown by adults, who primarily engaged in sniffing at the remains (Hawley et al., 2017). The only systematic and intentional observations of elephants' reactions to the remains of their dead has been so far conducted only for African savanna elephants, where wild subjects were presented with the bones and tusks of familiar and unfamiliar individuals as well as those of other species (McComb et al., 2006). In this study as well as other documented cases, elephants show greater interest in their own species than in others, regardless of whether or not the bones belonged to familiar individuals (Goldenberg & Wittemyer, 2020). From this collection of reports, it seems that the living elephant species all have some interest in engaging with dying or dead conspecifics, regardless of their social relationship to them and even when the remains are no longer fresh. This suggests some attempt at understanding and processing these indicators of death among conspecifics, a homologous characteristic that persists despite differences in environments and social systems. However, I note that the bulk of observations involve groups of females and calves, rather than males.

Do elephants have some conception of death then? What is particularly intriguing about elephants is that their interest extends beyond their own offspring or even kin. Many mammalian species have been reported to interact with dead and dying conspecifics, and the most frequently recorded are those involving maternal relationships because, with few exceptions, strong mother–offspring relationships are a hallmark of mammals as a clade (Bercovitch, 2020). This might underlie why thanatological behavior seems to be so often expressed by females. A particularly poignant example was the widely reported case of an orca identified as J-35 that carried the body of her dead calf for 17 days in 2018. Scientists do not shy away from terming such displays as indicative of "grief" (Bearzi et al., 2018). Anthropomorphism, long resisted by scientists, is actually being embraced in some circles as being a useful means of generating empathy for species of concern and countering a very pointed mind-denial of nonhuman animals prevalent in Western modes of thought. This is fraught territory. Although such moments may be reminiscent of *Hamlet*, unsatisfyingly, we still cannot know what is going on in their minds.

Personally, I don't believe that any person who has spent any time observing elephants has any doubt as to their capacity for emotion – but if so, we must allow for both the positive *and* the negative. The label of "gentle giant," so often applied to elephants by the popular media, hardly does them justice. Nor does it help the poorly informed would-be social media stars that view elephants as living plush toys and thus take naïve risks for the sake of selfies. Elephants do kill, by accident or intent. The result is hundreds of human casualties around the world, from infants to the elderly. Complex cognition also opens the door to complex pathologies, likely associated with traumatic events, which we may ignore at our own risk. In circumstances of conflict (or sometimes abuse), for instance, people report what appears to be vengeful or retaliatory behavior on the part of elephants directed at humans. There are even accounts of Asian elephants that go on homicidal killing sprees that defy alternate interpretations. In his book *To the Elephant Graveyard*, Tarquin Hall paints a chilling account tracking one such individual male elephant in India, who not only killed his victims, but also reportedly drank their blood (Hall, 2001). Unlike carnivores, which may acquire a certain ill-advised taste for humans or livestock, elephants have no such dietary excuse. Is this mere exaggeration? Or is there truly such a thing as a "rogue" elephant, a potential for animus that extends beyond an individual's justifiable and basic need to forage or protect itself? If so, what formative experiences drive such behavior? In the case documented by Hall (spoiler alert!), it would appear the animal in question

(a)

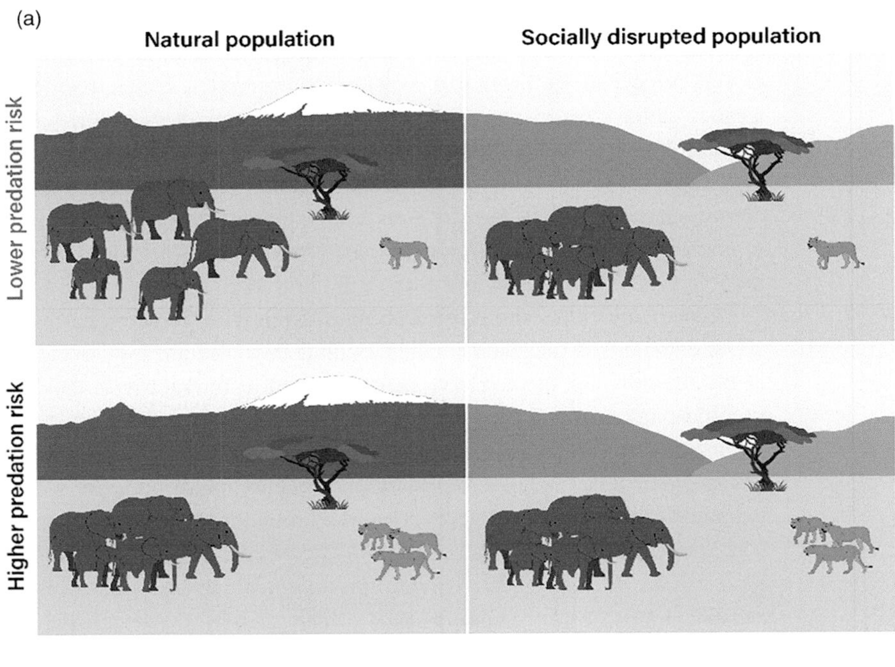

(b)

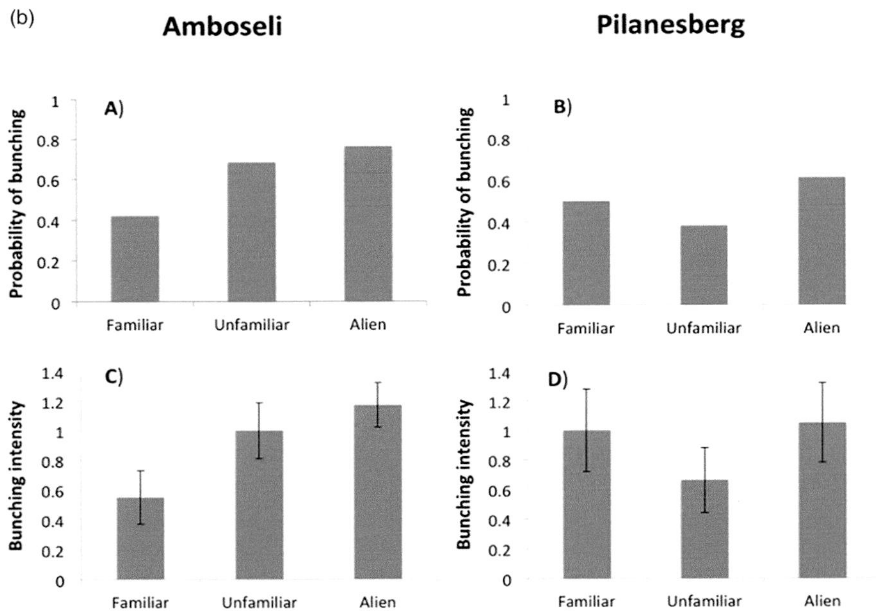

Figure 7.6 Possible effects of social disruption owing to culling of some savanna elephant populations. (a) A graphical summary of the study by Shannon et al., 2022. Intact groups respond (in Amboseli National Park, Kenya) more appropriately by showing a more pronounced defensive bunching response to playbacks of multiple lions vocalizing, which are a greater

had been subject to abuse by an alcoholic keeper while in captivity. Alternatively, as I suspect is frequently the case, are individual elephants offered up as scapegoats to appease the demands of a frustrated public? There are tensions here between trying to understand the animal as it really is, our inability to fully do so, the rigorous requirements of the scientific enterprise, the desire of conservationists to harness whatever means we have at our disposal to ultimately generate support for protecting the species of concern – and a very real need not to oversimplify the capacities of nonhuman minds. To walk this line, we must take seriously the evidence that nonhuman animals are capable of experiencing trauma, with downstream consequences for individuals and populations (Bradshaw et al., 2005). These species are in turn capable of inflicting trauma and emotional distress on people (Barua et al., 2013; de Silva et al., 2023; Pooley et al., 2021), which could persist intergenerationally.

We have already discussed what appears to be pathological behavior on the part of African savanna elephants in Chapter 4, with respect to the behavior of young males at Pilanesberg, South Africa, who killed white rhinoceros until the introduction of mature males into the population. The trauma they had experienced, the loss of their families during culling and their subsequent translocation to unfamiliar areas with unfamiliar individuals, may manifest in subtler ways as well. Using playback experiments, researchers evaluated the ability of elephants in the Pilanesberg population to appropriately classify and respond to vocalizations on the basis of social cues (i.e. identity and dominance status) and compared their responses with those of the Amboseli population, which has remained largely intact (Shannon et al., 2013). They found that while those in the Amboseli population showed responses varying systematically on the basis of social identity and dominance, those in the Pilanesberg population did not (Figure 7.6). In particular, elephants in Amboseli showed a stronger defensive response to individuals that were likely to be more socially dominant, whereas those in Pilanesberg seemed to fail to discriminate on the basis of features including age and identity, indicative of status. This despite the fact that the original culling events had taken place decades prior to the experiments. A similar pattern is also seen with their response to potential threat from predators (Shannon et al., 2022). Whereas elephants in the Amboseli population showed stronger defensive reactions in response to playbacks of multiple lions rather than those of a single lion, those in Pilanesberg showed heightened responses to both, thus they seemed to lack the knowledge and experience to react appropriately to the level of threat.

Now, it's possible that there is some other explanation, such as ecologically driven differences in social behavior or the degree of risk presented by lions, that underlies

Figure 7.6 caption (cont.)

threat than a solitary lion; disrupted groups (in Pilanesberg National Park, South Africa) show heightened defensiveness in both contexts. (b) Intact social groups show steadily more defensive responses to playbacks of vocalizations from increasingly unfamiliar individuals (A, C) whereas disrupted groups respond as strongly to both familiar and completely alien individuals and a somewhat weaker response to unfamiliar ones (B, D). It is proposed that social disruption impedes the individuals' ability to assess threats and potential competitors (panels (a) and (b) reproduced from Shannon et al., 2022, under Creative Commons license).

these differences. For instance, lions, which are ambush predators, may present more of a threat in more wooded locations, one site may be more woody than the other, and the amount of experience with lions may be different in the two populations/environments (Pilanesberg has areas of denser vegetation and a slightly higher density of lions than Amboseli). Elephant calves can also be more vulnerable to lion predation in drier periods (Loveridge et al., 2006), which may heighten vigilance, and it's not clear whether the different sets of experiments took place under similar conditions. Moreover, given the original traumatic experience of the founders of the Pilanesberg population, they may simply be more reactive overall – both with respect to conspecifics and predators. Despite these caveats, given the potential role of older individuals in identifying and responding appropriately to social and predatory cues, discussed in Section 7.2.2, it is plausible that the acquisition of these skills has been impeded in the disturbed population. Moreover, in a related but different comparison, another group of researchers studied the social tendencies of elephants in Kenya that had been orphaned though poaching relative to that of nonorphans (Goldenberg & Wittemyer, 2017). The clearest takeaway from this study was that orphans tended to interact less with matriarchs than nonorphans, who also of course could interact with their own mothers. This points to a mechanism by which social transmission of knowledge can be impeded, with orphans having reduced opportunities to learn from older females. These observations lead us into considerations of how behaviors are adopted and learned, and the implications for managing existing populations or recovering others.

7.3.2 Personality and Problem-Solving

The growing field of research exploring "personality" in nonhuman animals, characterized as "consistent among-individual differences in behavior" (Sih et al., 2015), reveals that some degree of standing variation in behavior is widespread among many species. These traits may fall along axes representing gradients such as gentle/aggressive, shy/bold, sociable/unsociable, and so on. Suites of traits may cluster together, forming particular behavioral "syndromes." Fundamentally, this standing variation in the suite of traits an individual possesses causes it to behave in a consistent manner with respect to various situations (contextual generality) that persists over time (temporal consistency) and involves some interaction of genes and environment (which is, of course, all of behavior; Stamps & Groothuis, 2010). It is suggested that the potential for such variation, and thus the evolution of personalities, may be related to there being more than one way to solve particular life history trade-offs (Wolf et al., 2007). Far from being inconsequential idiosyncrasies, such variation can matter for ecology and evolution (Wolf & Weissing, 2012).

To take one commonly used example, an organism has two possible choices for how to grow and reproduce: the "live fast, die young" strategy, which favors boldness and risk-taking so that it can grow fast at the expense of potentially earlier mortality, or a "slow and steady" strategy, which favors caution that may result in slower growth but longer lifespans. The fitness payoffs for both strategies may be similar, thus variation can be maintained. These two alternatives might be recognized as the r- versus K-selected ends of the life history spectrum, discussed in Chapter 4, except that now we focus on individual characteristics rather than reproductive tactics. Certain behavioral traits may correlate with others – for example, boldness may accompany neophilia or the tendency to experiment and innovate. Curious, exploratory individuals may be quick to colonize new environments, attributes that can either prove advantageous or detrimental. It is easy to see how this might play out, for instance, in terms of individual foraging tactics. These strategies can in turn influence how populations interact with their environment, including human artefacts and agriculture, and have cascading effects on linked organisms and throughout the ecosystem (Hunter et al., 2022). Seed dispersal, for instance, is a process by which foraging and movement decisions can ultimately affect plant distributions (Spiegel et al., 2017; Zwolak & Sih, 2020).

Elephants in zoos, both Asian and African, exhibit consistent differences in sociability as rated by their keepers (Barrett & Benson-Amram, 2021; Grand et al., 2012; Horback et al., 2013; Williams et al., 2019). Among the semicaptive Asian elephants of timber camps in Myanmar, males have been described as being on the whole more aggressive and less sociable than females (Seltmann et al., 2019), as one might expect given life history differences between the sexes. As discussed in Chapter 3, female Asian elephants also show individual variation in the degree of sociability or gregariousness, although it's not clear how much this has to do with personality as opposed to the particulars of their demographic circumstances (e.g. the number of living relatives she has). These observations do suggest, however, that there are different *social phenotypes*, with some individuals being more likely to interact with conspecifics than others. Some of these differences may be attributable to something like personality differences, while others, as in the case of orphans, may be tied to life events that position individuals differently within communities. Among wild savanna elephants, personality traits contribute to how individuals interact within and among social groups and could possibly have longer-term consequence for the success of these groups (Lee & Moss, 2012).

One of the domains in which these phenotypic differences might be relevant is problem-solving. As briefly discussed with respect to the mirror tests of self-awareness, individuals usually vary in how they tackle cognitive tasks or innovate solutions to problems. But it is difficult to assess the relationship between personality and problem-solving because sample sizes in captivity are usually quite small. For example, the apparent use of insight in problem solving (one-shot learning, often described as an "Aha!" moment) has been demonstrated by a single Asian elephant housed in a zoo (Foerder et al., 2011). Kandula, a male Asian elephant that was seven years old at the time, spontaneously figured out how to position and stand on a block in order to reach food items that had been hung out of reach. Two adult females that were also tested did not develop this solution, although one of them, his 33-year-old mother, had been observed to move objects and stand on them to reach items during her adolescence, but not since then. Kandula also demonstrated novel solutions to variations of the test, by using first a ball (an unstable object that the experimenters had not anticipated being used for the task) and then stacking multiple objects. One wonders if Kandula's success was driven by a particularly curious and innovative nature, or whether this is not unusual in elephants his age (perhaps the older females were simply more cautious and averse to falling in this particular task). In the absence of other age/sex-mates to be tested in a similar manner, one cannot say.

In a separate set of studies, experimenters directly evaluated 20 different personality traits in relation to problem-solving skills among 15 Asian elephants and 3 African elephants housed at three different North American zoos (Barrett & Benson-Amram, 2021). Personality traits were measured both in terms of keepers' ratings, as well as subjects' responses to various novel objects and fell along four axes which can roughly be characterized as how affectionate and sociable they were (the primary axis), energetic and curious (second axis), aggressive and protective (third axis), and excitable or mischievous (fourth axis). Then they were given three different puzzles to solve, with their adeptness at doing so measured in terms of how long it took them to solve the puzzle as well as the number of challenges they were able to solve. The results were mixed; there was no clear association between particular suites of traits and the various measures of success. Instead, particular suites of traits predicted success in some contexts but not others. Another group tested a set of captive Asian elephants with a puzzle-box apparatus and evaluated their success at solving the problem in various possible ways in relation to the behavioral attributes of neophilia, persistence, motivation, and exploratory diversity (Jacobson et al., 2022). They stop short of referring to these as personality traits, as they did not assess how consistently and reliably individuals expressed these characteristics. Eight elephants arrived at all three possible solutions, three of them managed two, and two solved only one. They found no significant effects of age or sex, but those that were more persistent (i.e. spent a greater proportion of time plugging away at the challenge) enjoyed greater success, but contrary to expectations, there was an inverse relationship between persistence and exploratory diversity (i.e. the number of unique actions and body parts the subject used to investigate the puzzle). This is surprising because in other species, such as wild spotted hyenas, exploratory diversity is associated with greater problem-solving success (Benson-Amram & Holekamp, 2012), suggesting either that these disparate studies were not measuring similar things, or that the predictors of success vary across species and contexts. Neophilia (how quickly they interacted with the puzzle when it was first introduced) and motivation (how quickly they interacted with the box in each session) didn't matter as much. Do these observations carry over into semicaptive or wild contexts?

Another study by Jacobson and colleagues, this time focused on wild elephants at Salakpra Wildlife Sanctuary in Thailand, provides some answers (Jacobson et al., 2023). Similar to the zoo-based set-ups, experimenters used puzzle boxes that could be opened in three possible ways. The boxes were attached to trees, baited with pieces of jackfruit (a naturally occurring fruit tree species) as the reward, and monitored with three trail cameras that provided lateral and overhead views. They quantified the number of types of doors individual elephants were able to open ("innovation"), the proportion of doors opened ("success"), and how long it took them to open a particular door ("solving latency"). They additionally defined "neophilia" based on how long it took the individual to make contact with the puzzle box once it had approached within 5 m, "exploratory diversity" based on the number of unique motor actions with different body parts the animal used, and "persistence" based on the time spent interacting with the box relative to the total time spent within 5 m of it. Out of 77 elephants that approached within 5 m of the apparatus, 44 interacted with it and 24 managed to open at least one door type; of this subset, eight (18%) opened two, but only five (11%) opened all three. Therefore, there was wide variation not only in overall success rates but also in what the researchers had defined as "innovation." Exploratory diversity and persistence had significant positive effects on success during the first interaction with the box as well as successive attempts. Individuals that interacted more with the box also opened doors more quickly on subsequent attempts. Persistence and the number of interactions per individual were also significantly associated with innovation score, as one would expect, but exploratory diversity was not. Age, sex, and being in a group did not seem to have any influence on whether subjects interacted with the boxes at all or on any of the tested outcomes; however, as the sample sizes for each category were not given, it is difficult to know if these effects could be evaluated robustly. The key takeaway nevertheless, in keeping with the zoo-based study, seems to be that the propensity to keep trying (which I would view as a combination of the proportion of time spent on the problem and the number of attempts made to solve it) is an important mechanism for achieving success, more so than measures of neophilia.

In the preceding experiments, the "innovative" behavior individuals could actually show was quite constrained by the design of the puzzle, unlike in the examples of insightful behavior discussed earlier. But we might ask whether varied explorations by particular individuals can lead to other, more creative, types of innovations such as tool use. Simply put, a tool is any object that is used (sometimes with modification) for a clear purpose. There was initially much ado about the use of sticks by chimpanzees for fishing out termites. But as with so many skills, tool use is now known to be far more widespread among not just primates and corvids but many other species, including, yes, fish (Coyer, 1995; Millot et al., 2014). Elephants, for their part, have been known to use branches to swat at flies on their own bodies (Hart et al., 2001; Hart & Hart, 1994), but my own observations indicate this behavior seems to be exceedingly rare (though seen in both sexes). They more readily use sticks to scratch themselves, as I've witnessed many times. Among the more interesting tools people have examined is water, inspired by Aesop's fable of the crow and the pitcher, wherein one must float an out-of-reach food reward by either filling the container with water or placing objects in it to raise the water level (Cheke et al., 2012; Jelbert et al., 2014). Classically tested on a range of corvid species, Aesop's original tale seems to suggest a stroke of insight bound with causal reasoning, but it now appears that trial-and-error learning may play an important part in this task as well (Hennefield et al., 2018; Jelbert et al., 2015). Given elephants' affinity for water and their possession of some handy anatomical tubing, one might expect them to be naturals at this task. And, indeed, a study of 12 captive Asian elephants found they could succeed somewhat unintentionally, simply out of their predisposition to put water into things, perhaps just for fun (Barrett & Benson-Amram, 2020). But

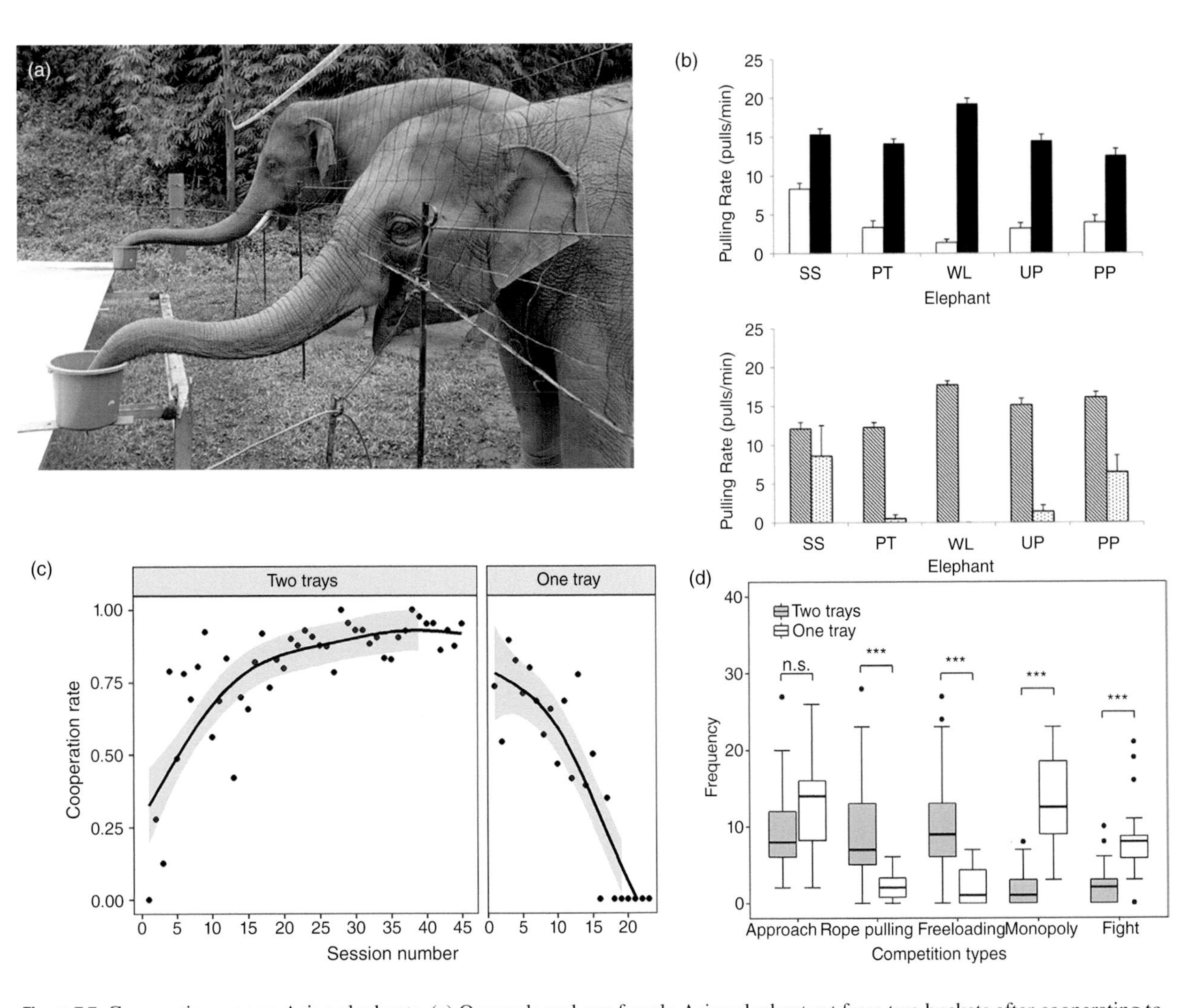

Figure 7.7 Cooperation among Asian elephants. (a) One male and one female Asian elephant eat from two buckets after cooperating to pull in an out-of-reach table using the two ends of a rope threaded through and around the table. The volleyball net in the picture was

only one individual seemed to solve the task in a goal-directed manner, though experimenters could not rule out that this was through some trial-and-error learning either throughout the experiment itself or during prior experiences. The evidence for causal reasoning in elephants remains rather sparse, however (Irie-Sugimoto et al., 2008; Nissani, 2006), and needs far more exploration, especially given their behavior in cooperative tasks.

This brings us to the point that individual capacities are only part of the story. Though many paradigms of cognition focus on individual learning abilities through solitary testing, social creatures have the benefit of opportunities to cooperate as well as learn from one another. In closing this section, let's consider explicit cooperation, which one expects to find among social species. Here I limit my discussion specifically to cooperative tasks and problem-solving, as opposed to behaviors associated with breeding or offspring-rearing. Of particular interest is cooperation among nonkin, which cannot be explained by kin-selection theory. Such cooperation must be explained in terms of direct benefits, where the individual has something to gain. Cooperative problem-solving has been experimentally demonstrated in a range of vertebrates including social carnivores (Drea & Carter, 2009; Schmelz et al., 2017), birds (Péron et al., 2011; Seed et al., 2008), and primates (Chalmeau et al., 1997; Drea, 2006; Hare et al., 2007). Captive Asian elephants have joined these ranks, by showing teamwork in pulling at a pair of ropes in order to bring a food reward within reach (Figure 7.7; Plotnik et al., 2011). Interestingly, not only did they cooperate, but little variations in tactics (including cheating) suggest that at least some individuals are capable of figuring out the mechanism of rope-pulling well enough to understand when it was appropriate to do so and even take shortcuts. They refrained from pulling if the partner could not also do so, and one individual spontaneously realized that they could avoid the actual work of pulling the rope by simply standing on it while their partner pulled, achieving the same end result (I would actually view this as another demonstration of insightfulness, though it was not the intent of the experiment). This again begs the question of how much causal reasoning they might actually be capable of.

However, cooperation need not be a universal response – especially when there is a limited resource over which individuals must compete. A follow-up study, which used

Figure 7.7 caption (cont.)

the starting position at which the elephants picked up the rope ends, which were laid on the ground near the base of the table (photo by Joshua Plotnik, taken at the Golden Triangle Asian Elephant Foundation, Chiang Rai, Thailand). (b) Individual elephants (x-axis) pull on the rope at a higher rate when their partners can also pull. Upper panel: pulling rate is lower before the partner enters (white bars) compared with after (black bars). Lower panel: pulling rate is lower when partner is present but cannot reach the rope (light bars) compared with when partner has rope access (dark bars) (reproduced from Plotnik et al., 2011, PNAS, Open Access). (c) When tested as a group, elephants cooperate when two food trays are present, allowing both participants to access the food but breaks down when only one tray is present. (d) Rope-pulling as well as freeloading are significantly more common in the two tray condition than the one tray condition, whereas monopolization and fighting show the opposite pattern ((c) and (d) reproduced from Li et al., 2021, PLOS, under Creative Commons license).

the same rope-pulling set-up to probe the subtle but important question of when cooperation among Asian elephants actually *breaks down*, found that it was mediated by the degree of affiliation and relative ranks of the participants (Li et al., 2021). Unlike the previous study, nine elephants were tested as a group rather than in pairs and under two conditions: first with two trays of food, then with one. In the latter condition one expects more competition that might interfere with cooperative efforts, and indeed the increasingly high cooperative behavior seen in the first 45 sessions was extinguished by the last session, with a concurrent rise in competitive, interfering interactions (Figure 7.7). There are noteworthy similarities and differences with similar experiments conducted with captive-born hyenas (Drea & Carter, 2009), which are highly social carnivores with strong dominance hierarchies. Like elephants, rank differences impeded performance wherein pairings containing the most dominant animals were less efficient and successful (owing to more aggression being expressed on the part of the dominant) than pairings containing lower-ranked individuals. But despite this, and unlike elephants, the overall effect of having more individuals present was positive, as inexperienced individuals could learn to master the task from those who were more experienced (this was not explicitly tested in the elephant study). Wild elephants, unlike hyenas, generally do not need to cooperate to obtain food, nor do they regularly face direct competition over food items, even if they may sometimes compete over other scarce resources such as water. Hyenas (and other cooperative hunters) may therefore have a higher predisposition to cooperate in the tested scenarios.

However, it is interesting to consider whether there might also be a distinction between species which normally exhibit clear dominance hierarchies and those that do not. As discussed in Chapter 3, one of the primary functions of dominance hierarchies is to resolve disputes more efficiently, be it over mates or resources (Hand, 1986). Comparing captive-housed bonobos and chimpanzees, it was found that although both species cooperated similarly when food was not monopolizable, bonobos did better than chimpanzees when food was highly monopolizable (Hare et al., 2007). This seems to be because bonobos are more tolerant of cofeeding and are even observed to share food with unfamiliar individuals (Tan & Hare, 2013). Asian elephants do not normally exhibit rigid hierarchies (de Silva et al., 2017), though researchers might well assign ranks to individuals confined in a restricted social-spatial environment for purposes of study. This lack of a conflict resolution mechanism might interfere with cooperation over an artificially limited resource. African savanna elephants, on the other hand, do naturally evidence strong dominance hierarchies as discussed in Chapter 3 (Archie et al., 2006a; Chiyo et al., 2011b; Wittemyer & Getz, 2007), and in at least some populations of males, such as at Etosha, they are more evident during resource-limited (drier) conditions (O'Connell-Rodwell et al., 2011). So if the hypothesis holds, savanna elephants should be better at maintaining cooperation even when there is potential for competition because lower-ranked individuals would be less likely to challenge those that are higher-ranked, and more likely to tolerate competitive behaviors directed at them. This remains to be seen. An intriguing implication is that resource scarcity, through their promotion of hierarchies and social mechanisms of conflict resolution, may wind up favoring greater levels of cooperation (within the limits of population size). As anthropogenic activities limit resources for elephants and other wildlife, this points to a possible means of adaptation in the future, which I will come back to in Section 7.4.

7.3.3 Observational Learning and Culture

Individuals can learn by observing others, with social species naturally having more opportunities to do so. They can learn not just about the world, but also about how to do things or even of certain preferences (or biases) that are perceived to be beneficial. All forms of social learning carry a potential for transmission vertically, horizontally, or obliquely. Such transmission may be intentional (i.e. taught) or not. By the same token, different populations that are not in contact with one another may acquire different ways of doing or favoring things that set them apart from one another, creating divergent traditions. The persistence of these attributes within a population over an extended period of time (even intergenerationally) results in *culture*, and there are by now a wealth of studies showing it to be present in a great diversity of species, from insects to whales (Allen, 2019; Whiten, 2021). While one may immediately think of domains such as communication, other behaviors such as foraging are also subject to cultural influences. For instance, vervet monkeys can be experimentally induced by researchers to arbitrarily prefer food dyed a particular color, and then pass on this preference to naïve individuals (van de Waal et al., 2013). Capuchin monkeys have been experimentally shown to adopt foraging tactics exhibited by the alpha male, by training the alpha males to open an artificial food source in specific ways (Dindo et al., 2009). These manipulations thus demonstrate the possibility of establishing foraging "traditions" in the studied primate groups. In a striking example of naturally occurring behavior, a single humpback whale in 1980 pioneered a novel variation on bubble-net fishing known as "lobtail feeding," where it first struck the water's surface with its tail flukes before following through with a bubble-net. This behavior was thought to improve the ability to prey on a different species of fish over the original technique. Over a 30-year period, the innovation was observed to spread to hundreds of other individuals, amounting to around 40% of the population (Allen et al., 2013).

The presence of culture among relatively large-brained cetaceans and primates, aspects of which have already been discussed, is perhaps less than surprising. But insects have brains the size of a pinhead or smaller, and their behavior has long been taken to be hard-wired genetically. Not so it seems. Take the honeybee. Honeybees have been model organisms for the study of social behavior and communication, being the chosen subjects of Karl von Frisch, one of the Nobel-prize winning founders of the field of ethology. von Frisch first described and decoded the dance "language" of honeybees in the 1940s, yet even after decades of study, we continue to learn more about the nuances of these behaviors. Perhaps surprisingly, it turns out that the skill of dancing is itself a socially learned rather than innate behavior: Naïve honeybees (*Apis mellifera*) deprived of experienced models produce messier error-prone dances (Dong et al., 2023). Some of these errors (accuracy of angle and direction) could be corrected with later experience, but others (distance) become fixed. On the flipside, though, various honeybee populations can show systematic differences in how they encode this distance information, referred to as dialects. Unlike whale song or human linguistic dialects, which show no obvious adaptive function, dance dialects may actually be adaptive responses to the foraging conditions bees experience. Comparing three different species of honeybees (*A. cerana*, *A. florea*, and *A. dorsata*), it was found that species foraging within a smaller radius showed a steeper relationship between waggle duration and distance to food than those foraging over longer ranges (Kohl et al., 2020). The relationship can sometimes vary even within species depending on the density of vegetation (George et al., 2021). Thus the propensity for flexibility of certain signaling components may be adaptive for bees overall, but at the cost of potentially making bees liable to errors under certain circumstances.

Why dedicate a paragraph to bees in a book about elephants? For one thing, it forces us to acknowledge that relatively large-brained birds and mammals don't necessarily exhibit greater sophistication than the rest of the animal kingdom (who knows what the mysterious gigantic cephalopods are doing when we're not looking?). Although it is commonly assumed and stated that elephants must have various cultural capacities owing to their social and cognitive attributes (Allen, 2019), the actual evidence is rather thin. These studies of bees and other species highlight a number of outstanding questions one might pose about Proboscideans as well. All of the domains of communication and cognition I have discussed in previous sections show potential for the manifestation of social learning and culture. Though the strongest case might be made for its role in vocal communication, with a little imagination we might ask whether there are any socially learned components to other communication modalities (do elephants learn from one another which odors are associated with friends or enemies?), how individual attributes originating from personality differences might manifest in behavior transmitted to conspecifics (can a problem-solving technique or food preference get copied?), whether adaptive or maladaptive behaviors might "fix" in populations (do populations react differently to human presence?), and many other questions that have not yet been thought of. The facility with which experiments can be conducted on bees provides insight into these complex behaviors to a degree that unfortunately cannot easily be replicated in experimental studies of elephants, but the space of possibilities is vast. Far from being inconsequential curiosities, the implications for evolution and conservation may be significant (Brakes et al., 2019, 2021).

For the moment, we are constrained to what can be observed in the wild, and that too under a limited set of conditions and sites. Savanna elephants appear to "simulate" oestrous signals through posture and gait even when they are not themselves reproductive, effectively a "false oestrous" (Bates et al., 2010). The behavior is quite rare (19 false oestrus events versus 980 genuine oestrus events observed across 480 months of daily data collection) but seems to occur disproportionately more often when there is a naïve young female relative who is herself entering oestrus for the first time. Observers therefore suggest that this may be helping inexperienced individuals to learn the appropriate thing to do and whom to target, effectively a type of behavioral modeling. In a completely different example, Fishlock and colleagues argue

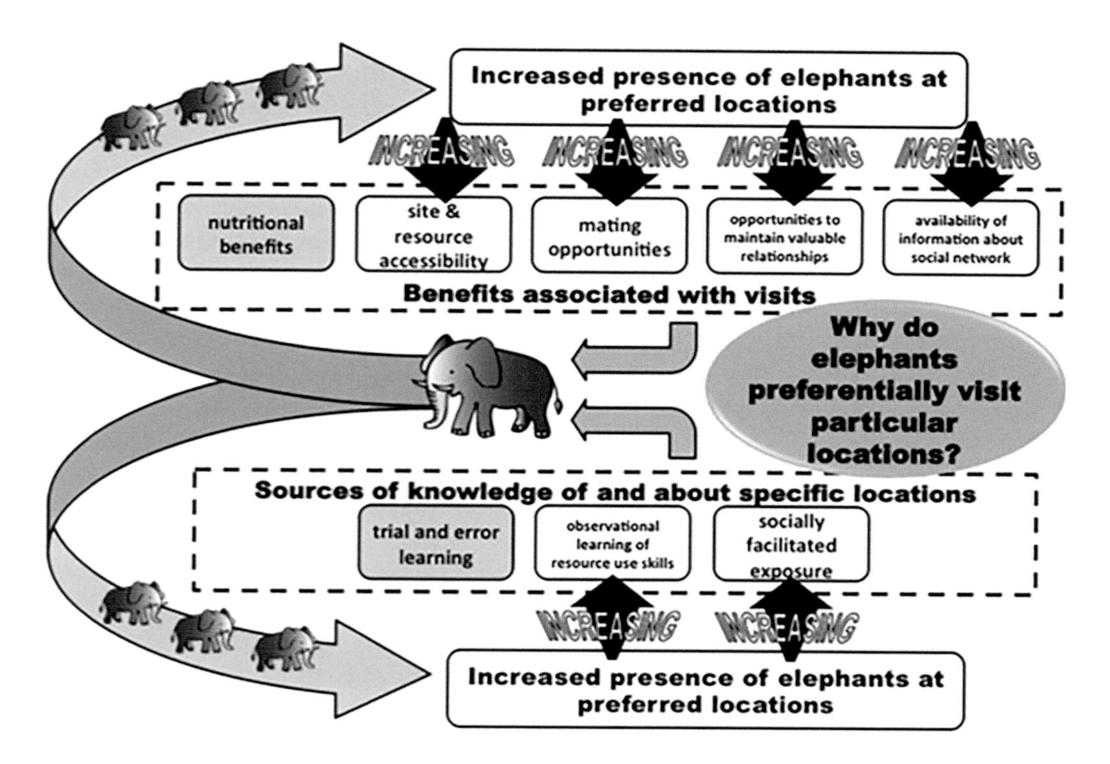

Figure 7.8 Possible dynamics establishing elephant resource-use traditions. Resources may be discovered through trial-and-error learning initially, but preferences for particular locations become preserved through self-reinforcing mechanisms owing to the presence of conspecifics (lower loop). Likewise, although there may be multiple sites with similar nutritional benefits, the presence of conspecifics at particular sites can be associated with social benefits, further reinforcing preferences (upper loop) (reproduced from Fishlock et al., 2016, with permission from Springer Nature BV).

that the preferential use of particular resources with great fidelity by African savanna and forest elephants constitutes traditions (Fishlock et al., 2016). We've already seen some specific examples of this in Chapter 6 with respect to waterhole use in arid environments. More generally, elephants may make use of particular swamps, forest clearings, caves (created by elephants themselves over many years), riverbeds, and so on. The used sites do not appear to differ substantially in the concentration of whatever resource they provide over unused sites, but their use is facilitated by the presence of well-trodden trails as well as incentivized by the fact that other animals also used them. Therefore, however a location becomes initially established, its use can become continually reinforced through these positive feedbacks (Figure 7.8). Reliance on such traditions can be greatly beneficial, as it spares individuals the costly process of exploration and trial and error that they might otherwise need to undertake; however, as I suggest in Chapter 6, it can also present a potential liability if these traditions reduce their ability to flexibly respond to change, trapping populations in a precarious environment. On the flipside, transmission of problematic behaviors through social learning can establish novel cultural traditions that bring elephants into greater conflict with people. Waiting for decades to establish with scientific rigor that such behaviors can be socially transmitted seems ill-advised. But unfortunately, they often cannot be documented in ethological or ecological journals owing to the small sample sizes and opportunistic nature of observations. The standards of rigor within these disciplines can thus act as an impediment to knowledge exchange, as Pooley and colleagues observe: "Dismissing anecdotes about singular animals renders ethologists unable to perceive unusual and singular animal behavior and protects their notions of typical species behavior, generalized from statistically significant sample sizes of observations…" (Pooley et al., 2021). I therefore take this opportunity to document and discuss case studies that signal worrying trends with respect to elephants.

7.4 Case Studies and Real-World Considerations

Rambo and Rivaldo are two wild adult male elephants in Sri Lanka and India respectively who grew accustomed to receiving food hand-outs from tourists. Rambo, who was estimated to be at least in his 40s by the year 2020, pioneered the tactic of "begging" at the electric fence delineating the southern border of Udawalawe National Park in Sri Lanka from at least the early 2000s, when he would have been in his 20s (Figure 7.9). At the time, the electric fence was perceived to be so effective in deterring elephants that the electricity was turned on only at night. Passing travelers and tourists, unaware of this, stopped to feed Rambo and be photographed with him. Although old enough to enter musth regularly, Rambo remained consistently at the fence throughout the year. Over time, other males copied this behavior, positioning themselves at various points along the fence. Between 2007 and 2021 we observed at least 66 different male elephants of varying ages seeking food at the fence. This is likely to be only a small subset of the number visiting the fence. Most were seen only over a single year, but nearly half were present in multiple years (Figure 7.9); thus begging had turned into a regular feeding strategy.

In 2011 a so-called problem elephant, a one-tusked adult male, was translocated into the park. As tusks do not conduct electricity, this individual was adept at breaking fences. We concurrently recorded an increase in the reports of the electric fences being broken, not only by this distinctively identifiable animal, but also multiple other elephants. One subadult male was struck and killed by a bus upon breaking the fence and attempting to cross the road. A second subadult male was shot when he commenced breaking the electric fence and feeding in the adjacent sugar cane plantation on multiple occasions. The latter individual never begged at the fence but had foraged near it as it was part of his natal family's range, likely observing fence-breaking techniques. A third male was killed by accidentally falling into a well in a village, though we could not confirm his identity and thus cannot know whether this individual engaged in begging. Elephant dung observed near the fence also contained plastic bags, evidencing ingestion. Although deaths attributable directly to feeding may be

Figure 7.9 Wild elephant feeding tactics. (a) A single individual known as Rambo pioneered the tactic of "begging" at the electric fence along the southern boundary of Udawalawe National Park, which was subsequently adopted by other young males, (b)–(c) Of N = 66 individuals, 43 were seen only in a single year and 4 were seen every year for 13 years, with the rest intermediate (photos by the Udawalawe Elephant Research Project).

rare, these examples show that merely remaining in close proximity to the fence can have negative impacts on individual animals through the social acquisition of one or more risky behaviors. Although we cannot definitively establish a link, the timing suggests social acquisition of fence-breaking behavior. Although authorities tried deterring elephants by means of chasing them away and adding additional layers of electric fencing, these efforts lost their efficacy over time. In 2016, Rambo was seen for the first time in consortship with females, leaving his customary position along the fence, possibly reflecting a shift in life history strategy with maturity (Madsen et al., 2022). Though Rambo was not known to approach adjacent sugarcane fields in well over a decade of highly visible "begging," he did nonfatally injure a person who crossed the fence right in front of him while drunk and was narrowly spared translocation into a problem elephant holding facility thanks to public opinion. However, in 2020 and 2021, Rambo commenced breaking the electric fence to feed on sugarcane.

Whether this was because of the drastic decrease in tourism owing to the COVID-19 pandemic or increased caloric requirements owing to his increasingly active musth period is difficult to say.

In the Sigur region bordering the Nilgiri Biosphere Reserve, observations likewise indicate negative consequences for 11 elephants that were observed to become food-habituated by tourism operators since 2007 (data courtesy of Davidar & Puyravaud). Four (36%) of these 11 elephants died unnatural deaths caused probably by people, while one, Rivaldo (Figure 7.10), survived an injury severing around 30 cm from the tip of his trunk in 2013. The remaining six are alive as of the time of this publication. Of these six, it seems that five no longer explore for human food. The COVID-19 lockdown, with its drastic reduction of tourism, appears to have decreased the animals' attempts to search for human food. In the case of Rivaldo, although he was initially released after treatment, he was again injured by a wild tusker and taken in for treatment in 2015, with additional food provisioning prior to release. Then in 2021 he was again taken back into custody for four months following misplaced concerns regarding his ability to forage for himself. When released again by court order, Rivaldo initially returned to the site where he had been food provisioned by tourists, but was driven back into the forest by authorities, who remained as escorts for an extended period of time, and subsequently reduced his visitation to villages. This demonstrates that dehabituation of individual animals is possible, but it requires substantial time investment and vigilance that are unlikely to be sustainable in the long term.

Elsewhere in Asia, male elephants are seen to block traffic in order to exact a food "toll" from passing vehicles (SdS personal observations). These behaviors, once they have become widespread in populations, will be difficult if not impossible to extinguish. Moreover, such behaviors may be adopted and therefore place at risk certain personality types – that is, individuals who are innovative, bold, and calm. These are attributes that might ordinarily serve their bearers well and should not be weeded out of elephant populations through the removal of so-called problem animals. Changing the human behavior that creates the problem animal in the first place, through the strict enforcement of penalties discouraging irresponsible practices such as feeding, would be a more practical and far-sighted approach. Likewise, the vast majority of research into techniques of conflict mitigation center on deterrent techniques of various kinds (Shaffer et al., 2019), none of which are universally applicable as they are likely to depend heavily on how habituated or desperate local elephant populations are. While particular deterrent techniques may work under specific conditions in one context or other, documenting each of these seems like the equivalent of attempting to plug the proverbial dam while largely ignoring the looming threat that lies behind it should the structure fail entirely. And the danger is this: In the face of ever-expanding agricultural lands and ever-depleting wild forage, elephant populations *may be changing their dietary preferences altogether*. For how long can we manage to deter them? There is an urgent need for conservationists, wildlife managers, and researchers to move beyond mere deterrence towards a better understanding of what elephant populations actually need and how these needs can be met alongside human activities. This

Figure 7.10 The rehabilitation of Rivaldo. (a) Rivaldo being treated for injuries. (b) Being escorted by forest guards. (c) Foraging after release (photos courtesy of Jean-Philippe Puyravaud and Priya Davidar, the Sigur Nature Trust).

also requires us to recognize and engage with the mechanisms by which elephants may be learning and acquiring their food preferences in the first place.

I suspect that elephants' foraging preferences are socially acquired – as generalist herbivores, they have many choices and the preferences of particularly influential individuals such as mothers or mentors may spread readily. Among humans, who are the ultimate generalist feeders, there is ample evidence that the food preferences of mothers are passed to offspring in utero (Ventura et al., 2021; Ventura & Worobey, 2013). Children's tastes are then continually shaped through subsequent exposure through breastfeeding and after weaning (Harris, 2008; Skinner et al., 2002). These preferences can cut both ways – predisposing the child to accept nutritious food or alternately to favor junk food. There is great interest in the mechanisms behind the acquisition of food preferences by children owing to the increasing rates of obesity and associated health problems in some countries. There should likewise be as much interest in the mechanisms behind the acquisition of food preferences for wildlife, in the face of increasing chances of conflict with humans. It is entirely possible that similar mechanisms for transmission of food preferences are present in other mammals, including elephants. A mother accustomed to a sugar-rich diet, for instance, may predispose her offspring to prefer such. After birth, the next critical phase is the juvenile period, where a calf learns its natal group's foraging areas and resources. I have often seen calves touching and presumably sniffing at the mouths of older individuals while they have bits of forage visibly protruding. Lee and Moss document that investigation of food constitutes a substantial fraction of the friendly contacts savanna elephant calves make, both with their mothers and other individuals (Lee & Moss, 1999). Some types of vegetation must not only be picked up or broken off, but also cleaned or otherwise processed and handled in some way. Finally, as individuals disperse, they may copy the foraging tactics of social affiliates. As yet there have been no studies of how younger individuals acquire preference for or learn to handle particular food items – perhaps they do so idiosyncratically, by trial and error, or perhaps they learn through observation. There is a growing awareness that behavioral traits, temperament, or personality types can potentially influence the success of animals that are translocated, especially where there is potential for conflict with people (de Azevedo & Young, 2021; Martínez-Abraín et al., 2022; May et al., 2016). Not only must animals learn where food and other resources are available, but dexterous generalists such as elephants and primates must possibly also need prior training and experience to know what is edible in the first place, as well as how to handle certain food items, much as predators do. "Easy" food cultivated by humans can become particularly appealing.

The established habits and expectations governing what constitutes acceptable behavior within a given population can be construed as analogous to the cultural norms (normative practices) familiar to humans, increasingly recognized as also present among nonhuman species (Vincent et al., 2018). This is referred to by de Waal as *natural normativity* (de Waal, 2014). Typically, the discussion centers around intraspecific social relationships – that is, how nonhuman species treat conspecifics, the evolutionary origins of nothing less than morality (Palao, 2021). In the early days of ethology, Konrad Lorenz proposed that group-living predators and species with

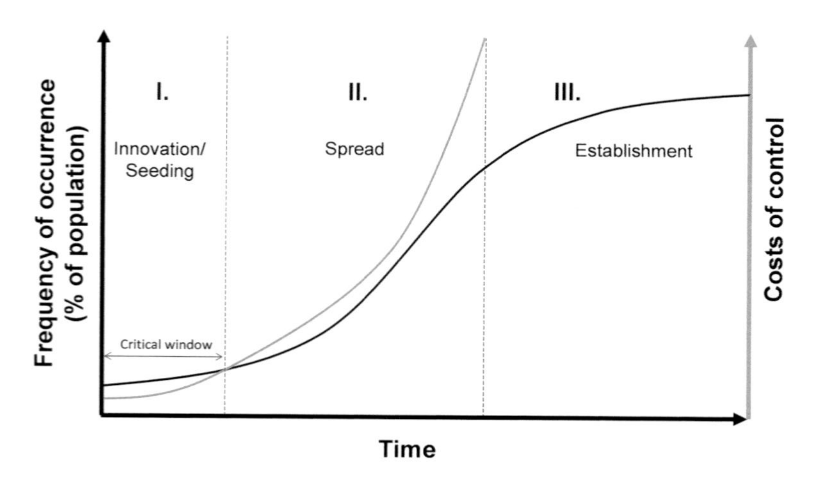

Figure 7.11 Behavior transmission as biological invasion. Problematic behavior begins as an innovation by one or a few individuals (Phase I), is spread to others by observational learning (Phase II), and may eventually saturate the entire population (Phase III). The costs of controlling such behavior is likely to increase exponentially as more individuals adopt the behavior; therefore the critical window for management will be in the earliest stage.

formidable weaponry evolved behavioral controls to mitigate bursts of aggression among social affiliates precisely for the reason that it would be all too easy for violence to turn fatal – perhaps much less so than in species that appear to us less capable (Lorenz, 1952). While Lorenz's proposition remains speculative, there is no reason why this concept of natural normativity should not be extended to interspecific interactions. Of special concern is how elephants react to human activity or presence. Indeed, I suggest that different elephant populations exhibit cultures with varying degrees of tolerance towards humans, just as humans do towards elephants and other wildlife. For example, in some populations it is extremely uncommon for female groups to forage on crops, while in other areas the practice is more widespread. Some elephant populations might be reluctant to enter villages or towns, whereas others do so freely. Some populations may be more given to direct confrontation than others. These tendencies may be driven not only by personality but also scarcity and necessity – for instance, the loss of forest cover which forces herds out into the open, then habituates subsequent generations to such exposure. There is a clear analogy between the social transmission of (problematic) behavior and classical biological invasion curves (Figure 7.11; see also Wright et al., 2010). The best chance of curbing problematic behavior will be in early stages, before it is widespread; as it reaches a larger fraction of the population, control will be both costlier and more difficult, if not impossible. This dynamic raises deep concerns about the future potential for conflict between people and a species such as the elephant, which is capable of holding preferences and traditions in memory over a multidecadal lifespan and transmitting acquired behaviors intergenerationally. So too can humans. Established norms of nonconfrontation between elephants and humans can be eroded over time, replaced by ever-escalating feedbacks of mutual aggression. May there be mutual tipping points of tolerance?

The preceding considerations regarding culture and tradition are salient not only for established populations but also when managers are considering the active relocation of wildlife. Individual elephants, moved between populations, can also potentially seed behaviors that create problems for the wider population as it spreads. Despite the frequent use of translocation as a management tool, as well as considerable interest in the rehabilitation and reintroduction of individual animals, the link between behavioral traits and the fates of animals has not been made. This might be because translocations are often highly time-sensitive operations that do not readily include a window of observation during which to assess these behavioral characteristics. The aspects of cognition we have been discussing might, regrettably, be the last thing on anyone's mind. However, given its potential importance for the outcomes and fates of translocated individuals, collaboration between behavioral researchers and wildlife managers would be useful for developing assessment techniques that can be applied with minimal interference. Likewise, those working with the aim of rescuing, rehabilitating, and releasing animals might also integrate behavioral and cognitive research into all phases of these programs to ensure that individuals possess the core competencies and skills required to survive in the wild (Goldenberg et al., 2022).

The living elephants represent species with sensory and cognitive capacities that allow them to make use of a breathtaking array of environments, from deserts to rainforests. Though no single species has conquered all environments, and although I have been emphasizing intra- and interspecific differences throughout, here it is worth stepping back to reflect on their similarities. All elephants live in dynamic social worlds and make use of varied communication signals. The core competencies an elephant needs in order to develop into a successful forager and breeder are very likely learned, although whether this is through passive observation and copying or active demonstration remains to be seen. Elephants have a capacity for insight and innovation, albeit with a high degree of variation among individuals, including attributes that may be termed personality characteristics. They are sophisticated, decision-making agents that respond to their own needs as well as those of their social companions. Behavioral innovations and preferences can spread in populations, be maintained through both horizontal and vertical transfer of knowledge, and possibly give rise to regional variations among populations that merit the term culture. While the combination of cognitive and behavioral flexibility of elephants may have served them well in the past, it can be a double-edged sword. It is allowing elephants to persist and even increasingly exploit heavily human-modified landscapes and resources and may be the key to finding ways of coping with a changing climate, but this brings new risks. Long-established sociocultural and behavioral norms, both elephant and human, may now be changing.

8 Conflict and Coexistence

Ecological, Political, or Psychological?

8.1 What Counts as an Elephant Ecosystem?

It is obvious, from looking at the fragmented distributions of all elephant species and subspecies, that populations that would have been connected have become cut off from one another at some point in the past (I have deliberately avoided providing maps of the current distributions for any species, as they are unlikely to be accurate or static enough to warrant placement in a book; it is better to refer to those that are compiled and periodically updated by the International Union for Conservation of Nature (IUCN) Specialist Groups and published in the Red List). In Asia, some populations were isolated by distant climatic and geological events, such as the inundation of the Sunda shelf in Southeast Asia (between 18,000 and 20,000 years ago), giving rise to the Bornean and Sumatran subspecies. Much more recently, the submersion of the land bridge between India and Sri Lanka (approximately in the fifteenth century) separated the populations in these countries. On the mainland, although Asian elephants are believed to have once ranged all the way from present-day Afghanistan to the eastern seaboard of China following the postglacial thaw of ice (Olivier, 1978),[1] this is no longer the case. More recent climatic and human elements are implicated in the patterns of fragmentation seen today.

Much has been written about the loss of "wilderness" areas (defined as areas with minimal human impacts rather than those that exclude people entirely), with conservationists rightfully alarmed by the rapidity of these changes. A review and summary of a report by the Intergovernmental Science-Policy Platform on Biodiversity and Ecosystem Services stresses that human impacts on life on Earth "have increased sharply since the 1970s" (Díaz et al., 2019). Where forests are concerned, this appears to coincide with the first state-run initiatives of road building and colonization that opened up new tracts of land for development, followed by various enterprises (Rudel, 2007), but may have diverse causes related to agricultural expansion into other biomes such as savannas and rangelands (Maitima et al., 2010). One tenth (approximately 3.3 million km^2)

[1] Olivier's outline is nevertheless rather crude, being based on sparse historical records and failing to account for ecological features that would have either excluded or favored elephant presence. It may be the best he could do at the time, but I expect that it vastly overrepresents the amount of habitat actually occupied by elephants. Newer modeling methods may allow for more reasonable characterizations of potential habitat.

of global wilderness areas was found to have been lost in twenty years since 1990s (Watson et al., 2016), with 14% of this occurring in Africa. This rate of loss was sufficiently alarming as to be termed "catastrophic" by the study's authors. In Asia, the study found there were no globally significant regions of contiguous wilderness (i.e. areas of at least 10,000 km^2) *remaining at all*, aside from central Borneo. Relevant to our species or interest, of the roughly 14 different major terrestrial biomes on Earth, the Tropical and Subtropical Dry Broadleaf forests had no wilderness areas at or above this extent remaining as of 2016. The rate of loss globally was twice the rate of protection. What happened?

While many studies focus on land use and land cover change over relatively recent decades owing to the availability of satellite data, they often overlook that all these changes continue a trend set in motion at least since the industrial revolution, and perhaps even earlier. About 8,000 years ago, forests covered around half of Earth's land area, making them the most widespread biome(s) on the planet, but they are now down to 30% or less (Ball, 2001); though Asian elephants thrive in nonforest habitat, as discussed below, this forest retreat might have extirpated some populations on both extremes of the range. Sometime during the intervening millennia, possibly concurrent with the rise of agricultural civilizations, elephants went into retreat across the continent. By the fourteenth century BCE they were gone from most of mainland China, and by the Roman era they were already regarded as strange curiosities in the western reaches beyond the Indian subcontinent (Williams et al., 2020, and references therein). The uptake and spread of intensive, permanently irrigated cultivations of rice in Southeast Asia may have taken place as early as the eighth century and is seen as being associated with the "retreat of large mammals, notably elephant and rhinoceros" (Reid, 1995: 93). In subsequent centuries following the industrial revolution, global forest cover is estimated to have decreased by 32% (FAO, 2010). On top of all this, it is unlikely that elephant populations were ever uniformly found throughout Asia given the intervening mountain ranges and deserts. African elephants, which are more populous and do not seem to have undergone range constriction that is nearly as dramatic as Asian elephants, nevertheless also evidence highly fragmented ranges (savanna elephants especially so), which would once have been more contiguous throughout the continent. Aside from the impact of hunting, populations would have also been displaced or extirpated by habitat modification. I now examine the history of the landscapes upon which elephants were found, and what it tells us about the requirements for populations that persist today.

8.1.1 Land-Use and Land-Cover Change

Bearing in mind this longer history, a team of collaborators and I set out to understand how land-use change within the past few centuries might have affected the availability of habitat for elephants in Asia (de Silva et al., 2023a). We constructed ecological niche models for Asian elephants to identify which areas might have actually contained suitable habitat in the past. Ecological niche models (also known as "species distribution models") relate data on species *actual* occurrence (and absence, if known) to environmental variables as a basis for modeling where a species *could* occur. This "prediction" of the model can be made for either a different area or time for which there are no data – including the past. There are a variety of techniques for doing so, and after evaluating several options, we used a machine-learning algorithm known as MAXENT (short for "maximum entropy"; Elith et al., 2011). Importantly, the same set of environmental variables have to be used in both model construction and prediction, to maintain consistency. This creates a challenge, because many datasets, including satellite imagery, are limited to just the twentieth century. Incredibly, researchers have modeled spatially explicit reconstructions of land use at annual increments between 850 and 2015 (Hurtt et al., 2016). Using 20 different land-use variables from the year 2000 as a reference, in combination with data on elephant occurrence in the 1990s and 2000s, we were able to identify areas that were more or less similar to areas in which wild elephants are today found in substantial numbers, at relatively coarse resolution (0.25° × 0.25°, which is approximately 30 km^2 at the equator but decreases toward the poles). This included areas outside the actual range, since a species is not necessarily found in all the areas that it could possibly survive.

The result was both striking and sobering. Up until the late seventeenth and early eighteenth centuries, the extent of elephant habitat bounces around a little but remains relatively stable. However, it then takes a nosedive, decreasing nearly two-thirds by the year 2015 (Table 8.1; a loss of more than 3.3 million km^2, comparable to the amount of global habitat loss over recent decades as reported by Watson et al., 2016). Whereas the entirety (i.e. 100%) of the area within at least 100 km of the present-day elephant range could have been considered "suitable" habitat for elephants, by the year 2015 less than half of the existing range could be considered as such (Figure 8.1a). In the year 1700, an elephant could have in principle traversed around 45% of the suitable habitat without interruption, since it formed a relatively large contiguous patch, but by 2015 it was down to just 7.5%. The average patch size, which represents the size of areas containing contiguous habitat, plummeted a whopping 83% from 99,000 km^2 to 16,000 km^2. If we believe these results (and there are some qualifications, discussed in a bit), they reveal two very important things. The first is that the extent of habitat loss is far greater than the extent of forest loss per se. Elephants, which can thrive in many different types of habitats (i.e. they have a wide "niche breadth"), can essentially stand in as surrogates for many biomes, including grass and scrublands that may not as valued as "forest" but which have nevertheless been settled and altered – notably, for agriculture. Second, it also shows that the long-term loss is far greater than what we can appreciate simply by looking at the past few decades. Indeed, it shows the timecourse over which "wilderness" has been eroded in Asia, to the extent that studies such as those by Watson et al. (2016) find no "globally significant" regions left.

India and China, which together contained the greatest share of elephant habitat, each lost over 80% (Table 8.1; interestingly, this suggests that India continued to have substantial elephant populations where there was insufficient habitat, whereas China lost most of its elephants before it lost the habitat). Because elephants are long-lived and demographic responses require time to manifest, it can take hundreds of years for populations to go extinct, even if they are in decline (Armbruster et al., 1999;

Table 8.1 Change in suitable habitat area by region, 1700–2015

	Suitable current area (km^2)			Suitable potential area (km^2)			Total suitable area (km^2)		
	1700	2015	% change	1700	2015	% change	1700	2015	% change
Mainland China	1,986	135	−93.2	1,119,258	65,054	−94.2	1,087,183	65,189	**−94.2**
India	216,207	82,793	−61.7	1,439,320	145,561	−89.9	1,655,527	228,354	**−86.2**
Bangladesh	6,322	1,770	−72	37,725	10,634	−71.9	44,046	12,405	**−71.8**
Thailand	40,449	31,303	−22.6	439,964	127,028	−71.2	480,413	158,331	**−67.0**
Vietnam	523	515	−1.4	196,259	80,923	−58.8	196,781	81,439	**−58.6**
Indonesia (Sumatra)	45,252	27,507	−39.2	317,636	123,278	−61.2	362,888	150,785	**−58.5**
Indonesia (Borneo)	826	928	12.3	428,709	282,061	−34.2	429,535	282,989	**−34.1**
Myanmar	32,026	36,591	14.3	289,533	181,179	−37.4	321,559	217,770	**−32.3**
Cambodia	7,904	12,508	58.3	147,795	102,867	−30.4	155,699	115,374	**−25.9**
Nepal	12,086	4,750	−60.7	44,333	37,905	−14.5	56,419	42,655	**−24.4**
Sri Lanka	31,654	22,603	−28.6	24,097	19,622	−18.6	55,750	42,225	**−24.3**
Bhutan	2,033	1,148	−43.6	5,560	5,126	−7.8	7,593	6,273	**−17.4**
Lao PDR	16,507	17,716	7.3	148,922	159,843	7.33	165,429	177,558	**7.3**
Malaysia (Peninsular)	7,796	10,682	37.0	95,029	105,418	10.9	102,825	116,100	**12.9**
Malaysia (Borneo)	7,216	12,007	66.4	100,023	161,316	61.3	107,239	173,323	**61.6**
Totals	**428,787**	**262,956**	**−38.7**	**4,834,163**	**1,607,815**	**−66.8**	**5,228,886**	**1,870,770**	**−64.2**

Areas are in km^2, ordered from ranges that experienced the greatest loss to those with the greatest gain in total suitable area. "Suitable current area" refers to the current range, that is, areas known to still contain elephants by the year 2000, whereas "Suitable potential area" refers to area that is outside the current range, where it is unknown whether elephants were ever present. The amount of suitable area within current range is only 14% of the total. Reproduced from de Silva et al. (2023b), CC-BY 4.0.

de Silva & Leimgruber, 2019). So, it is entirely possible that many populations are in fact persisting on suboptimal range – that is, areas that contained excellent elephant habitat several generations ago, but which have since deteriorated. Looking at the time course of habitat loss, another interesting observation is that the glowing patch of suitable habitat that was previously found in what is now central Thailand actually did not disappear until well into the 1950s and 1960s. This therefore is not directly tied to colonial expansion (and moreover Thailand maintains that it was never colonized). Instead, this period marks the beginning of ever more intensive logging that enabled the intensive form of cultivation associated with the "Green Revolution," which refers to the advent of modern, fertilizer-heavy, industrial agriculture. The threads that connect these developments to the colonial era here are more indirect – they are the result of the global economic transformation that linked together consumers and producers in disparate regions through an expansive supply chain (Lambin et al., 2001), a phenomenon sometimes referred to as "telecoupling" (Eakin et al., 2014; Liu et al., 2014). While some telecoupling relationships may be novel in the modern era, enabled by our fossil fuel-enabled transport capabilities, we must recognize that the process itself, and its devastating ecological impact, is not new.

One might ask whether these results are a reasonable representation of what took place over the past three centuries. While there is always bound to be some error in any modeling study and there's no way to go back in time to verify the truth, most of the uncertainty derives from whether the underlying land-use datasets are themselves accurate, given any assumptions upon which they are based. As a sanity check, we compared the result for the year 2000 to an alternate prediction generated using finer-scale variables based directly on satellite imagery, human census data, livestock densities, and other features. The two sets of results were in agreement for over 80% of the area. If anything, the coarse-resolution datasets may provide a more conservative estimate of land-use change (and therefore habitat loss) than fine-resolution datasets, which suggest the rate of conversions to be between two to four times *higher* (Winkler et al., 2021). This possibility is corroborated by an independent study which specifically modeled the amount of suitable habitat remaining in Sumatra, using entirely different fine-scale predictor variables alongside more than 2000 elephant occurrence locations (Imron et al., 2023). Being an island, Sumatra offers a clean, self-contained comparison. We found the total suitable area to be just over 150,000 km^2 or roughly 35% of the land mass, whereas Imron and colleagues found it to be a little over 135,000 km^2 or 32% of the island. Our pan-Asia estimates of suitable habitat and losses therefore seem reasonable, but are likely to be understatements, because our present-day sampling locations can only represent ecosystems in which Asian elephants remain, but not those from which they were entirely eliminated.

There are also a number of other studies that show similar trends. First, the period following the 1700s marks colonial influences on biomes throughout the world, and not only in Asia. One study found that in the year 1700 nearly half of terrestrial biomes could be characterized as "wild" (see Section 8.2) and much of the rest was in a seminatural state with little agriculture and settlement (Ellis et al., 2010).

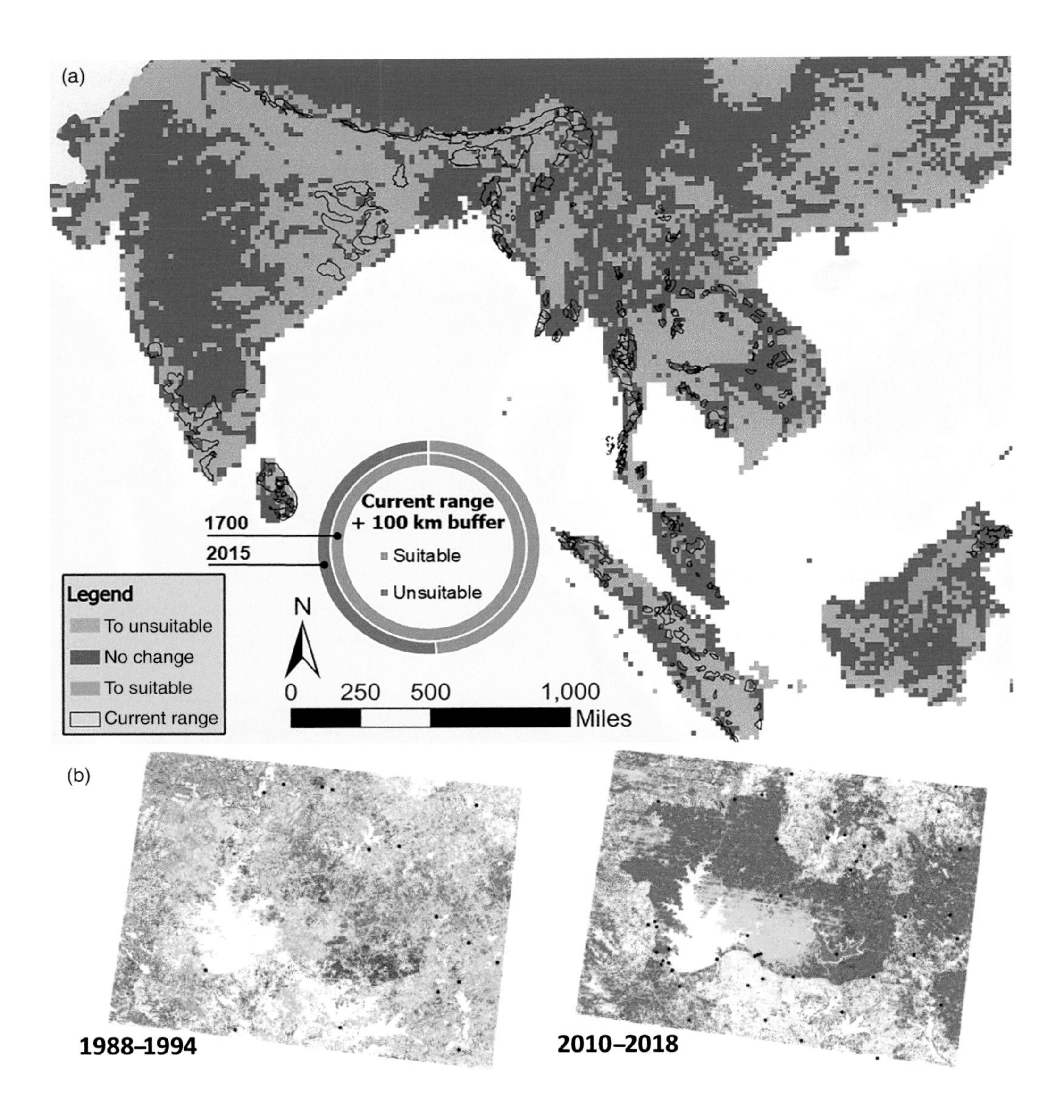

Figure 8.1 (a) Historic loss of elephant habitat, 1700–2015. The orange areas became unsuitable, the green areas became suitable, and the blue areas did not change. Black outlines show the known range of Asian elephants in the 2000s. Areas of note include central Thailand, Central and Southern India, northern Sri Lanka, and Sumatra, all of which lost extensive habitat. While elephants in Thailand appear restricted to the remaining areas of habitat, in India and Sri Lanka elephants continue to persist in what appears to be suboptimal habitat. Whereas all the range within 100 km of the extant range could be considered suitable habitat in the 1700s, by 2015 this was down to less than half. Such mismatch likely underlies so-called human–elephant conflict, in addition to habitat lost in more recent decades (adapted from de Silva et al., 2023a, under CC 4.0 license). (b) Land-use change in and around Udawalawe National Park, Sri Lanka. Colors represent land-use changes of different types and black dots

By 2000, only a quarter was characterized as wild and less than a fifth was semi-natural. Another study, which reviewed habitat suitability for the many mammals on IUCN's Red List, found that on average only 48.6% of a species' range could be classified as suitable for it, based on its ecological preferences (Crooks et al., 2017). Additional studies show that habitat adequate for elephants is being lost even inside protected areas (Clark et al., 2013; Neupane et al., 2020; Ram et al., 2021). The next question, then, is exactly what sorts of habitat do (and did) these areas represent? What were these past elephant ecosystems? While it may be tempting to think of these landscapes as pristine wilderness areas, there are many reasons to reevaluate such a proposition. Namely, that with the exception of extremely hot and cold deserts (or certain regions harboring endemic diseases) that made human habitation difficult if not impossible, the idea of such wilderness is turning out to be a myth, perpetuated by colonial settlers who were newly encountering these landscapes (Cronon, 1996).

The reality of these ecological changes and misrepresentations are perhaps especially evident on the African continent. While there has not yet been a similar study in terms of changes in habitat availability for elephants, there is of course a heavy colonial legacy with its accompanying policies of land use. In East Africa, large areas of forest were cleared for export crops such as cotton, coffee, tea, and sisal (a type of Agave native to the Americas, used for fiber production; Olson et al., 2004). Land-use intensification is linked to both land degradation and biodiversity loss (Maitima et al., 2010; explored more in Section 8.2). The colonial administrators also introduced a system of mixed cultivation and tree plantation known as the Shamba System, which appears to have gradually become widespread. At the same time, large areas of land were placed under some formal protected status: 12% in Kenya, 26% in Uganda, and up to 40% in Tanzania (Olson et al., 2004). Both the plantations and protected areas were, of course, imposed on top of many preexisting land-use traditions, setting up persistent tensions among local communities with one another, conservationists, and representatives of whatever governing authority that considered itself to be in charge. Olson and colleagues observe: "Particularly severe was the impact upon African livelihoods in areas of reliable rainfall, and on livestock based livelihoods in the semiarid regions as settlers occupied the best watered of their lands so vital in dry seasons and periods of droughts." Similar patterns are repeated throughout the continent, such as the widespread privatization of land in South Africa.

The major commonality throughout Asia and Africa (not to mention the Americas) is the subdivision of communally held and managed land into parcels of individual

Figure 8.1 caption (cont.)
show locations of "conflict" incidents recorded by the Department of Wildlife Conservation over the same period. These include elephant and human mortalities as well as injuries and property damage. In the period 1988–1994 there are many different land use changes creating a complex mosaic in which the protected areas are not clearly distinguishable. As the outlines of protected areas become more distinct from the background in later years (here, owing to forest regrowth), negative incidents largely seem to occur around forested edges (data visualized by Chithrangani Rathnayake).

ownership, or alternately nationalization, concurrent with the large-scale commodification and trade of resources derived from the land. As Eleanor Ostrom rightly recognized, the famous "Tragedy of the Commons" results not from some fundamental selfishness that prevents human agents from sustainably using common resources, but from the subversion of guiding principles that regulate fair use. This is a fundamental shift in the way that land is viewed and controlled that not only continues, but remains deeply embedded in scenario modeling, planning, and decision-making to date. Land-use change in more recent decades is inseparable from human population growth. But the relationship is far from straightforward. While expanding populations are indeed responsible for clearing, settling, and cultivating additional land, often the consumers who ultimately exploit many of these resources are not local – there is a continuous transfer of food and material products from these ecosystems to populations living distantly, whether from rural to urban or from one country to another. Historical context and telecoupling govern the proximate mechanisms (such as land-use concessions to international corporations and the participation of smallholders in a globalized economy) that directly drive habitat loss. The problem runs even deeper – it is becoming evident that not only were the precolonial traditions of resource management a means of maintaining sustainable *human* use of landscapes, but they were also essential to maintaining ecological regimes that actively sustained *wildlife*, which I explore in Section 8.2.

8.2 A Deeper History of Coexistence

In preceding sections and chapters, I have largely focused on the putatively antagonistic relationship between humans and elephants, both in terms of the ancient relationship between predator and prey and in terms of our modern-day struggles to share the landscape. The visible trends in land use and deforestation attest to the additional destructive effect our species can have on our surroundings, therefore it is no surprise that scientists and conservationists have largely tended to focus on the negative impacts of human activities on the various habitats where elephants occur (e.g. Sukumar, 2003, pp. 326–331, Shaffer et al., 2019). Nevertheless, underlying and predating these activities are older ecosystem management practices that went hand in hand with a diversity of distinctly local and place-centered worldviews. Equally fascinating to consider is how humans may have shaped landscapes in a manner that might have been beneficial for elephants and other wildlife, and possibly continues to be the case today. For instance, 36% of intact forest landscapes are found on lands managed by indigenous people, and although they are ever at risk of conversion owing to the vulnerability of indigenous communities themselves, they tend to experience less forest loss globally (Fa et al., 2020).

Recent movements to recognize indigenous land stewardship practices and divest conservation from its colonial baggage (often referred to "decolonizing" conservation in policy and practice) is increasingly revealing that the assumption that human influence is necessarily detrimental is not just false, but rooted in viewpoints originating from European settlers who were strangers to these systems. Yet these views have remained entrenched both in science and practice. While this discussion of human activities may seem a digression, conservation scientists, practitioners, and land managers must reckon with the long-overlooked anthropogenic role in shaping ecosystems of the past. The foraging behavior and food preferences of species living in what remains of these ecosystems today cannot be understood without including this historical and modern human dimension. For instance, Barnes and colleagues found that forest elephants in Gabon show preference for abandoned villages and plantations such that "elephant distribution is governed by the distribution of both past and present human settlement, even in the remotest and least disturbed forests of equatorial Africa" (Barnes et al., 1991: 54).

Using spatially explicit global models based on archeological and palaeoecological evidence, a large international team of researchers have mapped out what many local communities around the world already knew, namely that humans were widespread and shaping landscapes at least as early as 10,000 BCE (Ellis et al., 2021). These authors forcefully argue that the dominant portrayal of human impacts on the landscape as "mostly recent and inherently destructive" is incorrect as it ignores the many systems of sustainable land management practiced by societies predating the colonial era. This included more than 90% of tropical woodlands (together with 95% of temperate ones). The authors distinguish among three types of "anthromes" or cultural land-use practices with varying levels of intensity: *wildlands*, from which humans and intensive land uses are entirely absent, *cultured anthromes*, where less than 20% are covered by intensive land uses, and *intensive anthromes*, where more than 20% are covered by intensive land uses. Notably, of the current protected areas around the world, only 36.4% of the lands considered as wildlands were uninhabited 12,000 years ago. This means that *nearly two-thirds* of the land designated today as protected areas has some history of human influence in the distant past. They sum it up: "the current extinction crisis is better explained by the displacement of species-rich cultural natures sustained by past societies than the recent conversion and use of uninhabited Wildlands." How did elephants exist alongside so many human populations for so long – was so-called human–elephant conflict, just as prevalent and inevitable? While humans may never have accepted elephants being in our croplands, I argue that present-day conflict with elephants is the result of two major driving forces: land-use change, which deprives elephants and other wildlife of suitable habitat, as well as the loss of traditional knowledge systems and mindsets concurrent with dramatic changes to social systems and land rights. I next tie these two processes together with respect to both forests and rangelands.

8.2.1 Cultured Anthromes: Fire

Humans have altered many aspects of the biosphere since before recorded history, but here I would like to specifically focus first on the phenomenon of fire. While there has been extensive scrutiny of the relationship between elephants, woodlands, and fire (see also Sukumar, 2003), ecologists have tended to neglect fire regimes that involve humans other than the land managers themselves. Managers have generally focused on concerns regarding how to preserve woodlands and associated ecosystems within protected areas, rather than holistically in terms of the larger ecosystem (Nieman et al., 2021). In the scientific literature, this is reflected in the volumes of literature devoted to identifying alternative stable states and the processes that maintain them (Beisner et al., 2003; Schröder et al., 2005). Human influence, as represented by officially sanctioned management, is often seen as acceptable or even necessary for preserving some perceived "natural" or otherwise desired state (e.g. those that harbor greater biodiversity), whereas the influence of local communities is viewed as antithetical to the preservation of such a state.

While alternative stable states may well exist, the categorical exclusion of traditional human activities from the discussion seems a massive oversight. This blind spot is now being noticed. If one considers the dynamics of fire more completely, it cannot be denied that on many of the landscapes now designated for wildlife, it included people who intentionally used fire as a tool (Bowman et al., 2011; Shaffer, 2010). Indigenous communities have long used controlled burning techniques as part of culturally transmitted systems of land management (Lake et al., 2017), and the use of prescribed fire in the course of management has been credited with saving stands of ancient trees from present-day wildfires. It is now increasingly recognized that strict fire suppression policies in the temperate forests and grasslands of North America contribute to a build-up of fuel that can result in larger and more serious wildfires, while impeding nutrient cycling, seedling recruitment, and regeneration processes (Bowman et al., 2011; Costa & Thomaz, 2021; Lake et al., 2017). Nevertheless, the human application of fire continues to be a contentious topic. In tropical savanna regions in South America, Southern Africa, and Australia, the suppression of fire causes conflicts between local communities and authorities, and results either in the encroachment of woodlands or else facilitates late dry season wildfires (Moura et al., 2019). Globally, these policies have largely been carried over from the colonial era. Although I cannot do justice to this topic on a global scale, here I provide examples of the application of anthropogenic fires on landscapes occupied by elephants on both continents in the context of a diverse array of lifestyles. These include hunter-gatherers, pastoralists, and agrarian communities. In each case, fire has played a central role in both the lives of the human communities and the ecosystems to which they belong.

We begin with Asia, where contemporary debates and discourse unfortunately highlight the challenges of reconciling past and present. A compelling example comes from the many issues faced by indigenous Adivasi communities of India. "Adivasi" is a singular term that collectively refers to an extremely ethnolinguistically diverse group of indigenous peoples, many of whom have been displaced by the creation of exclusionary protected areas (Jolly et al., 2022). Their traditional use of fire, which continues to be banned, is described in detail by an ethnographic study of indigenous Katunayakan hunter-gatherer communities in the Wayanad district of Kerala by Jolly (2022: 83). She quotes an elder, who describes the role of fire in facilitating mobility as well as provisioning food for wildlife:

> Earlier times, the forest used to burn. That was when the forest was fresh and energetic. Now it is not easy to go through the thick forest. Now it isn't easy to move in the forest. Due to ponda [undergrowth], we cannot detect wild animals like elephants, bears, or tigers. Even it is difficult to cross the rivers or reach wayals [marshy wetlands]. In forests that do not burn, even the animals will not have enough food. Earlier, every year, there were small fires that turned the forest bright and clear with plenty of grasses. The animals had enough food back then. Such forests you will find more animals and more food.

Jolly goes on to quote additional community members, who remark on the decline of endemic species and increase in invasive nonnative vegetation, (especially those that are fire-adapted and perhaps even fire-dependent), which they also attribute to fire suppression. Traditionally set fires, which would have been timed over the course of the year such that they were followed by rain, were employed systematically to facilitate regeneration and promote the growth of native vegetation that is palatable, medicinal, or has other uses, which in turn supported the greater ecological community including the people themselves. Elephants, as I have seen and no doubt many others have as well, are especially attracted to the fresh growth of regenerating burn patches. This is known as patch-burn grazing and is also true of deer and other herbivores (including wild and domestic cattle) that benefit from the same vegetation. In north America, deer have a complicated response to fire in that they are attracted to the forage in smaller low-intensity burn patches, but avoid areas with recent fires and large-scale burns, especially when rearing young, owing to the higher risk of predation (Cherry et al., 2017; Wan et al., 2014). Although the association between elephants (or other herbivores) and fires in Asia have not received nearly as much research attention as it has elsewhere in the world (very likely precisely because intentional burning is largely still considered illegal), it is plausible that such an association may have existed and was mediated by humans. Hypotheses concerning the role of fire in land management, inspired and informed by the knowledge of indigenous elders, stand in great need of evaluation.

One source of contention may be the difficulty of distinguishing what might have actually maintained past ecosystem conditions and functionality in the face of present-day disturbances. While anthropogenic fire may be acknowledged as having long been part of the landscape, others argue that fires degrade forest ecosystems under present-day conditions, as forests are both more fragmented and subject to more intensive uses. Under this view, conservationists worry that fires have become *more* frequent, not less. Kodandapani and colleagues (2004) used remote-sensing and ground maps of fires within Mudumalai Wildlife Sanctuary in South India between 1982 and 2002 together with vegetation maps to calculate the average fire-return interval by vegetation type and for the sanctuary in its entirety. They found an average interval of less than seven years for all vegetation types and an interval of 3.3 years in all. They then compared this interval with those documented between 1909 and 1921 and noted a threefold increase in fire frequency within the 80-year span of time.

That a 3.3-year interval is considered more frequent than historic levels is odd, given the recollection of indigenous communities that fires occurred (i.e. were set) *annually* – alluded to in an account published in 1882 and cited by Hiremath and Sundaram (2005: 26–27):

> By January, the grass has all seeded and become dry, and it is then fired by the jungle-people. The hitherto impenetrable jungles are now reduced to clear forests of trees, interspersed with separate evergreen thickets. Moving about in such forests is rendered easy, but warm work ... The jungle people burn the grass to admit of their gathering certain fruits and jungle-products, especially the gall-nut, used in tanning. This burning insures a supply of sweet grass as soon as showers fall on the fertilizing ash.[2]

Nevertheless, citing the study by Kodandapani et al. (2004), Hiremath and Sundaram (2005: 27) maintain that whatever the past history of fire in India's forests, which may indeed have been and continue to be intimately associated with humans, present-day pressures create different dynamics: "Regardless of whether or not fires were part of the natural disturbance regime, or whether their arrival coincided with the arrival of the first hominids, it is probably fair to say that, with shrinking forests and growing human populations, the frequency of fires in Indian forests today is far greater than at any time in the past."

They describe accounts in which fires in India are attributed primarily to humans as opposed to lightning (which is likely to be accompanied by rain), the majority of which are set intentionally as part of a regime of shifting cultivation or to promote the growth of fodder for grazing livestock (both of these avenues are discussed shortly). They cite examples of a number of ways in which these fires may be qualitatively different and more damaging than fires of the past, owing to the stresses on remaining forests – which, they note, have also been subjected to logging. In this context, frequent fires may further reduce species diversity by impeding seedling recruitment and altering species composition in favor of those that are able to withstand fires (e.g. by being thicker-skinned) or regenerate more readily (e.g. clonally). But even more, they highlight the tendency of present-day fires to promote the growth of invasive nonnative grasses and woody perennials that are fire-tolerant. Extending this idea, they propose that fire facilitates the spread of *Lantana camara*. The authors acknowledge both the important role of fires for forest-dwelling communities, as well as their necessary role in other ecosystems such as grasslands. Rather than argue whether fires should be natural or anthropogenic, beneficial or detrimental, they instead encourage thinking of the specific management goals to be achieved through the use of fire.

Those working on African landscapes offer complementary perspectives. In Northern Tanzania, the pastoral Maasai have historically managed grasslands by setting a progression of small fires promoting regrowth for grazing, with the fragmented burn patterns curbing the potential for larger "catastrophic" accidental fires (Butz, 2009). As a consequence, the vegetation is also nonuniform. This practice goes hand in hand with the tradition of moving livestock around different parts of the rangeland

[2] The original quote is from G.P. Sanderson (1882). *Thirteen years among the wild beasts of India: Their haunts and habits from personal observation; With an account of the modes of capturing and taming elephants*, pp. 9–10. Reprinted by Asian Educational Services, New Delhi, 2000.

and would appear to be a human-managed ecological regime long embedded in these ecosystems. The effects of course extend beyond elephants, to the great diversity of other species in these systems. Bird populations, including long-distance migrants, have been found to congregate more densely in burn patches as opposed to recently abandoned *bomas* or undisturbed control plots (Gregory et al., 2010). Such associations between fire and grassland birds are widespread, being mirrored for instance on North American grasslands, where fire- and herbivore-induced spatial heterogeneity enhances both the diversity and stability of avian communities (Fuhlendorf et al., 2006; Hovick et al., 2015). These types of functional interaction between herbivores and fire have been referred to as "pyric herbivory," defined as "the spatial and temporal interaction of fire and grazing, where positive and negative feedbacks promote a shifting pattern of disturbance across the landscape" (Fuhlendorf et al., 2009: 588).

Experience from West Africa indicates that many decades of efforts by third parties attempting to run counter to traditional systems of nomadic pastoralism have worsened the productivity of these ecosystems quite independently of whatever climatic stressors they may now be under. Rangeland management practices derived from colonial-era views rested upon assumptions and experiments of the time (Laris & Wardell, 2006), which were in turn reflective of experience from wetter (by and large temperate) parts of the globe (Oba et al., 2000). Oba and colleagues characterize the difference as representing two ends of a spectrum, with drier landscapes being systems fundamentally at disequilibrium and wetter landscapes being closer to equilibrium (though I shall challenge the latter as well, at least insofar as the tropics are concerned). The two corresponding management paradigms are characterized as follows (Vetter, 2005: 321):

> The equilibrium model stresses the importance of biotic feedbacks such as density on vegetation composition, cover and productivity. Range management under this model centres on carrying capacity, stocking rates, and range condition assessment. In contrast, non-equilibrium rangeland systems are thought to be driven primarily by stochastic abiotic factors, notably variable rainfall, which result in highly variable and unpredictable primary production. Livestock populations are thought to have negligible feedback on the vegetation as their numbers rarely reach equilibrium with their fluctuating resource base.

Although the debate largely revolves around livestock, wild herbivores often tend to be viewed either explicitly or implicitly in quite the same way; both may share the very same landscapes. Hence, when managers try to manage elephant population sizes relative to their perceptions of "carrying capacity" for a given landscape, they are tacitly assuming an equilibrium model. This tends to cause discomfort among different stakeholders. As Vetter remarks, it is exceedingly difficult to infer whether or not there are density-dependent feedbacks on livestock population sizes, as posited by equilibrium models, owing to the time lags in population dynamics. One can expect that this would be even more difficult to infer for elephants, given the slow pace of demographic processes (de Silva & Leimgruber, 2019; de Silva, 2010a). Moreover, the studies reviewed by Vetter (2005) show that drier systems show both equilibrium and nonequilibrium dynamics depending on the spatial and temporal scale being considered. Elephants can and do modify landscapes; but whether or not such modification can be considered "necessary," "essential," or "healthy" seems largely a matter of taste based not only on ecological considerations but also on economic and aesthetic ones (e.g. how pleasant the landscape appears for tourists).

What is equally clear is that livestock, like wildlife, must move. A common error made by earlier settlers encountering arid or semiarid environments in West Africa, as well as on the Indian subcontinent, appears to be a tendency to classify lands with naturally sparse vegetation as "degraded" (Boateng, 2017). Policies of fire suppression have been one aspect of attempts to counter such supposed "degradation." At the same time, centuries-old traditions of movement by wild and domesticated grazers have increasingly been replaced by more sedentary paradigms of range use, through changes in land rights and accompanying land use practices, such as fencing. Notions of "carrying capacity" feature centrally, in which rangelands are managed to accommodate the particular number of individuals of a desired species that it is believed that landscape can support. But such notions of carrying capacity are illusory, as they simply cannot be predicted in disequilibrium systems with their more variable patterns of rainfall together with their inherent sensitivities to interactions between herbivory and fire.

The central role of fire in creating shifting disturbance patterns is not unique to rangelands with large grazers, though the latter can certainly capitalize on the aftermath of fires once they occur. Throughout the tropics, the practice of swiddening or shifting agriculture, often also referred to using the negative expression of "slash and burn" agriculture, encompasses a diverse set of practices that were widespread throughout the globe, along with traditional silviculture practices. Although the details may differ, central to this practice is the clearing of patches of forest by means of fire, which are then planted with crops. After some period of time, these plots are abandoned and allowed to regenerate while cultivation moves to a different area. The area may be revisited again later. Crucially, the time gaps between cultivations, which can be considered "fallow" periods, allow many palatable understory vegetation as well as fruiting tree species to grow. Such vegetation can benefit herbivores that would otherwise find little to eat in primary forest where lower-story vegetation is shaded out. Grasses, for instance, are scarce in primary forest but plentiful in recovering agricultural landscapes. In environments where the growth of forests is favored by climatic conditions, people may nevertheless maintain a patchwork of clearings and vegetation in varying stages of succession, thereby potentially enhancing the capacity for supporting herbivores, including not only elephants but also cervids and bovines.

It is entirely plausible that seasonally managed anthropogenic fires promoted ecological regimes that became naturalized over many generations, and to which ecological communities, including wildlife, were themselves adapted. This process of naturalization of a sort of wilderness in which (some) human cultures are an integral component, thus it is a form of *cultured anthrome*, to use the term of Ellis and colleagues (2021). While conservationists view primary forests as the ultimate goal, it is entirely possible that forest systems have been subject to long-term disequilibrium dynamics thanks to humans – to which local species became adapted. If human-managed fire regimes are indeed an essential ingredient of forest and savanna ecology, as they are increasingly believed to be, reinstating these practices calls for experimental testing and evaluation of the actual effects of fire on these dry forest ecosystems in terms of vegetation productivity and species richness, which would be preferable to unquestioning acceptance of assumptions made by European settlers. The baseline cannot be easily inferred from present-day dynamics; further research is needed to grapple with the human role in processes maintaining rangelands and woodlands/forests alike.

Obviously, traditional cultivation and grazing regimes can only be maintained if the human population is sufficiently small enough and the forest sufficiently large enough that some cyclic movement is possible without doing great damage. Given the large-scale historic takeover of communal lands, deforestation, and the replacement of native species with species more highly valued economically, this is now difficult. It does not help that traditional practices such as shifting cultivation have been demonized by conservationists and governments while being viewed by others as a primitive and inferior form of agriculture compared with permanently settled crops (Fox et al., 2000). The bind: it is argued that humanity must produce ever-higher yields per acre in order to sustain its population, and shifting agriculture practices cannot meet these needs (Foley et al., 2005). I do not dispute this. However, not all agricultural land is devoted to food production. Commodities such as oil palm, tea, coffee, sugar, fruits, and a host of other cultivated species cater to nonessential products, even if they are lucrative. This acreage is not devoted to essential food products, but commodities that were once luxuries. Certain working landscapes may already be capable of supporting elephants and other wildlife and others could be made capable, provided there is a will. Whether or not they do is not a matter of ecology, but rather economics, politics, and psychology.

8.2.2 Mindsets

Some land-management practices derive from value systems and outlooks that result in “deep coexistence” (Jolly et al., 2022), where the sharing of land and other resources with wildlife is viewed as simply part of the natural course of affairs. In this paradigm, conservation (i.e. the preservation of wildlife at healthy population levels) is simply an incidental byproduct, not a goal in itself. The ecosystem of elephants and other wildlife, quite simply, *is also* the ecosystem of certain human communities. Such a mindset cannot easily be transferred outside the community or even intergenerationally – the cultural transfer of ecological knowledge, which includes the particular manner of relating to nonhuman species, is crucially rooted in one’s sense of place. The displacement and dispossession of people from ancestral lands, together with well-intentioned reeducation of new generations in largely Western modes of thought concerning the economics of natural systems, undermines value systems built up over millennia. As Sayialel and Moss (Moss et al., 2011a) describe from their experience in Amboseli, sometimes disagreements over how to view and treat wildlife can create deep rifts even between age classes within a society.

The "deep coexistence" mindset contrasts with the somewhat lesser notion of coexistence pursued by conservation practitioners, which envisions resource-sharing through a "tolerance" of wildlife that is positioned as the means to overcome so-called conflict (Kansky et al., 2016; Pooley et al., 2021; Saif et al., 2019). Proponents may argue in terms of personal or cultural affinities, economic benefits, or ecosystem services, all of which position human interests fundamentally in opposition to those of elephants (and other wildlife). In many ways, this approach is largely pragmatic, precisely because the deep-rooted cultural norms and practices that give rise to the mindsets of deep coexistence are not necessarily translatable or easily conveyed across contexts. Indeed, it may be more appealing to younger generations, who may have economic aspirations divergent from previous ones. This does not mean such factors can be ignored; the calculus of economic costs and benefits alone is insufficient by far in motivating human attitudes and perceptions even among farming communities. However, the ignorance of traditional knowledge comes at great cost, unfortunately borne by the same communities.

The process of "desertification" in sub-Saharan Africa is one example. Boateng (2017) outlines how indigenous communities have been blamed for the supposed degradation of land in northern Ghana, according to the ecological characterizations of colonial administrators in the 1920s and 1930s. However, the land-use practices of the many different groups of people occupying the region prior to colonial settlement appear to have been quite fluid and responsive to environmental as well as social conditions. He discusses a group, the Tindaana, where one set of oral traditions contends that:

> the direct family ancestors of the present Tindaanas are the primordial residents of their respective communities. According to those beliefs, those primordial residents were the first to establish some sort of rapport with the nature spirits… If they were able to build rapport with the nature spirits of the territory, then the primordial inhabitants could permanently settle there. But if this failed to materialize, then they died or had to migrate elsewhere.

Other oral traditions and archeological findings suggest that these agrarian communities were not necessarily the first settlers, but that earlier human inhabitants may have been hunter-gatherer communities that did not persist (Boateng, 2017). For their part, the Tindaana viewed their survival and success as tied to having a favorable relationship with the natural environment and its spirits. Land was allocated for use among community members but could not be sold, as it placed nature spirits of the land into servitude. The Tindaana, like many communities around the world (Bhagwat & Rutte, 2006), also maintained sacred groves and sacred trees. Sacred groves are often important reservoirs of biodiversity and act as refuges for wildlife, in addition to being integral manifestations of cultural beliefs and heritage (Baker et al., 2018; Bhagwat & Rutte, 2006; Khan et al., 2008). The origins of these groves in drier ecosystems has been a matter of debate. While some have argued them to be the remains of larger forests, contrary evidence suggests that the cultural spiritual practices might actually be *encouraging* relatively recent tree growth. Boateng goes on to describe how a system of rule through local chieftainships became established with the takeover of these territories by the British and other colonial powers in the 1800s, together with land disputes meticulously documented with historic records. Striving to modernize the region and integrate it with the international economy, the spiritual leadership of these decentralized communities was substituted with more centralized secular leadership. Crucially, the spiritually guided authority of Tindaanas over land rights was sidelined, ushering in the economically driven use policies that persist today. This case study illustrates a pattern repeated throughout the continents, resulting in something like ossification in how land is managed, with people and wildlife becoming more sedentary. Ecologically and socially dynamic systems are forced to conform to static mindsets and frames of reference.

Summarizing two decades of research on human–elephant conflict mitigation throughout Africa, Hoare and colleagues identify 10 different types of measures that are routinely taken (Hoare, 2015):

1. Traditional deterrence and disturbance methods employed by communities
2. Disturbance and chasing of elephants by wildlife authorities
3. Killing of problem elephants by wildlife authorities
4. Translocation of problem elephants by wildlife authorities
5. Fencing options
6. Use of olfactory and auditory deterrents alongside fences
7. Systemic data collection and research
8. Compensation and insurance schemes
9. Wildlife utilization and benefit programs (including "community conservation")
10. Land-use planning, changes, and zonation.

They distinguish between measures 1–6, which are short term and applied against animals, and measures 7–10, which involve working with people in the longer term. Hoare and colleagues go on to discuss the research, practicalities, and drawbacks of each, which I will not reproduce here. However, I highlight one of the most important recommendations, which is the need for "Vertical Integration" in decision-making to reconcile the needs of local actors such as communities in which conflicts occur, and external actors, namely government policymakers. It is central governments that make decisions on how land, water, and other resources are to be used – or, indeed, exploited. The underlying causes of conflict therefore fall under no. 10, whereas responses to conflict largely fall under nos. 1–6, followed by the remaining options. This twenty-first-century disconnect can be traced directly back to its colonial roots, which removed decision-making authority and agency from local communities and replaced a model of respectful land-sharing with one in which wildlife are expected to be excluded from human spaces and vice versa.

Arguably, the degradation of land is, if anything, a consequence of such mismanagement (Maitima et al., 2010). Groom and colleagues compared the densities of livestock and wildlife along with grass biomass and ground cover on Maasai pastoral rangelands that were subdivided (i.e., individually owned and possibly fenced) with those that were not (Groom & Western, 2013). There was no significant difference in the densities of livestock between the two types of management system but the densities of wildlife were significantly higher in the unsubdivided areas. Despite this, the unsubdivided lands had a greater percentage of grass biomass and groundcover and evidenced less grazing pressure than the subdivided lands. Conservation advocates therefore now emphasize the need for starting with a recognition of community aspirations and pastoral management practices that are threatened by the processes of "land fragmentation, alienation and degradation" (Western et al., 2020: 279). These practices essentially rest upon requirements of open space and mobility that in turn require social and institutional arrangements that facilitate the management of common-pool resources.

The need to reverse centuries of top-down land use regulations with more just and equitable bottom-up governance is of course not limited to pastoralist communities. The distinction between human and "beastly" spaces is an idea that has been repeatedly introduced through colonial management practices (de Silva & Srinivasan, 2019), now perversely reinforced by human population pressures that require explicit protection of natural areas lest they be converted to some other use. The result is what known as fortress conservation (Hartter & Goldman, 2011; Sirua, 2006), where not only are (certain) human activities forbidden in "wilderness" areas, but also wildlife are prohibited from entering human areas. Many cultures that once viewed wildlife as part of a shared landscape have internalized this distinction. Protected areas turn into outdoor museums, their outlines coming into ever sharper relief against the background of human-dominated landscapes (Figure 8.1b). The proliferation of hard edges dividing nature and people makes ripe frontiers for negative interactions between people and wildlife. These edges themselves then become fundamentally part of the problem.

Given the dark origins of this separation, a new generation equipped with fresh eyes and knowledge of history must critically evaluate under what circumstances such a separation makes sense. Doing so entails engagement with extremely thorny questions: Which humans and which activities should be permitted in areas that are intended to conserve nature? The creeping loss of nature-centered spiritual, cultural, and practical value systems is cause for concern, for such a tide may not be easily (or ethically) reversed. It must also be acknowledged that even traditional sustainable uses of forest resources can give way to overexploitation of these same resources when economic forces create demand that outstrips the capacity of systems to provide it (Davidar et al., 2008; Nadal, 2008). And likewise, how could wildlife be safely accommodated in the human-dominated landscapes of the present? Certainly out of justice to both, we must consider not only the potential for physical harm but also economic or psychological harm. Aldo Leopold's land ethic remains extremely relevant today and provides guidance in both directions: A thing is right when it tends to preserve the beauty, integrity, and functioning of natural systems. The key is to recognize that *both* human and nonhuman agents have roles to play in maintaining this "rightness," including and especially those who have been the most marginalized historically.

8.3 Dentition and Diet Revisited

While the preceding sections are relevant to conservation, what does that have to do with elephant behavior? To answer this, we need to return to some aspects of elephant evolution we have briefly discussed elsewhere. Where teeth are concerned, elephants' tusks tend to get all the attention. But in fact, their molars do all the work, and potentially tell us a far more interesting and neglected story. In particular, they point to a seeming contradiction between the basic foraging preferences of elephants and their morphological adaptations that puzzled me as a graduate student. The lozenge-shaped lamellae characteristic *Loxodonta* molars, after which the genus is named, are evidently better adapted for browse than the straighter, narrower, and more numerous lamellae of *Elephas* molars, which are more appropriate for grazing. Dentition notwithstanding, both genera are found in ecosystems ranging the spectrum from grass-dominated to tree-dominated as we have already discussed. Still, would it not make more sense for savanna elephants, found in relatively open environments, to be the ones adapted for grazing while Asian elephants, inhabiting (on the whole) more forested environments, to be adapted to browse? Logical as it might be, evolution doesn't work that way.

An organism's adaptability is constrained by its generation time relative to environmental changes. While elephants may be very behaviorally flexible, morphology is slow to adapt given how long it takes them to reproduce. Recall that in Chapter 1 I discussed the ecological conditions and timing of divergence between the lineages, together with Lister's compelling case that proboscideans and other herbivores in Africa switched to a predominantly grazing diet before their dentition caught up to the fact (Lister, 2013, 2014), evidently driven by climatic changes favoring the spread of grasslands. Relative to earlier lineages, the molars of *Loxodonta* were indeed better adapted for processing grasses and the molars of *Elephas* (including those specimens previously labelled *Palaeoloxodon*) even more so (Figures 8.2 and 8.3). *Loxodonta* continued to forage on a mix of grass and browse (and fruits), whereas *Elephas* grew increasingly specialized on grasses, venturing further and further toward the edges of the continent and gradually coming to dominate the far eastern portion of it. By the time representatives of *Elephas* faded out in Africa between 205,000 and 130,000 years ago, their diets and dentition appear to have been well-aligned (Manthi et al., 2020). Fortunately, extinction in Africa wasn't the end of this genus, since they were on their way to spreading throughout Eurasia as discussed in Chapter 2.

This easterly dispersal history of *Elephas* means that their colonization of the wetter forested environments of South and Southeast Asia occurred relatively recently, compared with the evolution of their dentition, originally adapted to African savannas. In other words, some Asian elephant populations are now once again in a lag phase: Their teeth have yet to catch up to the fact that they inhabit environments in which grasses are less common. And although Asian elephants are generalist feeders, they evidently do still favor grasses and other herbaceous vegetation over browse – vegetation found in the greatest abundance in forest clearings, successional forests, and floodplains. However, in these environments was another species whose activities

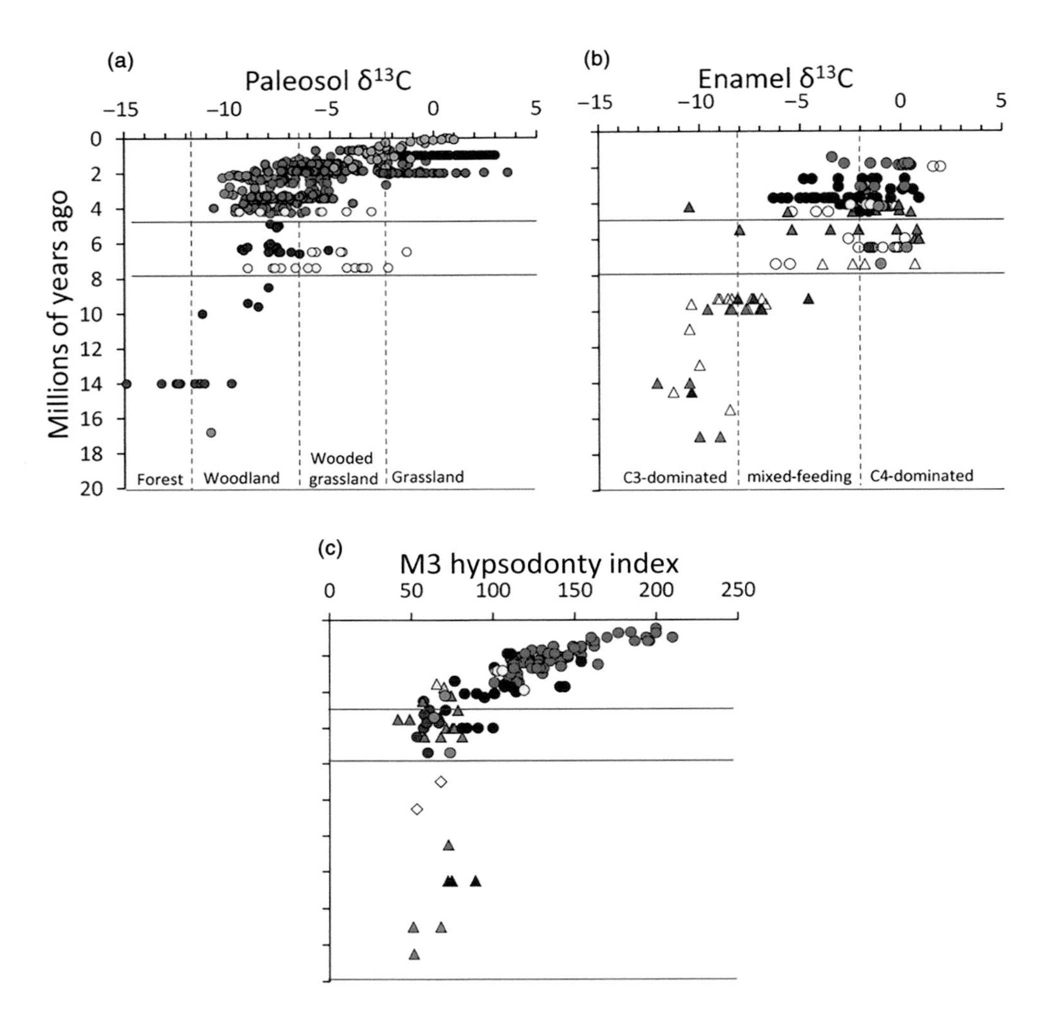

Figure 8.2 Evolution of hypsodonty (increased crown height) in East African proboscideans. Circles represent elephants, diamonds represent tetralophodont/elephantid intermediates, and triangles represent stegodonts and "gomphotheres." (a) Palaeosol δ13C is a proxy for vegetation type measured in terms of the ratio of Carbon 12 to Carbon 13 isotopes in ancient soil layers; (b) The same measured in tooth enamel, a proxy for diet; (c) Hypsodonty index in the upper third molar. Horizontal lines between approximately 8-5 mya when the diet became more grass dominated (b) with little change in crown height (c) (reproduced from Lister, 2014 with permission from Oxford University Press. Copyright © 2014, © 2013 The Linnean Society of London).

have perhaps maintained a greater abundance of such vegetation than would ordinarily occur. That species, of course, was us. Hunter-gatherers, shifting cultivators, livestock herders, and agriculturalists with their burning practices (as well as carefully managed irrigation, a subject I do not discuss in this book) promoted the very regimes that elephants would have benefited from. Hunted for food during the Paleolithic, elephants came to be venerated as the very embodiment of natural and spiritual power

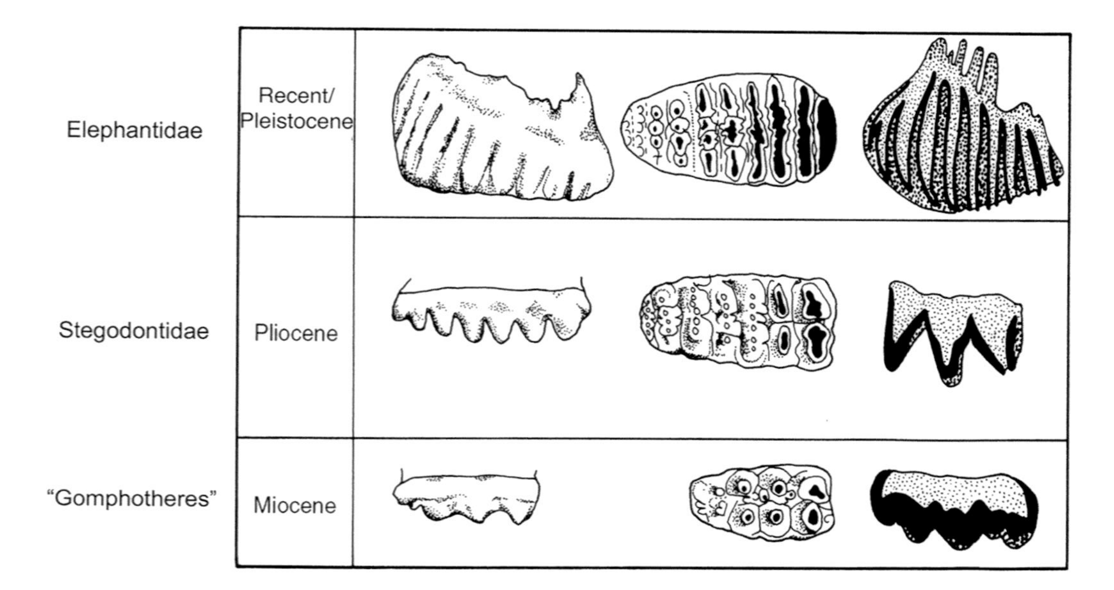

Figure 8.3 Evolution of proboscidean molars in the late Cenozoic. Upper molars shown in lateral (left), occlusal (centre) and cross-section views (right). The composition of tooth sections includes enamel (black), cementium (heavy stipple) and dentine (light stipple). The more recent Elephantidae (including both Elephas and Loxodonta) show taller crowns (i.e. the part of the tooth that protrudes beyond the gum) and a greater number of ridges, accompanied by more cementium as compared with the preceding lineages (reproduced from Zhang et al., 2017, with permission from Elsevier).

throughout many Asian cultures (Barkai, 2019; Lev & Barkai, 2016). Isn't it possible then that Asian elephants reached such great abundance in certain landscapes not only owing to the natural productivity with which these landscapes were already endowed, but the dynamic nonequilibrium regimes within these systems that humans actively facilitated? Likewise, throughout Africa, it is far more plausible to me that the thoughtful management of rangelands, wetlands, and forests through a mixture of sociocultural *as well as* environmentally mediated traditions would have been contributed to the maintenance of its incredible suite of diverse megaherbivores, rather than to their demise.

Now, in the twenty-first century, elephants are also encountering more new species in their ranges thanks to us. Some of these species present a potentially rich food source. The tall grasses *Cenchrus purpureus* (cf. *Pennisetum purpureum*), fittingly and colloquially referred to as "elephant grass" or napier grass, is a species native to African grasslands that was introduced throughout Asia as fodder for livestock. It thrives in fallow fields, at roadsides, and in other areas of disturbance. Asian elephants, with their savanna-adapted teeth, have thrived on such vegetation reminiscent of their distant past. Likewise, oil palms, native to West Africa, are now cultivated in great abundance in Southeast Asia, especially in Indonesia and Malaysia. Not only the palms themselves, but also the regenerating forests of abandoned palm concessions are rich in the herbaceous vegetation elephants prefer. Their consumption of economically valuable palms, however, creates tension with people. A similar story could be told about many of the nonnative fruit species that are grown alongside remnant forests in Southeast Asia and the maize fields that now form a staple crop in parts of sub-Saharan Africa. These agricultural fields and plantations become potential ecological traps and sinks for the elephant populations nearby.

Meanwhile, other introduced species such as *Lantana camara* threaten to overrun protected areas as they are not (as yet) consumed by elephants or other herbivores native to Asia. An important area of future research is to try to understand whether elephants are drawn to human cultivations, despite associated risks of injury or death, because of the inherent attractiveness of the food crops themselves, or else because the nearby forests, sanctuaries, and other uncultivated lands no longer contain sufficient food resources (or both). A study of forest elephant diets at Kibale National Park found that this population was sodium-limited in their wild diet, but crops had both higher concentrations of sodium and lower concentrations of fiber and secondary compounds. There is, understandably, extensive interest in the degree to which introduced species directly impact human societies. A recent study estimated that globally, the average cost of biological invasions ranged between USD $46.8 billion and $162.7 billion (Diagne et al., 2021). This may be an underestimate, as agriculture is one of the hardest hit sectors and regional studies show even greater losses. For example, invasive species across Africa alone can cost an estimated loss of USD $65.58 billion per year in the agriculture and livestock sectors (Eschen et al., 2021). But aside from this, introduced species can have unquantified negative impacts on wildlife by reducing forage in nonagricultural lands, which in turn has knock-on effects on human communities by adding more pressure on wildlife to forage on croplands. Given sufficient time, it is possible that native herbivores will learn to consume these nonnative yet superabundant plant species, but can we manage the new normal in the meantime? It seems prudent to *anticipate* rather than *react* to these processes by rethinking the functions of agricultural landscapes that now form the dominant human land-use category, especially when they are interspersed with critical wildlife habitat (Nyhus & Tilson, 2004; Obura et al., 2022; Plieninger et al., 2012; Western et al., 2020).

With present-day human densities and governance systems, we cannot turn back the clock in favor of fully returning to traditional systems of resource management. We are in need of a new paradigm. Fortunately, there is evidence that diverse agro-ecosystems can support biodiversity while also being economically productive. For example, in Southern India, the cultivation of Areca nut palms alongside broadleaf forest with the intercropping of woody species enables high levels of avian diversity (Ranganathan et al., 2008). The already megadiverse landscapes that elephants inhabit have a high potential for multiuse approaches that provide both economic and ecological benefits (Jackson et al., 2012), provided these benefits can be equitably shared rather than monopolized by those who already control and benefit from the status quo. This seems difficult, but achievable if the goals of conservationists and communities can be better aligned. A hitch is that the proposition of having elephants on one's land, especially if the relevant crops are susceptible to damage, might not be viewed with quite as much enthusiasm as that of more innocuous species, as we discuss in Chapter 6. Nevertheless, I believe the diversification of agro-ecological systems is one of the most promising ways forward, if not increasingly necessary as a means of buffering farming communities against oncoming global changes in climate, water supply, and the transport of nonnative species that result in costly biological invasions. Research is needed to assess both the economic and ecological viability of agro-ecological approaches, which will no doubt be highly context-specific for both social and ecological reasons.

When I earlier considered the possible causes of extinction that eliminated other megafauna in Chapter 2, I focused on the combination of hunting pressure and climate change. But there may be an additional reason. On the whole, ecological specialists tend to have greater risk of extinction because they are gambling on putting all their eggs in one basket along some dimension or other (Gallagher et al., 2015). This seems especially problematic for species with long generation times, as they cannot adapt and evolve as rapidly. It is possible that the taxa that were driven extinct were more specialized in some way, narrowing the niches they could occupy – either in terms of feeding habits, climate, or both. Support for this hypothesis comes from several observations. A study by Cantalapiedra and colleagues reviewed the palaeontological record for 185 proboscidean species on the basis of over 2,000 fossils encompassing 17 ecomorphological traits (Cantalapiedra et al., 2021). Niche differentiation among these taxa at the height of their diversity, which occurred approximately 15 million years ago, allowed multiple proboscidean species to co-occur in the same ecosystems. Following this period, at least eight different proboscidean functional types, with corresponding feeding strategies, have been described. However, after the shift to C_4 grass-dominated environments 8 million years ago, proboscidean diversity started to decline. Another pair of studies suggests a clue as to why, by examining the Pleistocene-era foraging ecology of species of proboscideans that evidently coexisted at the time in Southern China: *Stegodon*, *Sinomastodon*, and *Elephas*. Stable isotope analyses show that *Elephas maximus* already had a more diverse and mixed feeding strategy than *Stegodon orientalis*, which relied on a narrower range of plants in a more densely forested landscape (Ma et al., 2019). Corroborating these observations, patterns of dental microwear indicate that the *Stegodon* and *Sinomastodon* species were more specialized on browse, than *Elephas* (Zhang et al., 2017). As climatic changes favored the spread of more grass-dominated ecosystems, proboscideans that were more specialized on browse would have found themselves at a disadvantage. As we have seen, both African and Asian elephants have proven remarkably versatile, able to conquer astonishingly varied climatic zones and habitat types. Given their long generation times, elephants' behavioral flexibility (aided by social adaptations and certain cognitive capacities) provides a buffer against changing conditions much more quickly than physical attributes themselves. Reiterating statements at the close of Chapter 7, this adaptability, which was possibly the key to their success despite being such slow breeders, is also a potential liability when confronting the novel conditions humans create.

8.4 Future Predictions

One can never set foot in the same forest twice. Day by day, nature is ever in flux. The pace of some changes is slow to our eyes, mistaken for stability; others happen rapidly, even without our assistance. In mental contrast to so-called alternative stable states, I much prefer the characterization of alternative transient states (Fukami & Nakajima, 2011). As it should be clear by now, I am firmly in the disequilibrium camp when it comes to ecological dynamics. I view the phenomenon termed human–wildlife "conflict" as largely a byproduct of a seismic shift in perspective regarding how natural systems "should" behave, imposed upon landscapes and people largely by external actors with misguided baselines and value systems, causing profound ecological impacts. So-called human–elephant conflict derives not only from the propensity of elephants to be attracted to human crops, but also the ever-increasing share of land area devoted to static, large-scale agricultural fields and settlements at the expense of the dynamically shifting practices which preceded them for thousands of years, and perhaps in no small part to the sweeping dominance of newly introduced species. Unfortunately, dynamic land management seems fundamentally at odds with current economic and governance paradigms, as well as human population growth. But the dynamism inherent in the Earth system, subdued for much of the Holocene/Anthropocene, is becoming increasingly difficult to ignore thanks to another anthropogenically driven phenomenon: climate change.

The world might appear to be an exceedingly stable place if one is gnat or mayfly, but not if one is an elephant or human. Our long lifespans mean that the vicissitudes of climate change will be felt within a single generation, experienced by individuals who are now alive. Static paradigms of management that aim to contain wildlife (by means of fences and all manner of barriers) in specific locations (protected areas of various descriptions) are ill-suited to the changes that are even now occurring. Entire climatic zones will move. Equilibria, if they exist, will go out the window. Behavior is where the rubber meets the road – where ecology and evolution interact, with profound consequences. And elephants will not accept change quietly. As discussed earlier, it is possible that many elephant populations currently find themselves in suboptimal or unsuitable habitat. As these populations attempt to disperse to more suitable areas, they will find many human endeavors in their path. The intervening landscapes will thus have great potential for conflict. In this manner, elephants make us aware of the uncomfortable realities and consequences of ecological change.

Between 2020 and 2021, the world watched in astonishment as a small group of elephants embarked on an adventurous and unprecedented 500 km excursion out of their protected home in MengYang reserve, XiShuangBanNa National Nature Reserve located in Yunnan province, China (Jiang et al., 2023). MengYang is not small, an area encompassing 998 km^2. The group, consisting of both sexes and mixed ages including young calves, passed by roads, towns, and plantations over a period of several months – perhaps made easier by the vigilance of government authorities, who closely monitored the path of the group and evacuated over 150,000 people for the safety of all concerned. Although such a trek is unprecedented in recent memory, it is not difficult to imagine that the same exploratory spirit is what led the ancestors of elephants (and many other species besides) to disperse throughout the continents. It is rendered unusual today by the many barriers our species has newly created that impede such long-range movements of wildlife. It was rendered even more newsworthy by virtue of the fact that the elephants seemed to be headed directly toward KunMing, where the United Nations Biodiversity Conference was scheduled to be held (Figure 8.4). Journalists were quick to point out that it was as though they were on a mission to plead their case. It could not have been planned any better if the elephants had been advised by a savvy PR specialist!

Of course, the elephants could not know this. But they did drive scientists to investigate the reasons for such movement and make the case on their behalf (Wang et al., 2021). First, it appears that the annual mean temperature in their native range has been on the rise, increasing by 1.6°C in 2019–2020 compared with 1981–2010 (Figure 8.4b). It was also drier, with a decrease in mean precipitation. In March 2020, when the elephants initiated their journey, the drought index was at an "unusually low level" (Wang et al., 2021: 1), with a correspondingly steep drop in vegetation productivity (Figure 8.4c–e). Researchers further report that although the elephant population has virtually doubled from 150 to 300 individuals since the 1960s thanks to conservation efforts, between the years 1975 and 2014 the amount of forest cover and farmland in southwestern Yunnan decreased by 11.5% and 14.9% respectively, whereas plantations of rubber and tea increased by 25.4%. These observations suggest that a confluence of factors having to do with changes in land use as well as climate precipitated the sudden movement. Was this a one-off event or simply the first drop in a greater torrent still to come? The case underscores again that these movement decisions can be highly idiosyncratic: Out of the several hundred elephants inhabiting the reserve system, all of whom would have been affected by the same climatic and land-use changes, only this little group opted for such a remarkable adventure. Personality traits and social dynamics as much as environmental variables might have played a part.

Around the same time, halfway around the world in Botswana, another disturbing set of events was taking place. In May and June 2020, at least 350 elephants were found dead around the Okavango Delta ecosystem, some near water sources and others not (Azeem et al., 2020; van Aarde et al., 2021). They occurred in landscapes also occupied by people and livestock, rather than protected areas. More than 30 carcasses were likewise found between January and March 2021. Poaching was ruled out as tusks were intact, as was poisoning of water sources, since only elephants were affected (Azeem et al., 2020). One popular hypothesis was that the deaths were caused by blooms of toxic cyanobacteria in water holes, which are also (confusingly) referred to as a type of blue-green algae (Veerman et al., 2022). This again seems to be unlikely as no other species was affected, but it is possible elephants could have consumed toxic plants (e.g. introduced species) that other herbivores did not. But why the sudden spike in deaths? One intriguing early idea was that there may have been disease transfer between rodents (or livestock) and elephants, in the context of croplands,

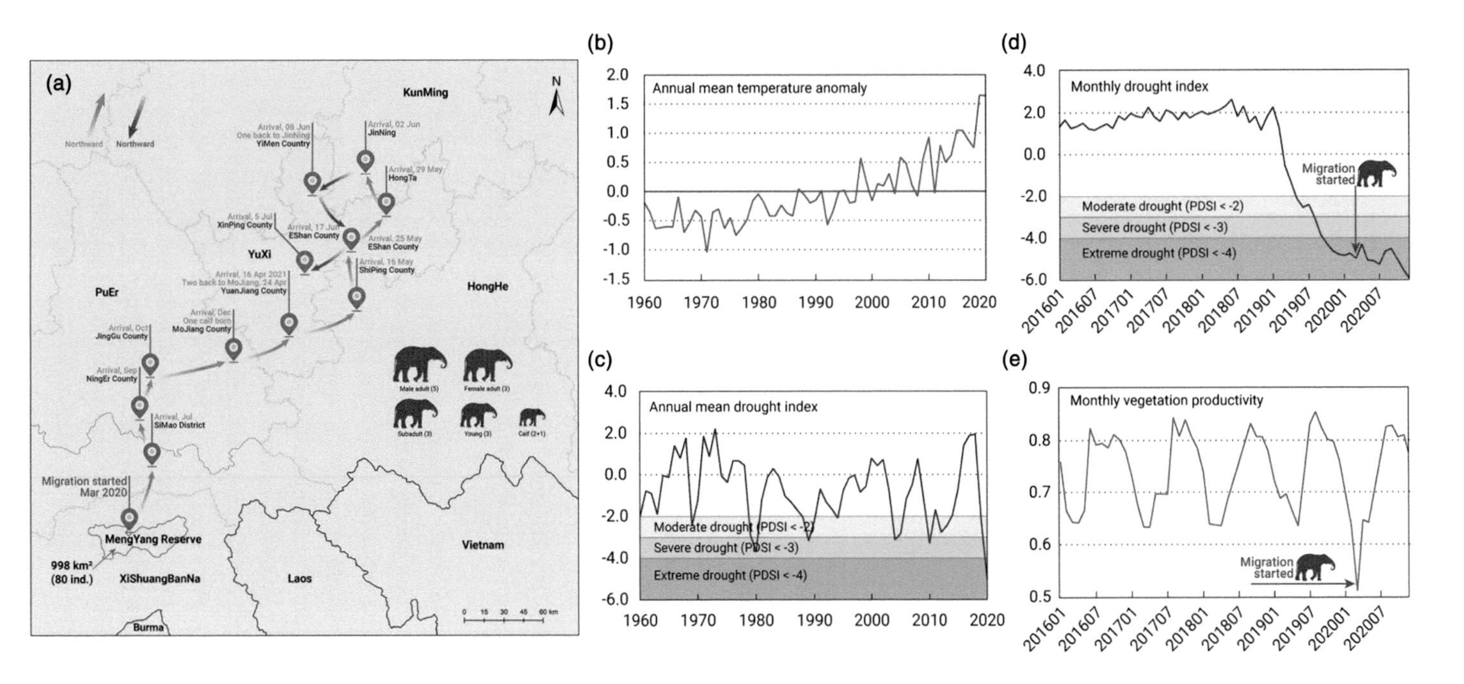

Figure 8.4 The unusual movement of an elephant group in Southern China. (a) The group consisting of sixteen individuals of mixed ages and both sexes (including a calf born en route) travelled well beyond their home reserve. The beginning of the journey appears to coincide with unusually high temperatures (b) and drought conditions (c)–(d), and low vegetation productivity (e). These conditions may have added to other chronic stressors such as land-use change (reproduced from Wang et al., 2021, with permission from Elsevier. Copyright © 2021 Wang et al.).

owing to late rains in the region resulting in bumper crops; February to May are also peak times when elephants forage on crops (Azeem et al., 2020). Other researchers point out that the effect of a disease outbreak would be compounded by the prevalence of fences and deep water, limiting the dispersal of the increasingly dense population of elephants (van Aarde et al., 2021). While this whodunnit remains unsolved as of the time of writing (December 2023), it suggests again that land use and climatic events, together with impediments to movement, may be conspiring together to create a "perfect storm" of dangerous conditions for wildlife (Azeem et al., 2020).

Picture a clear blue sky with wisps of clouds floating by. Now imagine that we place some arbitrarily shaped insubstantial boundary around such a cloud and call it protected air. We can no more contain and protect the cloud within this artificial boundary than we can contain the ecosystems and their many species that now make use of protected patches of the planet. This is not to say that protected spaces are unnecessary – but we must see them clearly for what they are. The ecosystems they are representing – be they pine forests, rainforests, or even kelp forests (and everything else besides) – are transient states of nature maintained by a combination of climate, biology and human activity. They have been so in the past and they will be so in the future. They may persist long enough to appear to us to be "stable," but this is relative to the pace of our lives and the pace of change. Like clouds blowing past our bounding box, the ecosystems within any arbitrary human-made boundary ("Protected Area," "Nation State") will change. Elephants force us to contend with how our own species will contend with such changes. They have demonstrated a remarkable adaptive capacity that is primarily behavioral rather than morphological, sufficient thus far to ensure their survival and dominance as the last remaining giants on their respective land masses. Will we reward this tenacity and our deep debt to the Proboscidean lineage for what they have provided us by way of sustenance and spiritual enrichment, so that they can continue to evolve into something else beyond the transient state represented by the taxonomic label "elephant"? Or will we continue to persecute them, as we have done all of our competitors? How we respond to this challenge will inevitably teach us more about ourselves and the prospects for our own longevity on the planet.

References

Abernethy, K. A., Coad, L., Taylor, G., Lee, M. E., & Maisels, F. (2013). Extent and ecological consequences of hunting in Central African rainforests in the twenty-first century. *Philosophical Transactions of the Royal Society B: Biological Sciences*, *368*(1625). https://doi.org/10.1098/rstb.2012.0303

Abrahams, P. W. (2012). Involuntary soil ingestion and geophagia: A source and sink of mineral nutrients and potentially harmful elements to consumers of Earth materials. *Applied Geochemistry*, *27*(5), 954–968. https://doi.org/10.1016/j.apgeochem.2011.05.003

Adamescu, G. S., Plumptre, A. J., Abernethy, K. A., Polansky, L., Bush, E. R., Chapman, C. A., Shoo, L. P., Fayolle, A., Janmaat, K. R. L., Robbins, M. M., Ndangalasi, H. J., Cordeiro, N. J., Gilby, I. C., Wittig, R. M., Breuer, T., Hockemba, M. B. N., Sanz, C. M., Morgan, D. B., Pusey, A. E., … Beale, C. M. (2018). Annual cycles are the most common reproductive strategy in African tropical tree communities. *Biotropica*, *50*(3), 418–430. https://doi.org/10.1111/BTP.12561

Adams, T. S. F., Mwezi, I., & Jordan, N. R. (2021). Panic at the disco: Solar-powered strobe light barriers reduce field incursion by African elephants *Loxodonta africana* in Chobe District, Botswana. *Oryx*, *55*(5), 739–746. https://doi.org/10.1017/S0030605319001182

Adams, T. S. F., Leggett, K. E. A., Chase, M. J., & Tucker, M. A. (2022). Who is adjusting to whom? Differences in elephant diel activity in wildlife corridors across different human-modified landscapes. *Frontiers in Conservation Science*, *3*, 872472. https://doi.org/10.3389/FCOSC.2022.872472/BIBTEX

Agrawal, A. A., & Rutter, M. T. (1998). Dynamic anti-herbivore defense in ant-plants: The role of induced responses. *Oikos*, *83*(2), 227–236.

Ahlering, M. A., Millspaugh, J. J., Woods, R. J., Western, D., & Eggert, L. S. (2011). Elevated levels of stress hormones in crop-raiding male elephants. *Animal Conservation*, *14*(2), 124–130. https://doi.org/10.1111/J.1469-1795.2010.00400.X

Alexander, K. A., Pleydell, E., Williams, M. C., Lane, E. P., Nyange, J. F. C., & Michel, A. L. (2002). Mycobacterium tuberculosis: An emerging disease of free-ranging wildlife. *Emerging Infectious Diseases*, *8*(6), 598–601. https://doi.org/10.3201/eid0806.010358

Alfred, R., Ahmad, A. H., Payne, J., Williams, C., & Ambu, L. N. (2010). Density and population estimation of the Bornean elephants (*Elephas maximus borneensis*) in Sabah. *Online Journal of Biological Sciences*, *10*(2), 92–102.

Alfred, R., Ahmad, A. H., Payne, J., Williams, C., Ambu, L. N., How, P. M., & Goossens, B. (2012). Home range and ranging behaviour of Bornean elephant (*Elephas maximus borneensis*) females. *PLOS One*, *7*(2). https://doi.org/10.1371/journal.pone.0031400

Allee, W. C., Emerson, A. E., Park, O., Park, T., & Schmidt, K. P. (1949). *Principles of animal ecology*. Saundere Co. Ltd.

Allen, C. R. B., Brent, L. J. N., Motsentwa, T., Weiss, M. N., & Croft, D. P. (2020). Importance of old bulls: Leaders and followers in collective movements of all-male groups in African savannah elephants (*Loxodonta africana*). *Scientific Reports*, *10*(1), 1–9. https://doi.org/10.1038/s41598-020-70682-y

Allen, C. R. B., Brent, L. J. N., Motsentwa, T., & Croft, D. P. (2021a). Field evidence supporting monitoring of chemical information on pathways by male African elephants. *Animal Behaviour*, *176*, 193–206. www.sciencedirect.com/science/article/pii/S0003347221001032

Allen, C. R. B., Croft, D. P., & Brent, L. J. N. (2021b). Reduced older male presence linked to increased rates of aggression to non-conspecific targets in male elephants. *Proceedings of the Royal Society B: Biological Sciences*, *288*(1965), 20211374. https://doi.org/10.1098/RSPB.2021.1374

Allen, J. A. (2019). Community through culture: From insects to whales. *BioEssays*, *41*(11), 1900060. https://doi.org/10.1002/BIES.201900060

Allen, J. A., Weinrich, M., Hoppitt, W., & Rendell, L. (2013). Network-based diffusion analysis reveals cultural transmission of lobtail feeding in humpback whales. *Science*, *340*(6131), 485–488. https://doi.org/10.1126/SCIENCE.1231976/SUPPL_FILE/ALLEN.SM.PDF

Amorim, T. O. S., Rendell, L., Di Tullio, J., Secchi, E. R., Castro, F. R., & Andriolo, A. (2020). Coda repertoire and vocal clans of sperm whales in the western Atlantic Ocean. *Deep Sea Research Part I: Oceanographic Research Papers*, *160*, 103254. https://doi.org/10.1016/J.DSR.2020.103254

Anderson, J. R. (2016). Comparative thanatology. *Current Biology*, *26*, R553–R556. https://doi.org/10.1016/j.cub.2015.11.010

Andersson, M., Michelsen, A., Jensen, M., & Kjøller, A. (2004). Tropical savannah woodland: Effects of experimental fire on soil microorganisms and soil emissions of carbon dioxide. *Soil Biology and Biochemistry*, *36*(5), 849–858.

Angeloni, L., Schlaepfer, M. A., Lawler, J. J., & Crooks, K. R. (2008). A reassessment of the interface between conservation and behaviour. *Animal Behaviour*, *75*(2), 731–737. https://doi.org/10.1016/j.anbehav.2007.08.007

Archie, E. A., Morrison, T. A., Foley, C. A. H., Moss, C. J., & Alberts, S. C. (2006a). Dominance rank relationships among wild female African elephants, *Loxodonta africana*. *Animal Behaviour*, *71*(1), 117–127. https://doi.org/10.1016/j.anbehav.2005.03.023

Archie, E. A., Moss, C. J., & Alberts, S. C. (2006b). The ties that bind: Genetic relatedness predicts the fission and fusion of social groups in wild African elephants. *Biological Sciences/The Royal Society*, *273*(1586), 513–522. https://doi.org/10.1098/rspb.2005.3361

Archie, E. A, Hollister-Smith, J. A, Poole, J. H., Lee, P. C., Moss, C. J., Maldonado, J. E., Fleischer, R. C., & Alberts, S. C. (2007). Behavioural inbreeding avoidance in wild African elephants. *Molecular Ecology*, *16*(19), 4138–4148. https://doi.org/10.1111/j.1365-294X.2007.03483.x

Archie, E. A., Moss, C. J., & Alberts, S. A. (2011). Friends and relations: Kinship and the nature of female elephant social relationships. In C. J. Moss, H. Croze, & P. C. Lee (Eds.), *The Amboseli Elephants: A long term perspective on a long-lived mammal* (pp. 238–245). University of Chicago Press.

Armbruster, P., Fernando, P., & Lande, R. (1999). Time frames for population viability analysis of species with long generations: An example with Asian elephants. *Animal Conservation*, *2*(1), 69–73. https://doi.org/10.1017/S1367943099000372

Ashley, G. M., Barboni, D., Dominguez-Rodrigo, M., Bunn, H. T., Mabulla, A. Z., Diez-Martin, F., Barba, R., & Baquedano, E. (2010). Paleoenvironmental and paleoecological reconstruction of a freshwater oasis in savannah grassland at FLK North, Olduvai Gorge, Tanzania. *Quaternary Research*, *73*(3), 333–342.

Aung, M. (1997). On the distribution, status and conservation of wild elephants in Myanmar. *Gajah*, *18*, 47–55.

Azeem, S., Bengis, R., Van Aarde, R., & Bastos, A. D. S. (2020). Mass die-off of African elephants in Botswana: Pathogen, poison or a perfect storm? *African Journal of Wildlife Research*, *50*(1), 149–156. https://doi.org/10.3957/056.050.0149

Babweteera, F., Savill, P., & Brown, N. (2007). *Balanites wilsoniana*: Regeneration with and without elephants. *Biological Conservation*, *134*, 40–47.

Baker, L. R., Tanimola, A. A., & Olubode, O. S. (2018). Complexities of local cultural protection in conservation: The case of an Endangered African primate and forest groves protected by social taboos. *Oryx*, *52*(2), 262–270. https://doi.org/10.1017/S0030605317001223

Bakker, E. S., Gill, J. L., Johnson, C. N., Vera, F. W. M., Sandom, C. J., Asner, G. P., & Svenning, J.-C. (2015). Combining paleo-data and modern exclosure experiments to assess the impact of megafauna extinctions on woody vegetation. *Proceedings of the National Academy of Sciences of the United States of America*, *113*(4), 847–855. https://doi.org/10.1073/pnas.1502545112

Baldwin, J. W., Garcia-Porta, J., & Botero, C. A. (2022). Phenotypic responses to climate change are significantly dampened in big-brained birds. *Ecology Letters*, *25*(4), 939–947. https://doi.org/10.1111/ELE.13971

Balenger, S. L., & Zuk, M. (2014). Testing the Hamilton–Zuk hypothesis: Past, present, and future. *American Zoologist*, *54*(4), 601–613.

Ball, J. B. (2001). Global forest resources: History and dynamics. In J. Evans (Ed.), *The forests handbook* (pp. 3–22). Blackwell Scientific Publications.

Ball, R., Jacobson, S. L., Rudolph, M. S., Trapani, M., & Plotnik, J. M. (2022). Acknowledging the relevance of elephant sensory perception to human-elephant conflict mitigation. *Animals*, *12*(8), 1018. https://doi.org/10.3390/ANI12081018

Baragli, P., Scopa, C., Maglieri, V., & Palagi, E. (2021). If horses had toes: Demonstrating mirror self recognition at group level in *Equus caballus*. *Animal Cognition*, *24*(5), 1099–1108. https://doi.org/10.1007/S10071-021-01502-7/TABLES/4

Barboza, P. S., Hartbauer, D. W., Hauer, W. E., & Blake, J. E. (2004). Polygynous mating impairs body condition and homeostasis in male reindeer (*Rangifer tarandus tarandus*). *Journal of Comparative Physiology B: Biochemical, Systemic, and Environmental Physiology*, *174*(4), 309–317. https://doi.org/10.1007/s00360-004-0416-6

Barkai, R. (2019). When elephants roamed Asia: The significance of Proboscideans in diet, culture and cosmology in Paleolithic Asia. In R. Kowner, G. Bar-Oz, M. Biran, M. Shahar, & G. Shelach-Lavi (Eds.), *Animals and human society in Asia* (pp. 33–62). Palgrave Macmillan. https://doi.org/10.1007/978-3-030-24363-0_2

Barnes, R. F. W., Barnes, K. L., Alers, M. P. T., & Blom, A. (1991). Man determines the distribution of elephants in the rain forests of northeastern Gabon. *African Journal of Ecology*, *29*(1), 54–63. https://doi.org/10.1111/J.1365-2028.1991.TB00820.X

Barnes, R. F. W., Dubiure, U. F., Danquah, E., Boafo, Y., Nandjui, A., Hema, E. M., & Manford, M. (2006). Crop-raiding elephants and the moon. *African Journal of Ecology*, *45*(1), 112–115.

Barnosky, A. D. (2004). Assessing the causes of late Pleistocene extinctions on the continents. *Science*, *306*(5693), 70–75. https://doi.org/10.1126/science.1101476

Barrett, L. P., & Benson-Amram, S. (2020). Can Asian elephants use water as a tool in the floating object task? *Animal Behavior and Cognition*, *7*(3), 310–326. https://doi.org/10.26451/abc.07.03.04.2020

Barrett, L. P., & Benson-Amram, S. (2021). Multiple assessments of personality and problem-solving performance in captive Asian elephants (*Elephas maximus*) and African savanna elephants (*Loxodonta africana*). *Journal of Comparative Psychology*, *135*(3), 406–419. https://doi.org/10.1037/COM0000281

Barrett, L. P., Stanton, L. A., & Benson-Amram, S. (2019). The cognition of 'nuisance' species. *Animal Behaviour*, *147*, 167–177. https://doi.org/10.1016/J.ANBEHAV.2018.05.005

Barua, M., Bhagwat, S. A., & Jadhav, S. (2013). The hidden dimensions of human-wildlife conflict: Health impacts, opportunity and transaction costs. *Biological Conservation*, *157*, 309–316. https://doi.org/10.1016/j.biocon.2012.07.014

Baruch-Mordo, S., Wilson, K. R., Lewis, D. L., Broderick, J., Mao, J. S., & Breck, S. W. (2014). Stochasticity in natural forage production affects use of urban areas by black bears: Implications to management of human-bear conflicts. *PLOS One*, *9*(1). https://doi.org/10.1371/journal.pone.0085122

Bates, L. A., Sayialel, K. N., Njiraini, N. W., Moss, C. J., Poole, J. H., & Byrne, R. W. (2007). Elephants classify human ethnic groups by odor and garment color. *Current Biology*, *17*(22), 1938–1942. https://doi.org/10.1016/J.CUB.2007.09.060

Bates, L. A., Sayialel, K. N., Njiraini, N. W., Poole, J. H., Moss, C. J., & Byrne, R. W. (2008). African elephants have expectations about the locations of out-of-sight family members. *Biology Letters*, *4*(1), 34–36. https://doi.org/10.1098/rsbl.2007.0529

Bates, L. A., Handford, R., Lee, P. C., Njiraini, N., Poole, J. H., Sayialel, K., Sayialel, S., Moss, C. J., & Byrne, R. W. (2010). Why do African elephants (*Loxodonta africana*) simulate oestrus? An analysis of longitudinal data. *PLOS One*, *5*(4), e10052. https://doi.org/10.1371/JOURNAL.PONE.0010052

Bateson, P., & Laland, K. N. (2013). Tinbergen's four questions: An appreciation and an update. *Trends in Ecology and Evolution*, *28*(12), 712–718.

Beale, C. M., Hauenstein, S., Mduma, S., Frederick, H., Jones, T., Bracebridge, C., Maliti, H., Kija, H., & Kohi, E. M. (2018). Spatial analysis of aerial survey data reveals correlates of elephant carcasses within a heavily poached ecosystem. *Biological Conservation*, *218*, 258–267. https://doi.org/10.1016/j.biocon.2017.11.016

Bearzi, G., Kerem, D., Furey, N. B., Pitman, R. L., Rendell, L., & Reeves, R. R. (2018). Whale and dolphin behavioural responses to dead conspecifics. *Zoology*, *128*, 1–15. https://doi.org/10.1016/J.ZOOL.2018.05.003

Beaune, D., Fruth, B., Bollache, L., Hohmann, G., & Bretagnolle, F. (2013). Doom of the elephant-dependent trees in a Congo tropical forest. *Forest Ecology and Management*, *295*, 109–117. https://doi.org/10.1016/j.foreco.2012.12.041

Beirne, C., Houslay, T. M., Morkel, P., Clark, C. J., Fay, M., Okouyi, J., White, L. J. T., & Poulsen, J. R. (2021). African forest elephant movements depend on time scale and individual behavior. *Scientific Reports*, *11*(1), 1–11. https://doi.org/10.1038/s41598-021-91627-z

Beisner, B. E., Haydon, D. T., & Cuddington, K. (2003). Alternative stable states in ecology. *Frontiers in Ecology and the Environment*, *1*(7), 376–382. https://doi.org/10.1890/1540-9295(2003)001[0376:ASSIE]2.0.CO;2

Ben-Dor, M., Gopher, A., Hershkovitz, I., & Barkai, R. (2011). Man the fat hunter: The demise of *homo erectus* and the emergence of a new hominin lineage in the middle Pleistocene (ca. 400 kyr) Levant. *PLOS One*, *6*(12). https://doi.org/10.1371/journal.pone.0028689

Bender, R., Tobias, P. V., & Bender, N. (2012). The Savannah Hypotheses: Origin, reception and impact on paleoanthropology. *History and Philosophy of the Life Sciences*, *34*(1/2), 147–184.

Benson-Amram, S., & Holekamp, K. E. (2012). Innovative problem solving by wild spotted hyenas. *Proceedings of the Royal Society B: Biological Sciences*, *279*(1744), 4087–4095. https://doi.org/10.1098/RSPB.2012.1450

Benson-Amram, S., Dantzer, B., Stricker, G., Swanson, E. M., & Holekamp, K. E. (2016). Brain size predicts problem-solving ability in mammalian carnivores. *Proceedings of the National Academy of Sciences of the United States of America, 113*(9), 2532–2537. https://doi.org/10.1073/PNAS.1505913113/SUPPL_FILE/PNAS.1505913113.SM01.MP4

Bercovitch, F. B. (2020). A comparative perspective on the evolution of mammalian reactions to dead conspecifics. *Primates, 61*(1), 21–28. https://doi.org/10.1007/S10329-019-00722-3/FIGURES/2

Berg, J. K. (1983). Vocalizations and associated behaviors of the African elephant (*Loxodonta africana*) in captivity. *Zeitschrift Für Tierpsychologie, 65*(1), 1983.

Berger-Tal, O., & Saltz, D. (2016). *Conservation behavior: Applying behavioral ecology to wildlife conservation and management* (Vol. 21). Cambridge University Press.

Berger-Tal, O., Polak, T., Oron, A., Lubin, Y., Kotler, B. P., & Saltz, D. (2011). Integrating animal behavior and conservation biology: A conceptual framework. *Behavioral Ecology, 22*(2), 236–239. https://doi.org/10.1093/beheco/arq224

Berglund, A., Pilastro, A., & Bisazza, A. (1996). Armaments and ornaments: An evolutionary explanation of traits of dual utility. *Biological Journal of the Linnean Society, 58*(4), 385–399.

Bergman, T. J. (2003). Hierarchical classification by rank and kinship in baboons. *Science, 302*(5648), 1234–1236. https://doi.org/10.1126/science.1087513

Bester, T., Schmitt, M. H., & Shrader, A. M. (2023a). The deterrent effects of individual monoterpene odours on the dietary decisions of African elephants. *Animal Cognition, 26*(3), 1049–1063. https://doi.org/10.1007/S10071-023-01755-4/METRICS

Bester, T., Schmitt, M. H., & Shrader, A. M. (2023b). Generalist dietary responses to individual versus combined plant toxin odors: An African elephant study. *Behavioral Ecology, 34*(5), 816–830. https://doi.org/10.1093/BEHECO/ARAD059

Bhagwat, S. A., & Rutte, C. (2006). Sacred groves: Potential for biodiversity management. *Frontiers in Ecology and the Environment, 4*(10), 519–524. https://doi.org/10.1890/1540-9295(2006)4[519:SGPFBM]2.0.CO;2

Bibi, F., Kraatz, B., Craig, N., Beech, M., Schuster, M., & Hill, A. (2012). Early evidence for complex social structure in Proboscidea from a late Miocene trackway site in the United Arab Emirates. *Biology Letters, 8*(4), 670–673. https://doi.org/10.1098/rsbl.2011.1185

Blake, S. (2002). *The ecology of forest elephant distribution and its implications for conservation*. University of Edinburgh.

Blake, S., & Hedges, S. (2004). Sinking the flagship: The case of forest elephants in Asia and Africa. *Conservation Biology, 18*(5), 1191–1202. https://doi.org/10.1111/j.1523-1739.2004.01860.x

Blake, S., & Inkamba-Nkulu, C. (2004). Fruit, minerals, and forest elephant trails: Do all roads lead to Rome? *Biotropica, 36*(3), 392. https://doi.org/10.1646/03215

Blake, S., Bouché, P., Rasmussen, H., Orlando, A., & Douglas-Hamilton, I. (2003). The last Sahelian elephants: Ranging behavior, population status and recent history of the desert elephants of Mali. In Save the Elephants (August). https://doi.org/11858/00-001M-0000-002D-0B5D-1

Blake, S., Deem, S. L., Strindberg, S., Maisels, F., Momont, L., Isia, I.-B., Douglas-Hamilton, I., Karesh, W. B., & Kock, M. D. (2008). Roadless wilderness area determines forest elephant movements in the Congo Basin. *PLOS One, 3*(10), e3546. https://doi.org/10.1371/journal.pone.0003546

Blumstein, D. T., & Fernández-Juricic, E. (2010). *A primer of conservation behavior*. Sinauer Associates Inc.

Boateng, P. K. (2017). Land access, agricultural land use changes and narratives about land degradation in the savannahs of northeast ghana during the pre-colonial and colonial periods. *Social Sciences*, *6*(1), 35. https://doi.org/10.3390/socsci6010035

Boehm, C. (1999). *Hierarchy in the forest: The evolution of egalitarian behavior*. Harvard University Press.

Boettiger, A. N., Wittemyer, G., Starfield, R., Volrath, F., Douglas-Hamilton, I., & Getz, W. M. (2011). Inferring ecological and behavioral drivers of African elephant movement using a linear filtering approach. *Ecology*, *92*(8), 1648–1657. https://doi.org/10.1890/10-0106.1

Bond, M. L., Lee, D. E., Ozgul, A., & König, B. (2019). Fission–fusion dynamics of a mega-herbivore are driven by ecological, anthropogenic, temporal, and social factors. *Oecologia*, *191*(2), 335–347. https://doi.org/10.1007/S00442-019-04485-Y

Bonenfant, C., Gaillard, J. M., Coulson, T. I. M., Bianchet, M. F., Loison, A., Garel, M., Loe, L. E., Blanchard, P., Pettorelli, N., & Smith, N. O. (2009). Empirical evidence of density-dependence in populations of large herbivores. *Advances in Ecological Research*, *41*(09), 313–357. https://doi.org/10.1016/S0065-2504(09)00405-X

Bouley, D. M., Alarcón, C. N., Hildebrandt, T., & O'Connell-Rodwell, C. E. (2007). The distribution, density and three-dimensional histomorphology of Pacinian corpuscles in the foot of the Asian elephant (*Elephas maximus*) and their potential role in seismic communication. *Journal of Anatomy*, *211*(4), 428–435. https://doi.org/10.1111/J.1469-7580.2007.00792.X

Bowell, R. J., Warren, A., & Redmond, I. (2008). Formation of cave salts and utilization by elephants in the Mount Elgon region, Kenya. *Geological Society, London, Special Publications*, *113*(1), 63–79. https://doi.org/10.1144/gsl.sp.1996.113.01.06

Bowman, D. M. J. S., Balch, J., Artaxo, P., Bond, W. J., Cochrane, M. A., D'Antonio, C. M., Defries, R., Johnston, F. H., Keeley, J. E., Krawchuk, M. A., Kull, C. A., Mack, M., Moritz, M. A., Pyne, S., Roos, C. I., Scott, A. C., Sodhi, N. S., & Swetnam, T. W. (2011). The human dimension of fire regimes on Earth. *Journal of Biogeography*, *38*(12), 2223–2236. https://doi.org/10.1111/J.1365-2699.2011.02595.X

Bracis, C., & Mueller, T. (2017). Memory, not just perception, plays an important role in terrestrial mammalian migration. *Proceedings of the Royal Society B: Biological Sciences*, *284*(1855). https://doi.org/10.1098/rspb.2017.0449

Bradshaw, G. A., Schore, A. N., Brown, J. L., Poole, J. H., & Moss, C. J. (2005). Elephant breakdown. *Nature*, *433*(7028), 807. https://doi.org/10.1038/433807a

Brakes, P., Dall, S. R. X., Aplin, L. M., Bearhop, S., Carroll, E. L., Ciucci, P., Fishlock, V., Ford, J. K. B., Garland, E. C., Keith, S. A., McGregor, P. K., Mesnick, S. L., Noad, M. J., di Sciara, G. N., Robbins, M. M., Simmonds, M. P., Spina, F., Thornton, A., Wade, P. R., … Rutz, C. (2019). Animal cultures matter for conservation. *Science*, *363*(6431), 1032–1034. https://doi.org/10.1126/science.aaw3557

Brakes, P., Carroll, E. L., Dall, S. R. X., Keith, S. A., McGregor, P. K., Mesnick, S. L., Noad, M. J., Rendell, L., Robbins, M. M., Rutz, C., Thornton, A., Whiten, A., Whiting, M. J., Aplin, L. M., Bearhop, S., Ciucci, P., Fishlock, V., Ford, J. K. B., Notarbartolo Di Sciara, G., … Garland, E. C. (2021). A deepening understanding of animal culture suggests lessons for conservation. *Proceedings of the Royal Society B: Biological Sciences*, *288*(1949). https://doi.org/10.1098/RSPB.2020.2718

Brandt, A. L., Ishida, Y., Georgiadis, N. J., & Roca, A. L. (2012). Forest elephant mitochondrial genomes reveal that elephantid diversification in Africa tracked climate transitions. *Molecular Ecology*, *21*(5), 1175–1189. https://doi.org/10.1111/j.1365-294X.2012.05461.x

Brent, L. J. N., Franks, D. W., Foster, E. A., Balcomb, K. C., Cant, M. A., & Croft, D. P. (2015). Ecological knowledge, leadership, and the evolution of menopause in killer whales. *Current Biology*, *25*(6), 746–750. https://doi.org/10.1016/j.cub.2015.01.037

Brittain, S., Ngo Bata, M., De Ornellas, P., Milner-Gulland, E. J., & Rowcliffe, M. (2020). Combining local knowledge and occupancy analysis for a rapid assessment of the forest elephant Loxodonta cyclotis in Cameroon's timber production forests. *Oryx*, *54*(1),90–100. https://doi.org/10.1017/S0030605317001569

Brown, S., Massilani, D., Kozlikin, M. B., Shunkov, M. V., Derevianko, A. P., Stoessel, A., Jope-Street, B., Meyer, M., Kelso, J., Pääbo, S., Higham, T., & Douka, K. (2021). The earliest Denisovans and their cultural adaptation. *Nature Ecology & Evolution*, *6*(1), 28–35. https://doi.org/10.1038/s41559-021-01581-2

Buchholtz, E. K., Redmore, L., Fitzgerald, L. A., Stronza, A., Songhurst, A., & McCulloch, G. (2019). Temporal partitioning and overlapping use of a shared natural resource by people and elephants. *Frontiers in Ecology and Evolution*, *7*(April), 1–12. https://doi.org/10.3389/fevo.2019.00117

Buniyaadi, A., Taufique, S. K. T., & Kumar, V. (2020). Self-recognition in corvids: Evidence from the mirror-mark test in Indian house crows (*Corvus splendens*). *Journal of Ornithology*, *161*(2), 341–350. https://doi.org/10.1007/S10336-019-01730-2/FIGURES/2

Bunney, K., Bond, W. J., & Henley, M. (2017). Seed dispersal kernel of the largest surviving megaherbivore – The African savanna elephant. *Biotropica*, *49*(3), 395–401. https://doi.org/10.1111/btp.12423

Buss, I. O., & Smith, N. S. (1966). Observations on reproduction and breeding behavior of the African elephant. *The Journal of Wildlife Management*, *30*, 375–388.

Butz, R. J. (2009). Traditional fire management: Historical fire regimes and land use change in pastoral East Africa. *International Journal of Wildland Fire*, *18*(4), 442–450. https://doi.org/10.1071/WF07067

Byrne, R. W., Bates, L. A., & Moss, C. J. (2009). Elephant cognition in primate perspective. *Comparative Cognition and Behavior Reviews*, *4*, 65–79. https://doi.org/10.3819/ccbr.2009.40009

Campbell-Staton, S. C., Arnold, B. J., Gonçalves, D., Granli, P., Poole, J., Long, R. A., & Pringle, R. M. (2021). Ivory poaching and the rapid evolution of tusklessness in African elephants. *Science*, *374*(6566), 483–487. https://doi.org/10.1126/SCIENCE.ABE7389

Campos-Arceiz, A. (2009). Shit happens (to be useful)! Use of elephant dung as habitat by amphibians. *Biotropica*, *41*(4), 406–407. https://doi.org/10.1111/j.1744-7429.2009.00525.x

Campos-Arceiz, A., & Blake, S. (2011). Megagardeners of the forest – The role of elephants in seed dispersal. *Acta Oecologica*, *37*(6), 542–553. https://doi.org/10.1016/j.actao.2011.01.014

Campos-Arceiz, A., Larrinaga, A. R., Weerasinghe, U. R., Takatsuki, S., Pastorini, J., Leimgruber, P., Fernando, P., & Santamaría, L. (2008). Behavior rather than diet mediates seasonal differences in seed dispersal by Asian elephants. *Ecology*, *89*(10), 2684–2691. www.ncbi.nlm.nih.gov/pubmed/18959306

Campos-Arceiz, A., Traeholt, C., Jaffar, R., Santamaria, L., & Corlett, R. T. (2011). Asian tapirs are no elephants when it comes to seed dispersal. *Biotropica*, *44*(2), 220–227. https://doi.org/10.1111/j.1744-7429.2011.00784.x

Canney, S., & Ganame, N. (2012). Impact of cattle on elephant use of key dry-season water in the Gourma of Mali. *Pachyderm*, *51*, 75–77.

Cant, M. A., & Johnstone, R. A. (2008). Reproductive conflict and the separation of reproductive generations in humans. *Proceedings of the National Academy of Sciences of the United States of America*, *105*(14), 5332–5336. https://doi.org/10.1073/pnas.0711911105

Cantalapiedra, J. L., Sanisidro, O., Zhang, H., Alberdi, M. T., Prado, J. L., Blanco, F., & Saarinen, J. (2021). The rise and fall of proboscidean ecological diversity. *Nature Ecology & Evolution*, *5*(9), 1266–1272.

Cappozzo, H. L., Túnez, J. I., & Cassini, M. H. (2008). Sexual harassment and female gregariousness in the South American sea lion, *Otaria flavescens*. *Naturwissenschaften*, *95*(7), 625–630. https://doi.org/10.1007/s00114-008-0363-2

Caro, T. (2007). Behavior and conservation: A bridge too far? *Trends in Ecology and Evolution*, *22*(8), 394–400. https://doi.org/10.1016/j.tree.2007.06.003

Caro, T., & Sherman, P. W. (2011). Endangered species and a threatened discipline: Behavioural ecology. *Trends in Ecology and Evolution*, *26*(3), 111–118. https://doi.org/10.1016/j.tree.2010.12.008

Caro, T., & Sherman, P. W. (2012). Vanishing behaviors. *Conservation Letters*, *5*(3), 159–166. https://doi.org/10.1111/j.1755-263X.2012.00224.x

Caro, T., & Sherman, P. W. (2013). Eighteen reasons animal behaviourists avoid involvement in conservation. *Animal Behaviour*, *85*(2), 305–312. https://doi.org/10.1016/j.anbehav.2012.11.007

Caro, T., Graham, C. M., Stoner, C. J., & Vargas, J. K. (2004). Adaptive significance of antipredator behaviour in artiodactyls. *Animal Behaviour*, *67*(2), 205–228. https://doi.org/10.1016/j.anbehav.2002.12.007

Carter, N. H., Shrestha, B. K., Karki, J. B., Pradhan, N. M. B., & Liu, J. (2012). Coexistence between wildlife and humans at fine spatial scales. *Proceedings of the National Academy of Sciences of the United States of America*, *109*(38), 15360–15365. https://doi.org/10.1073/pnas.1210490109

Cerling, T. E., Wittemyer, G., Rasmussen, H. B., Vollrath, F., Cerling, C. E., Robinson, T. J., & Douglas-Hamilton, I. (2006). Stable isotopes in elephant hair document migration patterns and diet changes. *Proceedings of the National Academy of Sciences of the United States of America*, *103*(2), 371–373. https://doi.org/10.1073/pnas.0509606102

Cerling, T. E., Wynn, J. G., Andanje, S. A., Bird, M. I., Korir, D. K., Levin, N. E., MacE, W., MacHaria, A. N., Quade, J., & Remien, C. H. (2011). Woody cover and hominin environments in the past 6 million years. *Nature*, *476*(7358), 51–56. https://doi.org/10.1038/nature10306

Chalmeau, R., Lardeux, K., Brandibas, P., & Gallo, A. (1997). Cooperative problem solving by orangutans (*Pongo pygmaeus*). *International Journal of Primatology*, *18*(1), 23–32. https://doi.org/10.1023/A:1026337006136/METRICS

Chamaillé-Jammes, S., Valeix, M., & Fritz, H. (2007). Managing heterogeneity in elephant distribution: Interactions between elephant population density and surface-water availability. *Journal of Applied Ecology*, *44*(3), 625–633. https://doi.org/10.1111/j.1365-2664.2007.01300.x

Chamaillé-Jammes, S., Fritz, H., Valeix, M., Murindagomo, F., & Clobert, J. (2008). Resource variability, aggregation and direct density dependence in an open context: The local regulation of an African elephant population. *Journal of Animal Ecology*, *77*, 135–144. https://doi.org/10.1111/j.1365-2656.2007.01307.x

Chandrajith, R., Kudavidanage, E., Tobschall, H. J., & Dissanayake, C. B. (2009). Geochemical and mineralogical characteristics of elephant geophagic soils in Udawalawe National Park, Sri Lanka. *Environmental Geochemistry and Health*, *31*(3), 391–400. https://doi.org/10.1007/s10653-008-9178-5

Chapman, C. A., Wrangham, R. W., & Chapman, L. J. (1995). Ecological constraints on group size: An analysis of spider monkey and chimpanzee subgroups. *Behavioral Ecology and Sociobiology*, *36*(1), 59–70.

Charif, R. A., Ramey, R. R., Langbauer, W. R., Payne, K. B., Martin, R. B., & Brown, L. M. (2004). Spatial relationships and matrilineal kinship in African savanna elephant (*Loxodonta africana*) clans. *Behavioral Ecology and Sociobiology*, *57*(4), 327–338. https://doi.org/10.1007/s00265-004-0867-5

Chase, M. J., Schlossberg, S., Griffin, C. R., Bouché, P. J. C., Djene, S. W., Elkan, P. W., Ferreira, S., Grossman, F., Kohi, E. M., Landen, K., & Omondi, P. (2016). Continent-wide survey reveals massive decline in African savannah elephants. *PeerJ*, *4*(e2354), 1–24. https://doi.org/10.7717/peerj.2354

Chave, E., Edwards, K. L., Paris, S., Prado, N., Morfeld, K. A., & Brown, J. L. (2019). Variation in metabolic factors and gonadal, pituitary, thyroid, and adrenal hormones in association with musth in African and Asian elephant bulls. *General and Comparative Endocrinology*, *276*(February), 1–13. https://doi.org/10.1016/j.ygcen.2019.02.005

Cheah, C., & Yoganand, K. (2022). Recent estimate of Asian elephants in Borneo reveals a smaller population. *Wildlife Biology*, *2022*(2), e01024. https://doi.org/10.1002/wlb3.01024

Cheke, L. G., Loissel, E., & Clayton, N. S. (2012). How do children solve Aesop's fable? *PLOS One*, *7*(7). https://doi.org/10.1371/JOURNAL.PONE.0040574

Chelliah, K., & Sukumar, R. (2013). The role of tusks, musth and body size in male–male competition among Asian elephants, *Elephas maximus*. *Animal Behaviour*, *86*(6), 1207–1214. https://doi.org/10.1016/j.anbehav.2013.09.022

Chelliah, K., & Sukumar, R. (2015). Interplay of male traits, male mating strategies and female mate choice in the Asian elephant, *Elephas maximus*. *Behaviour*, *152*(7–8), 1113–1144. https://doi.org/10.1163/1568539X-00003271

Cheney, D. L. (1977). The acquisition of rank and the development of reciprocal alliances among free-ranging immature baboons. *Behavioral Ecology and Sociobiology*, *2*(3), 303–318.

Cheney, D. L., & Seyfarth, R. (1990). *How monkeys see the world: Inside the mind of another species*. University of Chicago Press.

Cheney, D. L., & Seyfarth, R. (2008). *Baboon metaphysics*. University of Chicago Press.

Cherney, M. D., Fisher, D. C., Auchus, R. J., Rountrey, A. N., Selcer, P., Shirley, E. A., Beld, S. G., Buigues, B., Mol, D., Boeskorov, G. G., Vartanyan, S. L., & Tikhonov, A. N. (2023). Testosterone histories from tusks reveal woolly mammoth musth episodes. *Nature*, 1–7. https://doi.org/10.1038/s41586-023-06020-9

Cherry, M. J., Warren, R. J., & Conner, L. M. (2017). Fire-mediated foraging tradeoffs in white-tailed deer. *Ecosphere*, *8*(4), e01784. https://doi.org/10.1002/ECS2.1784

Chiyo, P. I., Archie, E. A., Hollister-Smith, J. A., Lee, P. C., Poole, J. H., Moss, C. J., & Alberts, S. C. (2011a). Association patterns of African elephants in all-male groups: The role of age and genetic relatedness. *Animal Behaviour*, *81*(6), 1093–1099. https://doi.org/10.1016/j.anbehav.2011.02.013

Chiyo, P. I., Lee, P. C., Moss, C. J., Archie, E. A., Hollister-Smith, J. A., & Alberts, S. C. (2011b). No risk, no gain: Effects of crop raiding and genetic diversity on body size in male elephants. *Behavioral Ecology*, *22*(3), 552–558.

Chiyo, P. I., Moss, C. J., Archie, E. A., Hollister-Smith, J. A., & Alberts, S. C. (2011c). Using molecular and observational techniques to estimate the number and raiding patterns of crop-raiding elephants. *Journal of Applied Ecology*, *48*(3), 788–796. https://doi.org/10.1111/j.1365-2664.2011.01967.x

Chiyo, P. I., Moss, C. J., & Alberts, S. C. (2012). The influence of life history milestones and association networks on crop-raiding behavior in male African elephants. *PLOS One*, *7*(2), e31382. https://doi.org/10.1371/journal.pone.0031382

Choudhury, A. L., Choudhury, D. K., Desai, A., Duckworth, J. W., Easa, P. S., Johnsingh, A. J. T., Fernando, P., Hedges, S., Gunawardena, M., Kurt, F., Karanth, U., Lister, A., Menon, V., Riddle, H., Rübel, A., & Wikramanayake, E. (2008). Elephas maximus. IUCN Red List of Threatened Species. www.iucnredlist.org/details/7140/0

Chowdhury, S., Alam, S., Labi, M. M., Khan, N., Rokonuzzaman, M., Biswas, D., Tahea, T., Mukul, S. A., & Fuller, R. A. (2022). Protected areas in South Asia: Status and prospects. *Science of The Total Environment*, *811*, 152316. https://doi.org/10.1016/J.SCITOTENV.2021.152316

Clark, N. E., Boakes, E. H., McGowan, P. J. K., Mace, G. M., & Fuller, R. A. (2013). Protected areas in South Asia have not prevented habitat loss: A study using historical models of land-Use change. *PLOS One*, *8*(5), 1–7. https://doi.org/10.1371/journal.pone.0065298

Clift, P., & Guedes, J. (2021). *Monsoon rains, great rivers and the development of farming civilisations in Asia*. Cambridge University Press.

Clutton-Brock, T. H. (1984). Reproductive effort and terminal investment in iteroparous animals. *The American Naturalist*, *123*(2), 212–229. https://doi.org/10.1086/284198

Clutton-Brock, T. H. (1989). Female transfer and inbreeding avoidance in social mammals. *Nature*, *337*(6202), 70–72. https://doi.org/10.1038/337070a0

Clutton-Brock, T. H. (2017). Reproductive competition and sexual selection. *Philosophical Transactions of the Royal Society B*, *372*, 1–22. https://doi.org/10.1098/not

Clutton-Brock, T. H., & Lukas, D. (2012). The evolution of social philopatry and dispersal in female mammals. *Molecular Ecology*, *21*(3), 472–492. https://doi.org/10.1111/j.1365-294X.2011.05232.x

Cochrane, E. (2003). The need to be eaten: Balanites wilsoniana with and without elephant seed-dispersal. *Journal of Tropical Ecology*, *19*, 579–589.

Cohen, A. A. (2004). Female post-reproductive lifespan: A general mammalian trait. *Biological Reviews of the Cambridge Philosophical Society*, *79*(4), 733–750. www.ncbi.nlm.nih.gov/pubmed/15682868

Comstock, K. E., Georgiadis, N., Pecon-Slattery, J., Roca, A. L., Ostrander, E. A., O'Brien, S. J., & Wasser, S. K. (2002). Patterns of molecular genetic variation among African elephant populations. *Molecular Ecology*, *11*(12), 2489–2498. https://doi.org/10.1046/j.1365-294X.2002.01615.x

Corde, S. C., Hagen, R. L. Von, Kasaine, S., Mutwiwa, U. N., Amakobe, B., Githiru, M., & Schulte, B. A. (2024). Lunar phase as a dynamic landscape of fear factor affecting elephant crop raiding potential. *Ethology Ecology & Evolution*, *36*(3), 295-308. https://doi.org/10.1080/03949370.2023.2263406

Corson, C., Gruby, R., Witter, R., Hagerman, S., Suarez, D., Greenberg, S., Bourque, M., Grayh, N., & Campbell, L. M. (2014). Everyone's solution? Defining and redefining protected areas at the Convention on Biological Diversity. *Conservation and Society*, *12*(2), 190–202. www.jstor.org/stable/26393154

Costa, Y. T., & Thomaz, E. L. (2021). Management, sustainability and research perspective of prescribed fires in tropical parks. *Current Opinion in Environmental Science & Health*, *22*, 100257. https://doi.org/10.1016/J.COESH.2021.100257

Coyer, J. A. (1995). Use of a rock as an anvil for breaking scallops by the Yellowhead Wrasse, *Halichoeres garnoti* (*Labridae*). *Bulletin of Marine Science*, *57*(2), 548–549.

Croft, D. P., Brent, L. J. N., Franks, D. W., & Cant, M. A. (2015). The evolution of prolonged life after reproduction. *Trends in Ecology and Evolution*, *30*(7), 407–416. https://doi.org/10.1016/j.tree.2015.04.011

Croft, D. P., Johnstone, R. A., Ellis, S., Nattrass, S., Franks, D. W., Brent, L. J. N., Mazzi, S., Balcomb, K. C., Ford, J. K. B., & Cant, M. A. (2017). Reproductive conflict and the evolution of menopause in Killer Whales. *Current Biology*, *27*(2), 298–304. https://doi.org/10.1016/j.cub.2016.12.015

Cronon, W. (1996). The trouble with wilderness – Or, getting back to the wrong nature. *Environmental History*, *1*(1), 7–28. https://doi.org/10.2307/3985059

Crooks, K. R., Burdett, C. L., Theobald, D. M., King, S. R. B., Di Marco, M., Rondinini, C., & Boitani, L. (2017). Quantification of habitat fragmentation reveals extinction risk in terrestrial mammals. *Proceedings of the National Academy of Sciences of the United States of America*, *114*(29), 7635–7640. https://doi.org/10.1073/pnas.1705769114

Darden, S. K., & Watts, L. (2012). Male sexual harassment alters female social behaviour towards other females. *Biology Letters*, *8*(2), 186–188. https://doi.org/10.1098/rsbl.2011.0807

Darden, S. K, James, R., Ramnarine, I. W., & Croft, D. P. (2009). Social implications of the battle of the sexes: Sexual harassment disrupts female sociality and social recognition. *Proceedings of the Royal Society B: Biological Sciences*, *276*(1667), 2651–2656. https://doi.org/10.1098/rspb.2009.0087

Davidar, P., Arjunan, M., & Puyravaud, J. P. (2008). Why do local households harvest forest products? A case study from the southern Western Ghats, India. *Biological Conservation*, *141*(7), 1876–1884. https://doi.org/10.1016/j.biocon.2008.05.004

de Azevedo, C. S., & Young, R. J. (2021). Animal personality and conservation: Basics for inspiring new research. *Animals*, *11*(4), 1019. https://doi.org/10.3390/ANI11041019

de Beer, Y., & van Aarde, R. J. J. (2008). Do landscape heterogeneity and water distribution explain aspects of elephant home range in southern Africa's arid savannas? *Journal of Arid Environments*, *72*(11), 2017–2025. https://doi.org/10.1016/j.jaridenv.2008.07.002

de la Torre, J. A., Wong, E. P., Lechner, A. M., Zulaikha, N., Zawawi, A., Abdul-Patah, P., Saaban, S., Goossens, B., & Campos-Arceiz, A. (2021). There will be conflict – Agricultural landscapes are prime, rather than marginal, habitats for Asian elephants. *Animal Conservation*, *24*(5), 720–732. https://doi.org/10.1111/ACV.12668

de la Torre, J. A., Cheah, C., Lechner, A. M., Wong, E. P., Tuuga, A., Saaban, S., Goossens, B., & Campos-Arceiz, A. (2022). Sundaic elephants prefer habitats on the periphery of protected areas. *Journal of Applied Ecology*, *59*(12), 2947–2958. https://doi.org/10.1111/1365-2664.14286

De, R., Sharma, R., Davidar, P., Arumugam, N., Sedhupathy, A., Puyravaud, J. P., Selvan, K. M., Rahim, P. P. A., Udayraj, S., Parida, J., Digal, D. K., Kanagaraj, R., Kakati, K., Nigam, P., Williams, A. C., Habib, B., & Goyal, S. P. (2021). Pan-India population genetics signifies the importance of habitat connectivity for wild Asian elephant conservation. *Global Ecology and Conservation*, *32*, e01888. https://doi.org/10.1016/J.GECCO.2021.E01888

de Silva, S. (2010a). On predicting elephant population dynamics. *Gajah*, *33*, 12–16.

de Silva, S. (2010b). Acoustic communication in the Asian elephant, *Elephas maximus maximus*. *Behaviour*, *147*(7), 825–852. https://doi.org/10.1163/000579510X495762

de Silva, S., & Leimgruber, P. (2019). Demographic tipping points as early indicators of vulnerability for slow-breeding megafaunal populations. *Frontiers in Ecology and Evolution*, *7*. https://doi.org/10.3389/fevo.2019.171

de Silva, S., & Srinivasan, K. (2019). Revisiting social natures: People-elephant conflict and coexistence in Sri Lanka. *Geoforum*, *102*(March), 182–190. https://doi.org/10.1016/j.geoforum.2019.04.004

de Silva, S., & Wittemyer, G. (2012). A comparison of social organization in Asian elephants and African savannah elephants. *International Journal of Primatology*, *33*(5), 1125–1141. https://doi.org/10.1007/s10764-011-9564-1

de Silva, S., Ranjeewa, A. D. G., & Kryazhimskiy, S. (2011a). The dynamics of social networks among female Asian elephants. *BMC Ecology*, *11*, 17. https://doi.org/10.1186/1472-6785-11-17

de Silva, S., Ranjeewa, A. D. G., & Weerakoon, D. (2011b). Demography of Asian elephants (*Elephas maximus*) at Uda Walawe National Park, Sri Lanka based on identified individuals. *Biological Conservation*, *144*(5), 1742–1752. https://doi.org/10.1016/j.biocon.2011.03.011

de Silva, S., Webber, C. E., Weerathunga, U. S., Pushpakumara, T. V., Weerakoon, D. D. K., & Wittemyer, G. (2013). Demographic variables for wild Asian elephants using longitudinal observations. *PLOS One*, *8*(12), e82788. https://doi.org/10.1371/journal.pone.0082788

de Silva, S., Weerathunga, U., & Pushpakumara, T. (2014). Morphometrics and behavior of a wild Asian elephant exhibiting disproportionate dwarfism. *BMC Research Notes*, *7*(1), 1–8. https://doi.org/10.1101/001594

de Silva, S., Schmid, V., & Wittemyer, G. (2017). Fission-fusion processes weaken dominance networks among female Asian elephants in a productive habitat. *Behavioral Ecology*, *28*(1), 243–252.

de Silva, S., Ruppert, K., Knox, J., Davis, E. O., Weerathunga, U. S., & Glikman, J. A. (2023a). Experiences and emotional responses of farming communities living with Asian Elephants in Southern Sri Lanka: Living with elephants in Sri Lanka. *Trees, Forests and People*, *14*, 100441. https://doi.org/10.1016/j.tfp.2023.100441

de Silva, S., Wu, T., Nyhus, P., Weaver, A., Thieme, A., Johnson, J., Wadey, J., Mossbrucker, A., Vu, T., Neang, T., Chen, B. S., Songer, M., & Leimgruber, P. (2023b). Land-use change is associated with multi-century loss of elephant ecosystems in Asia. *Scientific Reports*, *13*(1), 1–14. https://doi.org/10.1038/s41598-023-30650-8

de Waal, F. B. M. (2014). Natural normativity: The "is" and "ought" of animal behavior. *Behaviour*, *151*(2–3), 185–204. https://doi.org/10.1163/1568539x-00003146

de Waal, F. B. M. (2019). Fish, mirrors, and a gradualist perspective on self-awareness. *PLOS Biology*, *17*(2), e3000112. https://doi.org/10.1371/JOURNAL.PBIO.3000112

Debruyne, R. (2005). A case study of apparent conflict between molecular phylogenies: The interrelationships of African elephants. *Cladistics*, *21*(1), 31–50. https://doi.org/10.1111/j.1096-0031.2004.00044.x

Deecke, V. B., Ford, J. K. B., & Spong, P. (2000). Dialect change in resident killer whales: Implications for vocal learning and cultural transmission. *Animal Behaviour*, *60*(5), 629–638. https://doi.org/10.1006/ANBE.2000.1454

Desert Elephant Conservation Report. (2015). www.desertelephantconservation.org/NewsAndReports.html

Diagne, C., Leroy, B., Vaissière, A. C., Gozlan, R. E., Roiz, D., Jarić, I., Salles, J. M., Bradshaw, C. J. A., & Courchamp, F. (2021). High and rising economic costs of biological invasions worldwide. *Nature*, *592*(7855), 571–576. https://doi.org/10.1038/s41586-021-03405-6

Díaz, S., Settele, J., Brondízio, E. S., Ngo, H. T., Agard, J., Arneth, A., Balvanera, P., Brauman, K. A., Butchart, S. H. M., Chan, K. M. A., Lucas, A. G., Ichii, K., Liu, J., Subramanian, S. M., Midgley, G. F., Miloslavich, P., Molnár, Z., Obura, D., Pfaff, A., … Zayas, C. N. (2019). Pervasive human-driven decline of life on Earth points to the need for transformative change. *Science*, *366*(6471). https://doi.org/10.1126/science.aax3100

Dindo, M., Whiten, A., & de Waal, F. B. M. (2009). In-group conformity sustains different foraging traditions in Capuchin monkeys (*Cebus apella*). *PLOS One*, *4*(11). https://doi.org/10.1371/journal.pone.0007858

Dobzhansky, T. (1937). *Genetics and the origin of species*. Columbia University Press.

Dobzhansky, T. (1973). Nothing in biology makes sense except in the light of evolution. *American Biology Teacher*, *35*(3), 125–129.

Dong, S., Lin, T., Nieh, J. C., & Tan, K. (2023). Social signal learning of the waggle dance in honey bees. *Science*, *379*(6636), 1015–1018.

Doughty, C. E., Roman, J., Faurby, S., Wolf, A., Haque, A., Bakker, E. S., Malhi, Y., Dunning, J. B., & Svenning, J.-C. (2016). Global nutrient transport in a world of giants. *Proceedings of the National Academy of Sciences of the United States of America*, *113*(4), 868–873. https://doi.org/10.1073/pnas.1502549112

Douglas-Hamilton, I. (1972). *On the ecology and behaviour of the African elephant*. Oxford University.

Douglas-Hamilton, I., Krink, T., & Vollrath, F. (2005). Movements and corridors of African elephants in relation to protected areas. *Naturwissenschaften*, *92*(4), 158–163. https://doi.org/10.1007/s00114-004-0606-9

Drea, C. M. (2006). Studying primate learning in group contexts: Tests of social foraging, response to novelty, and cooperative problem solving. *Methods*, *38*(3), 162–177. https://doi.org/10.1016/J.YMETH.2005.12.001

Drea, C. M., & Carter, A. N. (2009). Cooperative problem solving in a social carnivore. *Animal Behaviour*, *78*(4), 967–977. https://doi.org/10.1016/J.ANBEHAV.2009.06.030

Dublin, H. T. (1983). Cooperation and reproductive competition among female African elephants. In S. Wasser (Ed.), *Social behavior of female vertebrates* (pp. 291–313). Academic Press.

Ducatez, S., Clavel, J., & Lefebvre, L. (2015). Ecological generalism and behavioural innovation in birds: Technical intelligence or the simple incorporation of new foods? *Journal of Animal Ecology*, *84*(1), 79–89. https://doi.org/10.1111/1365-2656.12255

Dumonceaux, G. A. (2006). Digestive system. In M. E. Fowler, & S. K. Mikota (Eds.), *Biology, medicine, and surgery of elephants* (pp. 299–307). Blackwell Publishers.

Dunkin, R. C., Wilson, D., Way, N., Johnson, K., & Williams, T. M. (2013). Climate influences thermal balance and water use in African and Asian elephants: Physiology can predict drivers of elephant distribution. *Journal of Experimental Biology*, *216*(15), 2939–2952. https://doi.org/10.1242/jeb.080218

Dupke, C., Bonenfant, C., Reineking, B., Hable, R., Zeppenfeld, T., Ewald, M., & Heurich, M. (2017). Habitat selection by a large herbivore at multiple spatial and temporal scales is primarily governed by food resources. *Ecography*, *40*(8), 1014–1027. https://doi.org/10.1111/ECOG.02152

Duporge, I., Finerty, G. E., Ihwagi, F., Lee, S., Wathika, J., Wu, Z., Macdonald, D. W., & Wang, T. (2022). A satellite perspective on the movement decisions of African elephants in relation to nomadic pastoralists. *Remote Sensing in Ecology and Conservation*, *8*(6), 841–854. https://doi.org/10.1002/rse2.285

Duquette, C. A., Loss, S. R., & Hovick, T. J. (2021). A meta-analysis of the influence of anthropogenic noise on terrestrial wildlife communication strategies. *Journal of Applied Ecology*, *58*(6), 1112–1121. https://doi.org/10.1111/1365-2664.13880

Eakin, H., DeFries, R., Kerr, S., Lambin, E. F., Liu, J., Marcotullio, P. J., Messerli, P., Reenberg, A., Rueda, X., Swaffield, S. R., & Wicke, B. (2014). Significance of telecoupling for exploration of land-use change. In Karen Ching-Yee Seng, & Anette Reenberg (Eds.), *Rethinking global landuse in an urban era* (pp. 141–161). MIT Press. https://asu.pure.elsevier.com/en/publications/significance-of-telecoupling-for-exploration-of-land-use-change

Eggert, L. S., Rasner, C. A, & Woodruff, D. S. (2002). The evolution and phylogeography of the African elephant inferred from mitochondrial DNA sequence and nuclear microsatellite markers. *Proceedings of the Royal Society B: Biological Sciences*, *269*(1504), 1993–2006. https://doi.org/10.1098/rspb.2002.2070

Eisenberg, J. F, Mckay, G. M., & Jainudeen, M. R. (1971). Reproductive behavior of the Asiatic elephant (*Elephas maximus*). *Behaviour*, *38*(3), 193–225.

Eisenberg, J. F., McKay, G. M., & Seidensticker, J. (1990). *Asian elephants*. Friends of the National Zoo and National Zoological Park.

Elith, J., Phillips, S. J., Hastie, T., Dudík, M., Chee, Y. E., & Yates, C. J. (2011). A statistical explanation of MaxEnt for ecologists. *Diversity and Distributions*, *17*(1), 43–57. https://doi.org/10.1111/j.1472-4642.2010.00725.x

Ellis, E. C., Goldewijk, K. K., Siebert, S., Lightman, D., & Ramankutty, N. (2010). Anthropogenic transformation of the biomes, 1700 to 2000. *Global Ecology and Biogeography*, *19*(5), 589–606. https://doi.org/10.1111/j.1466-8238.2010.00540.x

Ellis, S., Franks, D. W., Nattrass, S., Currie, T. E., Cant, M. A., Giles, D., Balcomb, K. C., & Croft, D. P. (2018). Analyses of ovarian activity reveal repeated evolution of post-reproductive lifespans in toothed whales. *Scientific Reports*, *8*(1), 1–10. https://doi.org/10.1038/s41598-018-31047-8

Ellis, E. C., Gauthier, N., Goldewijk, K. K., Bird, R. B., Boivin, N., Díaz, S., Fuller, D. Q., Gill, J. L., Kaplan, J. O., Kingston, N., Locke, H., McMichael, C. N. H., Ranco, D., Rick, T. C., Rebecca Shaw, M., Stephens, L., Svenning, J. C., & Watson, J. E. M. (2021). People have shaped most of terrestrial nature for at least 12,000 years. *Proceedings of the National Academy of Sciences of the United States of America*, *118*(17), 1–8. https://doi.org/10.1073/pnas.2023483118

Emlen, S. T. (1982). The evolution of helping. I. an ecological constraints model. *American Naturalist*, *119*, 29–39.

Emlen, S. T. (1984). Cooperative breeding in birds and mammals. In J. R. Krebs, & N. B. Davies (Eds.), *Behavioural ecology* (pp. 305–339). Blackwell Scientific.

Emlen, S. T., & Oring, L. W. (1977). Ecology, sexual selection, and the evolution of mating systems. *Science*, *197*(4300), 215–223.

Eschen, R., Beale, T., Bonnin, J. M., Constantine, K. L., Duah, S., Finch, E. A., Makale, F., Nunda, W., Ogunmodede, A., Pratt, C. F., Thompson, E., Williams, F., Witt, A., & Taylor, B. (2021). Towards estimating the economic cost of invasive alien species to African crop and livestock production. *CABI Agriculture and Bioscience*, *2*(1), 1–18. https://doi.org/10.1186/s43170-021-00038-7

Evans, A. R., Jones, D., Boyer, A. G., Brown, J. H., Costa, D. P., Ernest, S. K. M., Fitzgerald, E. M. G., Fortelius, M., Gittleman, J. L., Hamilton, M. J., Harding, L. E., Lintulaakso, K., Lyons, S. K., Okie, J. G., Saarinen, J. J., Sibly, R. M., Smith, F. A., Stephens, P. R., Theodor, J. M., & Uhen, M. D. (2012). The maximum rate of mammal evolution. *Proceedings of the National Academy of Sciences of the United States of America*, *109*(11), 4187–4190. https://doi.org/10.1073/pnas.1120774109

Evans, K. E., & Harris, S. (2008). Adolescence in male African elephants, *Loxodonta africana*, and the importance of sociality. *Animal Behaviour*, *76*, 779–787. https://doi.org/10.1016/j.anbehav.2008.03.019

Evans, L. J., Asner, G. P., & Goossens, B. (2018). Protected area management priorities crucial for the future of Bornean elephants. *Biological Conservation*, *221*, 365–373. https://doi.org/https://doi.org/10.1016/j.biocon.2018.03.015

Evans, L. J., Goossens, B., Davies, A. B., Reynolds, G., & Asner, G. P. (2020). Natural and anthropogenic drivers of Bornean elephant movement strategies. *Global Ecology and Conservation*, *22*, e00906. https://doi.org/https://doi.org/10.1016/j.gecco.2020.e00906

Fa, J. E., Watson, J. E. M., Leiper, I., Potapov, P., Evans, T. D., Burgess, N. D., Molnár, Z., Fernández-Llamazares, Á., Duncan, T., Wang, S., Austin, B. J., Jonas, H., Robinson, C. J., Malmer, P., Zander, K. K., Jackson, M. V., Ellis, E., Brondizio, E. S., & Garnett, S. T. (2020). Importance of indigenous peoples' lands for the conservation of intact forest landscapes. *Frontiers in Ecology and the Environment*, *18*(3), 135–140. https://doi.org/10.1002/fee.2148

FAO. (2010). Global Forest Resources Assessment 2010.

Fernando, C., Weston, M. A., Corea, R., Pahirana, K., & Rendall, A. R. (2022). Asian elephant movements between natural and human-dominated landscapes mirror patterns of crop damage in Sri Lanka. *Oryx*, *57*(4), 481–488. https://doi.org/10.1017/S0030605321000971

Fernando, P., & Lande, R. (2000). Molecular genetic and behavioral analysis of social organization in the Asian elephant (*Elephas maximus*). *Behavioral Ecology and Sociobiology*, *48*(1), 84–91. https://doi.org/10.1007/s002650000218

Fernando, P., Wikramanayake, E. D., Janaka, H. K., Jayasinghe, L. K. a., Gunawardena, M., Kotagama, S. W., Weerakoon, D., & Pastorini, J. (2008). Ranging behavior of the Asian elephant in Sri Lanka. *Mammalian Biology – Zeitschrift Für Säugetierkunde*, *73*(1), 2–13. https://doi.org/10.1016/j.mambio.2007.07.007

Fernando, P., Leimgruber, P., Prasad, T., & Pastorini, J. (2012). Problem-elephant translocation: Translocating the problem and the elephant? *PLOS One*, *7*(12), e50917. https://doi.org/10.1371/journal.pone.0050917

Fernando, P., de Silva, M. K. C. R., Jayasinghe, L. K. A., Janaka, H. K., & Pastorini, J. (2021). First country-wide survey of the endangered Asian elephants: Towards better conservation and management in Sri Lanka. *Oryx*, *55*(1), 46–55. https://doi.org/10.1017/s0030605318001254

Festa-Bianchet, M. (2018). Learning to migrate. *Science*, *361*(6406), 972–973. https://doi.org/10.1126/science.aau6835

Fishlock, V. (2010). Bai use in forest elephants (*Loxodonta africana cyclotis*): Ecology, sociality & risk.

Fishlock, V., & Lee, P. C. (2013). Forest elephants: Fission–fusion and social arenas. *Animal Behaviour*, *85*(2), 357–363. https://doi.org/10.1016/j.anbehav.2012.11.004

Fishlock, V., Lee, P. C., & Breuer, T. (2008). Quantifying forest elephant social structure in Central African bai environments. *Pachyderm*, *44*, 19–28.

Fishlock, V., Caldwell, C., & Lee, P. C. (2016). Elephant resource-use traditions. *Animal Cognition*, *19*(2), 429–433. https://doi.org/10.1007/s10071-015-0921-x

Flint, B. F., Hawley, D. M., & Alexander, K. A. (2016). Do not feed the wildlife: Associations between garbage use, aggression, and disease in banded mongooses (*Mungos mungo*). *Ecology and Evolution*, *6*(16), 5932–5939. https://doi.org/10.1002/ece3.2343

Flower, S. S. (1943). Notes on age at sexual maturity, gestation period and growth of the Indian elephant, *Elephas maximus*. *Journal of Zoology*, *113*, 21–26.

Foerder, P., Galloway, M., Barthel, T., Moore, D. E., & Reiss, D. (2011). Insightful problem solving in an Asian elephant. *PLOS One*, *6*(8), e23251. https://doi.org/10.1371/journal.pone.0023251

Foley, A. M., Hewitt, D. G., DeYoung, R. W., Schnupp, M. J., Hellickson, M. W., & Lockwood, M. A. (2018). Reproductive effort and success of males in scramble-competition polygyny: Evidence for trade-offs between foraging and mate search. *Journal of Animal Ecology*, *87*(6), 1600–1614. https://doi.org/10.1111/1365-2656.12893

Foley, C., Pettorelli, N., & Foley, L. (2008). Severe drought and calf survival in elephants. *Biology Letters*, *4*(5), 541–544.

Foley, J. A., Defries, R., Asner, G. P., Barford, C., Bonan, G., Carpenter, S. R., Chapin, F. S., Coe, M. T., Daily, G. C., Gibbs, H. K., Helkowski, J. H., Holloway, T., Howard, E. A., Kucharik, C. J., Monfreda, C., Patz, J. A., Prentice, I. C., Ramankutty, N., & Snyder, P. K. (2005). Global consequences of land use. *Science*, *309*(5734), 570–574. https://doi.org/10.1126/science.1111772

Foster, E. A., Franks, D. W., Mazzi, S., Darden, S. K., Balcomb, K. C., Ford, J. K. B., & Croft, D. P. (2012). Adaptive prolonged postreproductive life span in killer whales. *Science*, *337*(6100), 1313. https://doi.org/10.1126/science.1224198

Found, R., & St. Clair, C. C. (2016). Behavioural syndromes predict loss of migration in wild elk. *Animal Behaviour*, *115*(5), 35–46. https://doi.org/10.1016/J.ANBEHAV.2016.02.007

Fowler, C. W., & Smith, T. (1973). Characterizing stable populations: An application to the African elephant population. *The Journal of Wildlife Management*, *37*(4), 513–523.

Fox, J., Truong, D. M., Rambo, A. T., Tuyen, N. P., Cuc, L. T., & Leisz, S. (2000). Shifting cultivation: A new old paradigm for managing tropical forests. *BioScience*, *50*(6), 521–528. https://doi.org/10.1641/0006-3568(2000)050[0521:SCANOP]2.0.CO;2

Freckleton, R. P., Watkinson, A. R., Green, R. E., William, J., Freckleton, R. P., Watkinson, A. R., & Greent, R. E. (2006). Census error and the detection of density dependence. *Journal of Animal Ecology*, *75*(4), 837–851.

Freeberg, T. M., Dunbar, R. I. M., & Ord, T. J. (2012). Social complexity as a proximate and ultimate factor in communicative complexity. *Philosophical Transactions of the Royal Society B: Biological Sciences*, *367*(1597), 1785–1801. https://doi.org/10.1098/RSTB.2011.0213

Fretwell, S. D., & Lucas, H. L. J. (1969). On territorial behavior and other factors influencing habitat distribution in birds. *Acta Biotheoretica*, *19*, 16–36.

Fuhlendorf, S. D., Harrell, W. C., Engle, D. M., Hamilton, R. G., Davis, C. A., & Leslie, D. M. (2006). Should heterogeneity be the basis for conservation? Grassland bird response to fire and grazing. *Ecological Applications*, *16*(5), 1706–1716. https://esajournals.onlinelibrary.wiley.com/doi/abs/10.1890/1051-0761(2006)016[1706:SHBTBF]2.0.CO;2

Fuhlendorf, S. D., Engle, D. M., Kerby, J., & Hamilton, R. (2009). Pyric herbivory: Rewilding landscapes through the recoupling of fire and grazing. *Conservation Biology*, *23*(3), 588–598. https://doi.org/10.1111/J.1523-1739.2008.01139.X

Fukami, T., & Nakajima, M. (2011). Community assembly: Alternative stable states or alternative transient states? *Ecology Letters*, *14*(10), 973–984. https://doi.org/10.1111/j.1461-0248.2011.01663.x

Furuichi, T., & Ihobe, H. (1994). Variation in male relationships in bonobos and chimpanzees. *Behaviour*, *130*, 211–228.

Galanti, V., Preatoni, D., Martinoli, A., Wauters, L. A., & Tosi, G. (2006). Space and habitat use of the African elephant in the Tarangire-Manyara ecosystem, Tanzania: Implications for conservation. *Mammalian Biology*, *71*(2), 99–114. https://doi.org/10.1016/j.mambio.2005.10.001

Galetti, M., Pedrosa, F., Keuroghlian, A., & Sazima, I. (2016). Liquid lunch–vampire bats feed on invasive feral pigs and other ungulates. *Frontiers in Ecology and the Environment*, *14*, 505–506.

Galetti, M., Moleón, M., Jordano, P., Pires, M. M., Guimarães, P. R., Pape, T., Nichols, E., Hansen, D., Olesen, J. M., Munk, M., de Mattos, J. S., Schweiger, A. H., Owen-Smith, N., Johnson, C. N., Marquis, R. J., & Svenning, J. C. (2018). Ecological and evolutionary legacy of megafauna extinctions. *Biological Reviews*, *93*(2), 845–862. https://doi.org/10.1111/brv.12374

Gallagher, A. J., Hammerschlag, N., Cooke, S. J., Costa, D. P., & Irschick, D. J. (2015). Evolutionary theory as a tool for predicting extinction risk. *Trends in Ecology and Evolution*, *30*(2), 61–65. https://doi.org/10.1016/j.tree.2014.12.001

Ganswindt, A., Rasmussen, H. B., Heistermann, M., & Hodges, K. J. (2005). The sexually active states of free ranging male African elephants (*Loxodonta africana*): Defining musth and non-musth using endocrinology, physical signals, and behvior. *Hormones and Behavior*, *47*, 83–91.

Gardner, S. L., & Campbell, M. L. (1992). Parasites as probes for biodiversity. *Journal of Parasitology*, *78*(4), 596–600.

Garland, E. C., Goldizen, A. W., Rekdahl, M. L., Constantine, R., Garrigue, C., Hauser, N. D., Poole, M. M., Robbins, J., & Noad, M. J. (2011). Dynamic horizontal cultural transmission of humpback whale song at the ocean basin scale. *Current Biology*, *21*(8), 687–691. https://doi.org/10.1016/j.cub.2011.03.019

Garstang, M. (2004). Long-distance, low-frequency elephant communication. *Journal of Comparative Physiology. A, Neuroethology, Sensory, Neural, and Behavioral Physiology*, *190*(10), 791–805. https://doi.org/10.1007/s00359-004-0553-0

Garstang, M., Davis, R. E., Leggett, K., Frauenfeld, O. W., Greco, S., Zipser, E., & Peterson, M. (2014). Response of African elephants (*Loxodonta africana*) to seasonal changes in rainfall. *PLOS One*, *9*(10), e108736. https://doi.org/10.1371/journal.pone.0108736

Gaynor, K. M., Branco, P. S., Long, R. A., Gonçalves, D. D., Granli, P. K., & Poole, J. H. (2018a). Effects of human settlement and roads on diel activity patterns of elephants (*Loxodonta africana*). *African Journal of Ecology*, *56*(4), 872–881. https://doi.org/10.1111/AJE.12552

Gaynor, K. M., Hojnowski, C. E., Carter, N. H., & Brashares, J. S. (2018b). The influence of human disturbance on wildlife nocturnality. *Science*, *360*(6394), 1232–1235.

Gaynor, K. M., Brown, J. S., Middleton, A. D., Power, M. E., & Brashares, J. S. (2019). Landscapes of Fear: Spatial patterns of risk perception and response. *Trends in Ecology and Evolution*, *34*(4), 355–368. https://doi.org/10.1016/j.tree.2019.01.004

Gaynor, K. M., Cherry, M. J., Gilbert, S. L., Kohl, M. T., Larson, C. L., Newsome, T. M., Prugh, L. R., Suraci, J. P., Young, J. K., & Smith, J. A. (2021). An applied ecology of fear framework: Linking theory to conservation practice. *Animal Conservation*, *24*(3), 308–321. https://doi.org/10.1111/ACV.12629

Geffroy, B., Samia, D. S. M., Bessa, E., & Blumstein, D. T. (2015). How nature-based tourism might increase prey vulnerability to predators. *Trends in Ecology & Evolution*, *30*(12), 755–765. https://doi.org/10.1016/J.TREE.2015.09.010

George, E. A., Thulasi, N., Kohl, P. L., Suresh, S., Rutschmann, B., & Brockmann, A. (2021). Distance estimation by Asian honey bees in two visually different landscapes. *Journal of Experimental Biology*, *224*(9), jeb242404. https://doi.org/10.1242/JEB.242404/238110

Gniadek, S. J., & Kendall, K. C. (1998). A summary of bear management in Glacier National Park, Montana, 1960–1994. In S. D. Miller, & H. V Reynolds (Eds.), *Ursus* (pp. 155–159). www.jstor.org/stable/3873122

Göbbel, L., Fischer, M. S., Smith, T. D., Wible, J. R., & Bhatnagar, K. P. (2004). The vomeronasal organ and associated structures of the fetal African elephant, *Loxodonta africana* (Proboscidea, Elephantidae). *Acta Zoologica*, *85*(1), 41–52. https://doi.org/10.1111/j.0001-7272.2004.00156.x

Gobetz, K. E., & Bozarth, S. R. (2001). Implications for late pleistocene mastodon diet from opal phytoliths in tooth calculus. *Quaternary Research2*, *55*(2), 115–122. https://doi.org/10.1006/qres.2000.2207

Gobush, K., Kerr, B., & Wasser, S. (2009). Genetic relatedness and disrupted social structure in a poached population of African elephants. *Molecular Ecology*, *18*(4), 722–734. https://doi.org/10.1111/j.1365-294X.2008.04043.x

Goldenberg, S. Z., & Wittemyer, G. (2017). Orphaned female elephant social bonds reflect lack of access to mature adults. *Scientific Reports*, *7*(1), 1–7. https://doi.org/10.1038/s41598-017-14712-2

Goldenberg, S. Z., & Wittemyer, G. (2018). Orphaning and natal group dispersal are associated with social costs in female elephants. *Animal Behaviour*, *143*, 1–8. https://doi.org/10.1016/j.anbehav.2018.07.002

Goldenberg, S. Z., & Wittemyer, G. (2020). Elephant behavior toward the dead: A review and insights from field observations. *Primates*, *61*(1), 119–128. https://doi.org/10.1007/S10329-019-00766-5

Goldenberg, S. Z., de Silva, S., Rasmussen, H. B., Douglas-Hamilton, I., & Wittemyer, G. (2014). Controlling for behavioural state reveals social dynamics among male African elephants, *Loxodonta africana*. *Animal Behaviour*, *95*, 111–119. https://doi.org/10.1016/j.anbehav.2014.07.002

Goldenberg, S. Z, Douglas-Hamilton, I., & Wittemyer, G. (2016). Vertical transmission of social roles drives resilience to poaching in elephant networks. *Current Biology*, *26*(1), 75–79. https://doi.org/10.1016/j.cub.2015.11.005

Goldenberg, S. Z., Douglas-Hamilton, I., & Wittemyer, G. (2018). Inter-generational change in african elephant range use is associated with poaching risk, primary productivity and adult mortality. *Proceedings of the Royal Society B: Biological Sciences*, *285*(1879). https://doi.org/10.1098/rspb.2018.0286

Goldenberg, S. Z., Parker, J. M., Chege, S. M., Greggor, A. L., Hunt, M., Lamberski, N., Leigh, K. A., Nollens, H. H., Ruppert, K. A., Thouless, C., Wittemyer, G., & Owen, M. A. (2022). Revisiting the 4 R's: Improving post-release outcomes for rescued mammalian wildlife by fostering behavioral competence during rehabilitation. *Frontiers in Conservation Science*, *3*, 910358. https://doi.org/10.3389/fcosc.2022.910358

Goossens, B., Sharma, R., Othman, N., Kun-Rodrigues, C., Sakong, R., Ancrenaz, M., Ambu, L. N., Jue, N. K., O'Neill, R. J., Bruford, M. W., & Chikhi, L. (2016). Habitat fragmentation and genetic diversity in natural populations of the Bornean elephants: Implications for conservation. *Biological Conservation*, *196*, 80–92. https://doi.org/10.1016/J.BIOCON.2016.02.008

Gough, K. F., & Kerley, G. I. H. (2006). Demography and population dynamics in the elephants *Loxodonta africana* of Addo Elephant National Park, South Africa: Is there evidence of density dependent regulation? *Oryx*, *40*(4), 434–441. https://doi.org/10.1017/S0030605306001189

Graham, M. D., Douglas-Hamilton, I., Adams, W. M., & Lee, P. C. (2009). The movement of African elephants in a human-dominated land-use mosaic. *Animal Conservation*, *12*(5), 445–455. https://doi.org/10.1111/j.1469-1795.2009.00272.x

Grand, A. P., Kuhar, C. W., Leighty, K. A., Bettinger, T. L., & Laudenslager, M. L. (2012). Using personality ratings and cortisol to characterize individual differences in African Elephants (*Loxodonta africana*). *Applied Animal Behaviour Science*, *142*(1–2), 69–75. https://doi.org/10.1016/J.APPLANIM.2012.09.002

Greenwood, P. (1980). Mating systems, philopatry and dispersal in birds and mammals. *Animal Behaviour*, *28*(4), 1140–1162.

Greggor, A. L., Berger-Tal, O., Blumstein, D. T., Angeloni, L., Bessa-Gomes, C., Blackwell, B. F., St Clair, C. C., Crooks, K., de Silva, S., Fernández-Juricic, E., Goldenberg, S. Z., Mesnick, S. L., Owen, M., Price, C. J., Saltz, D., Schell, C. J., Suarez, A. V., Swaisgood, R. R., Winchell, C. S., & Sutherland, W. J. (2016). Research priorities from animal behaviour for maximising conservation progress. *Trends in Ecology and Evolution*, *31*(12). https://doi.org/10.1016/j.tree.2016.09.001

Gregory, N. C., Sensenig, R. L., & Wilcove, D. S. (2010). Effects of controlled fire and livestock grazing on bird communities in East African savannas. *Conservation Biology*, *24*(6), 1606–1616. https://doi.org/10.1111/J.1523-1739.2010.01533.X

Groom, R. J., & Western, D. (2013). Impact of land subdivision and sedentarization on wildlife in Kenya's Southern rangelands. *Rangeland Ecology & Management*, *66*(1), 1–9. https://doi.org/10.2111/rem-d-11-00021.1

Gross, E. M., Drouet-Hoguet, N., Subedi, N., & Gross, J. (2017). The potential of medicinal and aromatic plants (MAPs) to reduce crop damages by Asian Elephants (*Elephas maximus*). *Crop Protection*, *100*, 29–37. https://doi.org/10.1016/j.cropro.2017.06.002

Gross, E. M., Lahkar, B. P., Subedi, N., Nyirenda, V. R., Lichtenfeld, L. L., & Jakoby, O. (2018). Seasonality, crop type and crop phenology influence crop damage by wildlife herbivores in Africa and Asia. *Biodiversity and Conservation*, *27*(8), 2029–2050. https://doi.org/10.1007/s10531-018-1523-0

Guimarães, P. R., Galetti, M., & Jordano, P. (2008). Seed dispersal anachronisms: Rethinking the fruits extinct megafauna ate. *PLOS One*, *3*(3), e1745. https://doi.org/10.1371/journal.pone.0001745

Gulati, S., Karanth, K. K., Le, N. A., & Noack, F. (2021). Human casualties are the dominant cost of human–wildlife conflict in India. *Proceedings of the National Academy of Sciences of the United States of America*, *118*(8), e1921338118. https://doi.org/10.1073/pnas.1921338118

Guldemond, R. A. R., Purdon, A., & Van Aarde, R. J. (2017). A systematic review of elephant impact across Africa. *PLOS One*, *12*(6), e0178935. https://doi.org/10.1371/journal.pone.0178935

Gunn, J., Hawkins, D., Barnes, R. F. W., Mofulu, F., Grant, R. A., & Norton, G. W. (2014). The influence of lunar cycles on crop-raiding elephants; evidence for risk avoidance. *African Journal of Ecology*, *52*(2), 129–137. https://doi.org/10.1111/AJE.12091

Gunther, K. A., Haroldson, M. A., Frey, K., Cain, S. L., Copeland, J., & Schwartz, C. C. (2004a). Grizzly bear-human conflicts in the Greater Yellowstone ecosystem, 1992–2000. *Ursus*, *15*(1), 10–22. https://doi.org/10.2192/1537-6176(2004)015<0010:gbcitg>2.0.co;2

Günther, R. H., O'Connell-Rodwell, C. E., & Klemperer, S. L. (2004b). Seismic waves from elephant vocalizations: A possible communication mode? *Geophysical Research Letters*, *31*(11). https://doi.org/10.1029/2004GL019671

Hagenah, N., Prins, H. H. T., & Olff, H. (2009). Effects of large herbivores on murid rodents in a South African savanna. *Journal of Tropical Ecology*, *25*(5), 483–492. https://doi.org/doi.org/10.1017/S0266467409990046

Hall, T. (2001). *To the elephant graveyard*. Grove Press.

Hall-Martin, A. J. (1987). Role of musth in the reproductive strategy of the African elephant (*Loxodonta africana*). *South African Journal of Science*, *83*, 616–620.

Hamilton, W. D. (1971). Geometry of the selfish herd. *Journal of Theoretical Biology*, *31*(2), 295–311.

Hamilton, W. D., & Zuk, M. (1982). Heritable true fitness and bright birds: A role for parasites? *Science*, *218*(4570), 384–387.

Hand, J. L. (1986). Resolution of social conflicts: Dominance, egalitarianism, spheres of dominance, and game theory. *The Quarterly Review of Biology*, *61*(2), 201–220.

Hare, B., Melis, A. P., Woods, V., Hastings, S., & Wrangham, R. (2007). Tolerance allows bonobos to outperform chimpanzees on a cooperative task. *Current Biology*, *17*(7), 619–623. https://doi.org/10.1016/J.CUB.2007.02.040

Harris, G. (2008). Development of taste and food preferences in children. *Current Opinion in Clinical Nutrition and Metabolic Care*, *11*(3), 315–319. https://doi.org/10.1097/MCO.0b013e3282f9e228

Harrison, R. D. (2011). Emptying the forest: Hunting and the extirpation of wildlife from tropical nature reserves. *BioScience*, *61*(11), 919–924. https://doi.org/10.1525/bio.2011.61.11.11

Hart, B. L., & Hart, L. A. (1994). Fly switching by Asian elephants: Tool use to control parasites. *Animal Behaviour*, *48*, 35–45.

Hart, B. L., Hart, L. A., McCoy, M., & Sarath, C. R. (2001). Cognitive behaviour in Asian elephants: Use and modification of branches for fly switching. *Animal Behaviour*, *62*(5), 839–847. https://doi.org/10.1006/anbe.2001.1815

Hartter, J., & Goldman, A. (2011). Local responses to a forest park in western Uganda: Alternate narratives on fortress conservation. *Oryx*, *45*(1), 60–68. https://doi.org/10.1017/s0030605310000141

Hatchwell, B., & Komdeur, J. (2000). Ecological constraints, life history traits and the evolution of cooperative breeding. *Animal Behaviour*, *59*(6), 1079–1086. https://doi.org/10.1006/anbe.2000.1394

Hawkes, K., & Coxworth, J. E. (2013). Grandmothers and the evolution of human longevity: A review of findings and future directions. *Evolutionary Anthropology*, *22*(6), 294–302. https://doi.org/10.1002/evan.21382

Hawkes, K., O'Connell, J. F., Jones, N. G. B., Alvarez, H., & Charnov, E. L. (1998). Grandmothering, menopause, and the evolution of human life histories. *Proceedings of the National Academy of Sciences of the United States of America*, *95*(3), 1336–1339. https://doi.org/10.1073/pnas.95.3.1336

Hawlena, D., Strickland, M. S., Bradford, M. A., & Schmitz, O. J. (2012). Fear of predation slows plant-litter decomposition. *Science*, *336*(6087), 1434–1438. https://doi.org/10.1126/science.1220097

Hawley, C. R., Beirne, C., Meier, A., & Poulsen, J. R. (2017). Conspecific investigation of a deceased forest elephant (*Loxodonta cyclotis*). *Pachyderm*, *59*, 97–100.

Haynes, G. (2012). Elephants (and extinct relatives) as earth-movers and ecosystem engineers. *Geomorphology*, *157–158*, 99–107. https://doi.org/10.1016/j.geomorph.2011.04.045

Hayward, A. D., Mar, K. U., Lahdenperä, M., & Lummaa, V. (2014). Early reproductive investment, senescence and lifetime reproductive success in female Asian elephants. *Journal of Evolutionary Biology*, *27*(4), 772–783. https://doi.org/10.1111/jeb.12350

Head, J. S., Robbins, M. M., Mundry, R., Makaga, L., & Boesch, C. (2012). Remote video-camera traps measure habitat use and competitive exclusion among sympatric chimpanzee, gorilla and elephant in Loango National Park, Gabon. *Journal of Tropical Ecology*, *28*(6), 571–583. https://doi.org/10.1017/S0266467412000612

Hedwig, D., DeBellis, M., & Wrege, P. H. (2018). Not so far: Attenuation of low-frequency vocalizations in a rainforest environment suggests limited acoustic mediation of social interaction in African forest elephants. *Behavioral Ecology and Sociobiology*, *72*(3). https://doi.org/10.1007/s00265-018-2451-4

Hedwig, D., Poole, J., & Granli, P. (2021). Does social complexity drive vocal complexity? Insights from the two african elephant species. *Animals*, *11*(11), 3071. https://doi.org/10.3390/ANI11113071/S1

Heffner, R., & Heffner, H. (1980). Hearing in the elephant (*Elephas maximus*). *Science*, *208*(4443), 518–520.

Heffner, R. S., & Heffner, H. E. (1982). Hearing in the elephant (*Elephas maximus*): Absolute sensitivity, frequency discrimination, and sound localization. *Journal of Comparative Psychology*, *96*(6), 926–944.

Hennefield, L., Hwang, H. G., Weston, S. J., & Povinelli, D. J. (2018). Meta-analytic techniques reveal that corvid causal reasoning in the Aesop's Fable paradigm is driven by trial-and-error learning. *Animal Cognition*, *21*(6), 735–748. https://doi.org/10.1007/S10071-018-1206-Y/TABLES/9

Herbst, C. T., Stoeger, A. S., Frey, R., Lohscheller, J., Titze, I. R., Gumpenberger, M., & Fitch, W. T. (2012). How low can you go? Physical production mechanism of elephant infrasonic vocalizations. *Science*, *337*(6094), 595–599. https://doi.org/10.1126/science.1219712

Herculano-Houzel, S., Avelino-de-Souza, K., Neves, K., Porfírio, J., Messeder, D., Feijó, L. M., Maldonado, J., & Manger, P. R. (2014). The elephant brain in numbers. *Frontiers in Neuroanatomy*, *8*(JUN). https://doi.org/10.3389/fnana.2014.00046

Hernández, L., & Laundré, J. W. (2007). Foraging in the 'landscape of fear' and its implications for habitat use and diet quality of elk *Cervus elaphus* and bison *Bison bison*. *Wildlife Biology*, *11*(3), 215–220. https://doi.org/10.2981/0909-6396(2005)11[215:fitlof]2.0.co;2

Hill, C. M. (2018). Crop foraging, crop losses, and crop raiding. *Annual Review of Anthropology*, *47*(October 2018), 377–394. https://doi.org/10.1146/annurev-anthro-102317-050022

Hing, S., Othman, N., Nathan, S. K., Fox, M., Fisher, M., & Goossens, B. (2013). First parasitological survey of Endangered Bornean elephants Elephas maximus borneensis. *Endangered Species Research*, *21*(3), 223–230.

Hiremath, A. J., & Sundaram, B. (2005). The Fire-Lantana Cycle hypothesis in Indian forests. *Conservation & Society*, *3*(1), 26–42.

Hladký, V., & Havlíček, J. (2013). Was Tinbergen an Aristotelian? Comparison of Tinbergen's four whys and Aristotle's four causes. *Human Ethology Bulletin*, *28*(4), 3–11.

Hoare, R. (2015). Lessons from 20 Years of human–elephant conflict mitigation in Africa. *Human Dimensions of Wildlife*, *20*(4), 289–295. https://doi.org/10.1080/10871209.2015.1005855

Hohmann, G., & Fruth, B. (2003). Intra- and inter-sexual aggression by bonobos in the context of mating. *Behaviour*, *140*, 1389–1413. https://doi.org/10.1163/156853903771980648

Holdo, R. M., & McDowell, L. R. (2006). Termite mounds as nutrient-rich food patches for elephants. *Biotropica*, *36*(2), 231. https://doi.org/10.1646/03025-q1564

Holdo, R. M., Holt, R. D., & Fryxell, J. M. (2009). Grazers, browsers, and fire influence the extent and spatial pattern of tree cover in the Serengeti. *Ecological Applications*, *19*(1), 95–109. https://doi.org/10.1890/07-1954.1

Holdo, R. M., Galvin, K. A., Knapp, E., Polasky, S., Hilborn, R., & Holt, R. D. (2010). Responses to alternative rainfall regimes and antipoaching in a migratory system. *Ecological Applications*, *20*(2), 381–397. https://doi.org/10.1890/08-0780.1

Hollister-Smith, J. A., Poole, J. H., Archie, E. A., Vance, E. A., Georgiadis, N. J., Moss, C. J., & Alberts, S. C. (2007). Age, musth and paternity success in wild male African elephants, *Loxodonta africana*. *Animal Behaviour*, *74*(2), 287–296. https://doi.org/10.1016/j.anbehav.2006.12.008

Hollister-Smith, J. A., Alberts, S. C., & Rasmussen, L. E. L. (2008). Do male African elephants, *Loxodonta africana*, signal musth via urine dribbling? *Animal Behaviour*, *76*(6), 1829–1841. https://doi.org/10.1016/j.anbehav.2008.05.033

Hoppe, K. A. (2004). Late Pleistocene mammoth herd structure, migration patterns, and Clovis hunting strategies inferred from isotopic analyses of multiple death assemblages. *Paleobiology*, *30*(1), 129–145. https://doi.org/10.1666/0094-8373(2004)030<0129:LPMHSM>2.0.CO;2

Horback, K., Miller, L., Science, & Kuczaj II, S. (2013). Personality assessment in African elephants (*Loxodonta africana*): Comparing the temporal stability of ethological coding versus trait rating. *Applied Animal Behaviour Science*, *149*(1–4), 55–62.

Houston, D. C., Gilardi, J. D., & Hall, A. J. (2001). Soil consumption by elephants might help to minimize the toxic effects of plant secondary compounds in forest browse. *Mammal Review*, *31*(3–4), 249–254.

Hovick, T. J., Elmore, R. D., Fuhlendorf, S. D., Engle, D. M., & Hamilton, R. G. (2015). Spatial heterogeneity increases diversity and stability in grassland bird communities. *Ecological Applications*, *25*(3), 662–672. https://doi.org/10.1890/14-1067.1

Hunter, M. L., Boone, S. R., Brehm, A. M., & Mortelliti, A. (2022). Modulation of ecosystem services by animal personalities. *Frontiers in Ecology and the Environment*, *20*(1), 58–63. https://doi.org/10.1002/FEE.2418

Hurtt, G. C., Chini, L. P., Sahajpal, R., & Frolking, S. (2016). Harmonization of global land-use change and management for the period 850–2100. https://luh.umd.edu/.

Idani, G. (1991). Social relationships between immigrant and resident bonobo (*Pan paniscus*) females at Wamba. *Folia Primatologica*, *57*, 83–95.

Ihwagi, F. W., Thouless, C., Wang, T., Skidmore, A. K., Omondi, P., & Douglas-Hamilton, I. (2018). Night-day speed ratio of elephants as indicator of poaching levels. *Ecological Indicators*, *84*, 38–44. https://doi.org/10.1016/J.ECOLIND.2017.08.039

Imron, M. A., Glass, D. M., Tafrichan, M., Crego, R. D., Stabach, J. A., & Leimgruber, P. (2023). Beyond protected areas: The importance of mixed-use landscapes for the conservation of Sumatran elephants (*Elephas maximus sumatranus*). *Ecology and Evolution*, *13*(10), e10560. https://doi.org/10.1002/ECE3.10560

Irie-Sugimoto, N., Kobayashi, T., Sato, T., & Hasegawa, T. (2008). Evidence of means-end behavior in Asian elephants (*Elephas maximus*). *Animal Cognition*, *11*(2), 359–365. https://doi.org/10.1007/S10071-007-0126-Z/FIGURES/4

Jackson, J., Childs, D. Z., Mar, K. U., Htut, W., & Lummaa, V. (2019). Long-term trends in wild-capture and population dynamics point to an uncertain future for captive elephants. *Proceedings. Biological Sciences*, *286*(1899), 20182810. https://doi.org/10.1098/rspb.2018.2810

Jackson, L. E., Pulleman, M. M., Brussaard, L., Bawa, K. S., Brown, G. G., Cardoso, I. M., de Ruiter, P. C., Garcia-Barrios, L., Hollander, A. D., Lavelle, P., Ouedraogo, E., Pascual, U., Setty, S., Smukler, S. M., Tscharntke, T., & Van Noordwijk, M. (2012). Social-ecological and regional adaptation of agrobiodiversity management across a global set of research regions. *Global Environmental Change-Human and Policy Dimensions*, *22*(3), 623–639. https://doi.org/10.1016/j.gloenvcha.2012.05.002

Jacobson, S. L., & Plotnik, J. (2020). The importance of sensory perception in an elephant's cognitive world. *Comparative Cognition and Behavior Reviews*, *15*. https://doi.org/10.3819/CCBR.2020.150006E

Jacobson, S. L., Puitiza, A., Snyder, R. J., Sheppard, A., & Plotnik, J. M. (2022). Persistence is key: Investigating innovative problem solving by Asian elephants using a novel multi-access box. *Animal Cognition, 25*(3), 657–669. https://doi.org/10.1007/S10071-021-01576-3/FIGURES/4

Jacobson, S. L., Dechanupong, J., Horpiencharoen, W., Yindee, M., & Plotnik, J. M. (2023). Innovating to solve a novel puzzle: Wild Asian elephants vary in their ability to problem solve. *Animal Behaviour, 205*, 227–239. https://doi.org/10.1016/J.ANBEHAV.2023.08.019

Janik, V. M., & Knörnschild, M. (2021). Vocal production learning in mammals revisited. *Philosophical Transactions of the Royal Society B, 376*(1836), 20200244. https://doi.org/10.1098/RSTB.2020.0244

Janzen, D. H., & Martin, P. (1982). Neotropical anachronisms: The fruits the Gomphotheres ate. *Science, 215*(4528), 19–27.

Jelbert, S. A., Taylor, A. H., Cheke, L. G., Clayton, N. S., & Gray, R. D. (2014). Using the Aesop's fable paradigm to investigate causal understanding of water displacement by New Caledonian crows. *PLOS One, 9*(3), e92895. https://doi.org/10.1371/JOURNAL.PONE.0092895

Jelbert, S. A., Taylor, A. H., & Gray, R. D. (2015). Investigating animal cognition with the Aesop's Fable paradigm: Current understanding and future directions. *Communicative & Integrative Biology, 8*(4), 1–6. https://doi.org/10.1080/19420889.2015.1035846

Jesmer, B. R., Merkle, J. A., Goheen, J. R., Aikens, E. O., Beck, J. L., Courtemanch, A. B., Hurley, M. A., McWhirter, D. E., Miyasaki, H. M., Monteith, K. L., & Kauffman, M. J. (2018). Is ungulate migration culturally transmitted? Evidence of social learning from translocated animals. *Science, 361*(6406), 1023–1025. https://doi.org/10.1126/science.aat0985

Jiang, L., Liu, Y., Xu, H., Jiang, L., Liu, Y., & Xu, H. (2023). Losing the way or running off? An unprecedented major movement of Asian elephants in Yunnan, China. *Land, 12*(2), 460. https://doi.org/10.3390/LAND12020460

Jin, Z. W., Cho, K. H., Xu, D. Y., You, Y. Q., Kim, J. H., Murakami, G., & Abe, H. (2020). Pacinian corpuscles in the human fetal foot: A study using 3D reconstruction and immunohistochemistry. *Annals of Anatomy – Anatomischer Anzeiger, 227*, 151421. https://doi.org/10.1016/J.AANAT.2019.151421

Johnson, D. D. P., Kays, R., Blackwell, P. G., & Macdonald, D. W. (2002). Does the resource dispersion hypothesis explain group living? *Trends in Ecology & Evolution, 17*(12), 563–570. https://doi.org/10.1016/S0169-5347(02)02619-8

Johnson, E., & Rasmussen, L. (2002). Morphological characteristics of the vomeronasal organ of the newborn Asian elephant (*Elephas maximus*). *The Anatomical Record: An Official Publication of the American Association of Anatomists, 267*(3), 252–259. https://doi.org/10.1002/ar.10112

Jolly, H., Satterfield, T., Kandlikar, M., & TR, S. (2022). Indigenous insights on human–wildlife coexistence in southern India. *Conservation Biology, 36*(6), e13981. https://doi.org/10.1111/cobi.13981

Joppa, L. N., & Pfaff, A. (2009). High and far: Biases in the location of protected areas. *PLOS One, 4*(12), e8273. https://doi.org/10.1371/JOURNAL.PONE.0008273

Kalumanga, E., Mpanduji, D. G., & Cousins, S. A. O. (2017). Geophagic termite mounds as one of the resources for African elephants in Ugalla Game Reserve, Western Tanzania. *African Journal of Ecology, 55*(1), 91–100. https://doi.org/10.1111/aje.12326

Kansky, R., Kidd, M., & Knight, A. T. (2016). A wildlife tolerance model and case study for understanding human wildlife conflicts. *Biological Conservation, 201*, 137–145. https://doi.org/10.1016/j.biocon.2016.07.002

Karanth, K. K., & DeFries, R. (2011). Nature-based tourism in Indian protected areas: New challenges for park management. *Conservation Letters, 4*(2), 137–149. https://doi.org/10.1111/j.1755-263X.2010.00154.x

Katlam, G., Prasad, S., Aggarwal, M., & Kumar, R. (2018). Trash on the menu: Patterns of animal visitation and foraging behaviour at garbage dumps. *Current Science*, *115*(12), 2322–2326. https://doi.org/10.18520/cs/v115/i12/2322-2326

Kaufmann, L. V., Schneeweiß, U., Maier, E., Hildebrandt, T., & Brecht, M. (2022). Elephant facial motor control. *Science Advances*, *8*(43), 2789. https://doi.org/10.1126/SCIADV.ABQ2789/SUPPL_FILE/SCIADV.ABQ2789_SM.PDF

Kauffman, M. J., Brodie, J. F., & Jules, E. S. (2010). Are wolves saving Yellowstone's aspen? A landscape-level test of a behaviorally mediated trophic cascade. *Ecology*, *91*(9), 2742–2755.

Keerthipriya, P., Nandini, S., & Vidya, T. N. C. (2021). Effects of male age and female presence on male associations in a large, polygynous mammal in Southern India: The Asian elephant. *Frontiers in Ecology and Evolution*, *9*, 348. https://doi.org/10.3389/FEVO.2021.616666/BIBTEX

Kennedy, M., & Gray, R. D. (1993). Can ecological theory predict the distribution of foraging animals? A critical analysis of experiments on the Ideal Free Distribution. *Oikos*, *68*, 158–166.

Khan, M. L., Khumbongmayum, A. D., & Tripathi, R. S. (2008). The sacred groves and their significance in conserving biodiversity: An overview. *International Journal of Ecology and Environmental Sciences*, *34*(3), 277–291.

King, L. E., Douglas-Hamilton, I., & Vollrath, F. (2007). African elephants run from the sound of disturbed bees. *Current Biology*, *17*(19), R832–R833. https://doi.org/10.1016/J.CUB.2007.07.038

King, L. E., Soltis, J., Douglas-Hamilton, I., Savage, A., & Vollrath, F. (2010). Bee threat elicits alarm call in African elephants. *PLOS One*, *5*(4). https://doi.org/10.1371/JOURNAL.PONE.0010346

King, L. E., Pardo, M., Weerathunga, S., Pushpakumara, T., Jayasena, N., Soltis, J., & de Silva, S. (2018). Wild Sri Lankan Elephants (*Elephas maximus*) retreat from the sound of disturbed Asian honey bees (*Apis cerana indica*). *Current Biology*, *28*(2), R64–R65. https://doi.org/10.1016/j.cub.2017.12.018

Kinsella, J. M., Deem, S. L., Blake, S., & Freeman, A. S. (2004). Endoparasites of African forest elephants (*Loxodonta africana cyclotis*) from the Republic of Congo and Central African Republic. *Comparative Parasitology*, *71*(2), 104–110. https://doi.org/10.1654/4131

Kirkwood, T. B. L., & Rose, M. R. (1991). Evolution of senescence: Late survival sacrificed for reproduction. *Philosophical Transactions of the Royal Society B: Biological Sciences*, *332*, 15–24.

Kodandapani, N., Cochrane, M. A., & Sukumar, R. (2004). Conservation threat of increasing fire frequencies in the Western Ghats, India. *Conservation Biology*, *18*(6), 1553–1561. https://doi.org/10.1111/j.1523-1739.2004.00433.x

Koenig, A. (2002). Competition for resources and its behavioral consequences among female primates. *International Journal of Primatology*, *23*(4), 759–783.

Koenig, A., Scarry, C. J., Wheeler, B. C., & Borries, C. (2013). Variation in grouping patterns, mating systems and social structure: What socio-ecological models attempt to explain. *Proceedings of the Royal Society B: Biological Sciences*, *368*(1638), 20120348. https://doi.org/10.1098/rstb.2012.0348

Kohda, M., Bshary, R., Kubo, N., Awata, S., Sowersby, W., Kawasaka, K., Kobayashi, T., & Sogawa, S. (2023). Cleaner fish recognize self in a mirror via self-face recognition like humans. *Proceedings of the National Academy of Sciences of the United States of America*, *120*(7), e2208420120. https://doi.org/10.1073/PNAS.2208420120/SUPPL_FILE/PNAS.2208420120.SM01.MP4

Kohl, P. L., Thulasi, N., Rutschmann, B., George, E. A., Steffan-Dewenter, I., & Brockmann, A. (2020). Adaptive evolution of honeybee dance dialects. *Proceedings of the Royal Society B: Biological Sciences*, *287*(1922). https://doi.org/10.1098/RSPB.2020.0190

Koskei, M., Kolowski, J., Wittemyer, G., Lala, F., Douglas-Hamilton, I., & Okita-Ouma, B. (2022). The role of environmental, structural and anthropogenic variables on underpass use by African savanna elephants (*Loxodonta africana*) in the Tsavo Conservation Area. *Global Ecology and Conservation*, *38*, e02199. https://doi.org/10.1016/J.GECCO.2022.E02199

Kumar, M. A., Mudappa, D., & Raman, T. R. S. (2010). Asian elephant *Elephas maximus* habitat use and ranging in fragmented rainforest and plantations in the Anamalai Hills, India. *Tropical Conservation Science*, *3*(2), 143–158. https://doi.org/10.1177/194008291000300203

Kumar, M. A., Vijayakrishnan, S., & Singh, M. (2018). Whose habitat is it anyway? Role of natural and anthropogenic habitats in conservation of charismatic species. *Tropical Conservation Science*, *11*, 194008291878845. https://doi.org/10.1177/1940082918788451

Kummer, H. (1968). *Social organization of hamadryas baboons*. University of Chicago Press.

Kunc, H. P., & Schmidt, R. (2019). The effects of anthropogenic noise on animals: A meta-analysis. *Biology Letters*, *15*(11), 20190649. https://doi.org/10.1098/RSBL.2019.0649

Kurt, F. (1974). Remarks on the social structure and ecology of the Ceylon elephant in the Yala National Park. In V. Geist, & F. Walther (Eds.), *The behaviour of ungulates and its relation to management* (pp. 618–634). International Union for Conservation of Nature and Natural Resources.

Kurt, F., Hartl, G. B., & Tiedemann, R. (1995). Tuskless bulls in Asian elephant *Elephas maximus*. History and population genetics of a man-made phenomenon. *Acta Theriologica*, *3*, 125–143.

LaDue, C. A., Goodwin, T. E., & Schulte, B. A. (2018). Concentration-dependent chemosensory responses towards pheromones are influenced by receiver attributes in Asian elephants. *Ethology*, *124*(6), 387–399. https://doi.org/10.1111/ETH.12741

LaDue, C. A., Vandercone, R. P. G., Kiso, W. K., & Freeman, E. W. (2022). Behavioral characterization of musth in Asian elephants (*Elephas maximus*): Defining progressive stages of male sexual behavior in in-situ and ex-situ populations. *Applied Animal Behaviour Science*, *251*(April), 105639. https://doi.org/10.1016/j.applanim.2022.105639

Lahdenperä, M., Lummaa, V., Helle, S., Tremblay, M., & Russell, A. F. (2004). Fitness benefits of prolonged post-reproductive lifespan in women. *Nature*, *428*(6979), 178–181. https://doi.org/10.1038/nature02367

Lahdenperä, M., Mar, K. U., & Lummaa, V. (2014). Reproductive cessation and post-reproductive lifespan in Asian elephants and pre-industrial humans. *Frontiers in Zoology*, *11*(1), 1–14. https://doi.org/10.1186/s12983-014-0054-0

Lahdenperä, M., Mar, K. U., & Lummaa, V. (2015). Short-term and delayed effects of mother death on calf mortality in Asian elephants. *Behavioral Ecology*, *27*(1). https://doi.org/10.1093/beheco/arv136

Lake, F. K., Wright, V., Morgan, P., McFadzen, M., McWethy, D., & Stevens-Rumann, C. (2017). Returning fire to the land: Celebrating traditional knowledge and fire. *Journal of Forestry*, *115*(5), 343–353. https://doi.org/10.5849/JOF.2016-043R2

Laland, K. N., Sterelny, K., Odling-Smee, J., Hoppitt, W., & Uller, T. (2011). Cause and effect in biology revisited: Is Mayr's proximate-ultimate dichotomy still useful? *Science*, *334*(6062), 1512–1516. https://doi.org/10.1126/science.1210879

Lamarck, J. B. (1809). Philosophie zoologique, ou exposition des considérations relatives à l'histoire naturelle des animaux, Volume 1 (Oxford).

Lambin, E. F., Turner, B. L., Geist, H. J., Agbola, S. B., Angelsen, A., Folke, C., Bruce, J. W., Coomes, O. T., Dirzo, R., George, P. S., Homewood, K., Imbernon, J., Leemans, R., Li, X., Moran, E. F., Mortimore, M., Ramakrishnan, P. S., Richards, J. F., Steffen, W., … Veldkamp, T. A. (2001). The causes of land-use and land-cover change: Moving beyond the myths. *Global Environmental Change*, *11*(4), 261–269.

Laris, P., & Wardell, D. A. (2006). Good, bad or 'necessary evil'? Reinterpreting the colonial burning experiments in the savanna landscapes of West Africa. *Geographical Journal*, *172*(4), 271–290. https://doi.org/10.1111/J.1475-4959.2006.00215.X

Larramendi, A. (2016). Shoulder height, body mass and shape of proboscideans. *Acta Palaeontologica Polonica*, *61*(3), 1–109. https://doi.org/10.4202/app.2010.0047

Laundré, J. W., Hernández, L., & Altendorf, K. B. (2001). Wolves, elk, and bison: Reestablishing the "landscape of fear" in Yellowstone National Park, U.S.A. *Canadian Journal of Zoology*, *79*(8), 1401–1409. https://doi.org/10.1139/z01-094

Laundré, J. W., Hernandez, L., & Ripple, W. J. (2010). The landscape of fear: Ecological implications of being afraid. *The Open Ecology Journal*, *3*(3), 1–7. https://doi.org/10.2174/1874213001003030001

Laurance, W. F., Goosem, M., & Laurance, S. G. W. (2009). Impacts of roads and linear clearings on tropical forests. *Trends in Ecology and Evolution*, *24*(12), 659–669. https://doi.org/10.1016/j.tree.2009.06.009

Laurance, W. F., Campbell, M. J., Alamgir, M., & Mahmoud, M. I. (2017). Road expansion and the fate of Africa's tropical forests. *Frontiers in Ecology and Evolution*, *5*(JUL), 1–7. https://doi.org/10.3389/fevo.2017.00075

Lea, S. E. G., Chow, P. K. Y., Leaver, L. A., & McLaren, I. P. L. (2020). Behavioral flexibility: A review, a model, and some exploratory tests. *Learning and Behavior*, *48*(1), 173–187. https://doi.org/10.3758/S13420-020-00421-W/FIGURES/3

Leach, S., Sutton, R. M., Dhont, K., Douglas, K. M., & Bergström, Z. M. (2023). Are we smart enough to remember how smart animals are? *Journal of Experimental Psychology: General*, *152*(8), 2138–2159. https://doi.org/10.1037/XGE0001401

Lee, P. C., & Moss, C. J. (1999). The social context for learning and behavioural development among wild African elephants. In Hilary O. Box, & Kathleen R. Gibson (Eds.), *Mammalian social learning: Comparative and ecological perspectives* (pp. 102–125). Cambridge University Press.

Lee, P. C., & Moss, C. J. (2012). Wild female African elephants (*Loxodonta africana*) exhibit personality traits of leadership and social integration. *Journal of Comparative Psychology*, *126*(3), 224–232. https://doi.org/10.1037/a0026566

Lee, P. C., Poole, J. H., Njiraini, N., Sayialel, C., N., & Moss, C. J. (2011). Male social dynamics: Independence and beyond. In Moss, H. Croze, & P. C. Lee (Eds.), *The Amboseli Elephants: A long term perspective on a long-lived mammal* (pp. 260–271). University of Chicago Press.

Leggett, K. E. A. (2006). Home range and seasonal movement of elephants in the Kunene Region, northwestern Namibia. *African Zoology*, *41*, 17–36.

Leggett, K. E. A. (2008). Diurnal activities of the desert-dwelling elephants in northwestern Namibia. *Pachyderm*, *45*, 20–33.

Leggett, K. E. A., Fennessy, J., & Schneider, S. (2003). Seasonal distributions and social dynamics of elephants in Hoanib River catchment, northwestern Namibia. *African Zoology*, *38*(2), 305–316.

Leggett, K. E. A., MacAlister Brown, L., & Ramey II, R. R. (2011). Matriarchal associations and reproduction in a remnant subpopulation of desert-dwelling elephants in Namibia. *Pachyderm*, *49*, 20–32. https://cmsdata.iucn.org/downloads/pachy49small.pdf#page=24

Leighty, K. A., Soltis, J., Leong, K., & Savage, A. (2008a). Antiphonal exchanges in African elephants (*Loxodonta africana*): Collective response to a shared stimulus, social facilitation, or true communicative event? *Behaviour*, *145*(3), 297–312. https://doi.org/10.1163/156853908783402885

Leighty, K. A., Soltis, J., Wesolek, C. M., & Savage, A. (2008b). Rumble vocalizations mediate interpartner distance in African elephants, *Loxodonta africana*. *Animal Behaviour*, *76*, 1601–1608. https://doi.org/10.1016/j.anbehav.2008.06.022

Leimgruber, P., Gagnon, J. B., Wemmer, C., Kelly, D. S., Songer, M. A., & Selig, E. R. (2003). Fragmentation of Asia's remaining wildlands: Implications for Asian elephant conservation. *Animal Conservation*, *6*(4), 347–359. https://doi.org/10.1017/S1367943003003421

Leuthold, W. (1977). Spatial organization and strategy of habitat utilization of elephants in Tsavo National Park, Kenya. *Zeitschrift Fur Saugetierkunde-International Journal of Mammalian Biology*, *42*, 358–379.

Lev, M., & Barkai, R. (2016). Elephants are people, people are elephants: Human-elephant similarities as a case for cross cultural animal humanization in recent and Paleolithic times. *Quaternary International*, 406. https://doi.org/10.1016/j.quaint.2015.07.005

Levin, N. E., Brown, F. H., Behrensmeyer, A. K., Bobe, R., & Cerling, T. E. (2011). Paleosol carbonates from the Omo Group: Isotopic records of local and regional environmental change in East Africa. *Palaeogeography, Palaeoclimatology, Palaeoecology*, *307*(1–4), 75–89. https://doi.org/10.1016/J.PALAEO.2011.04.026

Levitis, D. A., & Lackey, L. B. (2011). A measure for describing and comparing postreproductive life span as a population trait. *Methods in Ecology and Evolution*, *2*(5), 446–453. https://doi.org/10.1111/j.2041-210X.2011.00095.x

Li, L. L., Plotnik, J. M., Xia, S. W., Meaux, E., & Quan, R. C. (2021). Cooperating elephants mitigate competition until the stakes get too high. *PLOS Biology*, *19*(9), e3001391. https://doi.org/10.1371/JOURNAL.PBIO.3001391

Li, Q. L., & Zhang, L. (2010). Parent-offspring recognition in Brandt's voles, *Lasiopodomys brandti*. *Animal Behaviour*, *79*(4), 797–801. https://doi.org/10.1016/j.anbehav.2009.12.001

Lister, A. M. (2013). The role of behaviour in adaptive morphological evolution of African proboscideans. *Nature*, *500*(7462), 331–334. https://doi.org/10.1038/nature12275

Lister, A. M. (2014). Behavioural leads in evolution: Evidence from the fossil record. *Biological Journal of the Linnean Society*, *112*(2), 315–331. https://doi.org/10.1111/BIJ.12173

Lister, A. M., & Bahn, P. (2007). *Mammoths: Giants of the Ice Age*. University of California Press.

Lister, A. M., Dirks, W., Assaf, A., Chazan, M., Goldberg, P., Applbaum, Y. H., Greenbaum, N., & Horwitz, L. K. (2013). New fossil remains of *Elephas* from the southern Levant: Implications for the evolutionary history of the Asian elephant. *Palaeogeography, Palaeoclimatology, Palaeoecology*, *386*, 119–130. https://doi.org/10.1016/J.PALAEO.2013.05.013

Liu, J., Hull, V., Moran, E., Nagendra, H., Swaffield, S. R., & Turner, B. L. (2014). Applications of the telecoupling framework to land-change science. In *Rethinking global landuse in an urban era* (pp. 119–140). MIT Press. https://asu.pure.elsevier.com/en/publications/applications-of-the-telecoupling-framework-to-land-change-science

Loarie, S. R., van Aarde, R. J., & Pimm, S. L. (2009). Elephant seasonal vegetation preferences across dry and wet savannas. *Biological Conservation*, *142*(12), 3099–3107. https://doi.org/10.1016/j.biocon.2009.08.021

Loizi, H., Goodwin, T., Whitehouse, A., & Schulte, B. (2009). Sexual dimorphism in the performance of chemosensory investigatory behaviours by African elephants (*Loxodonta africana*). *Behaviour*, *146*(3), 373–392. https://doi.org/10.1163/156853909X410964

Lorenz, K. (1952). *King Solomon's Ring*. Methuen & Co., London. https://doi.org/10.4324/9780203165966/KING-SOLOMON-RING-JULIAN-HUXLEY-KONRAD-LORENZ

Loveridge, A. J., Hunt, J. E., Murindagomo, F., & Macdonald, D. W. (2006). Influence of drought on predation of elephant (*Loxodonta africana*) calves by lions (*Panthera leo*) in an African wooded savannah. *Journal of Zoology*, *270*(3), 523–530. https://doi.org/10.1111/J.1469-7998.2006.00181.X

Ma, J., Wang, Y., Jin, C., Hu, Y., & Bocherens, H. (2019). Ecological flexibility and differential survival of Pleistocene *Stegodon orientalis* and *Elephas maximus* in mainland southeast Asia revealed by stable isotope (C, O) analysis. *Quaternary Science Reviews*, *212*, 33–44. https://doi.org/10.1016/j.quascirev.2019.03.021

MacDougall-Shackleton, S. A. (2011). The levels of analysis revisited. *Philosophical Transactions of the Royal Society B: Biological Sciences*, *366*(1574), 2076–2085. https://doi.org/10.1098/rstb.2010.0363

MacFadyen, S., Hui, C., Verburg, P. H., & Van Teeffelen, A. J. A. (2019). Spatiotemporal distribution dynamics of elephants in response to density, rainfall, rivers and fire in Kruger National Park, South Africa. *Diversity and Distributions*, *25*(6), 880–894. https://doi.org/10.1111/ddi.12907

Madsen, A. E., Minge, C., Pushpakumara, T. V, Weerathunga, U. S., Padmalal, U. K., Weerakoon, D. K., & Silva, S. De. (2022). Strategies of protected area use by Asian elephants in relation to motivational state and social affiliations. *Scientific Reports*, *12*, 18490. https://doi.org/10.1038/s41598-022-22989-1

Main, M. B., Roka, F. M., & Noss, R. F. (1999). Evaluating costs of conservation. *Conservation Biology*, *13*(6), 1262–1272. https://doi.org/10.1046/j.1523-1739.1999.98006.x

Mainguy, J., & Cote, S. D. (2008). Age- and state-dependent reproductive effort in male mountain goats, *Oreamnos americanus*. *Behavioral Ecology and Sociobiology*, *62*(6), 935–943. https://doi.org/10.1007/s00265-007-0517-9

Maisels, F., Strindberg, S., Blake, S., Wittemyer, G., Hart, J., Williamson, E. A., Aba'a, R., Abitsi, G., Ambahe, R. D., Amsini, F., Bakabana, P. C., Hicks, T. C., Bayogo, R. E., Bechem, M., Beyers, R. L., Bezangoye, A. N., Boundja, P., Bout, N., Akou, M. E., … Warren, Y. (2013). Devastating decline of forest elephants in Central Africa. *PLOS One*, *8*(3). https://doi.org/10.1371/journal.pone.0059469

Maitima, J. M., Mugatha, S. M., Reid, R. S., Gachimbi, L. N., Majule, A., Lyaruu, H., Pomery, D., Mathai, S., & Mugisha, S. (2010). The linkages between land use change, land degradation and biodiversity across East Africa. *African Journal of Environmental Science and Technology*, *3*(10), 310–325. https://doi.org/10.4314/ajest.v3i10.56259

Makecha, R., Fad, O., & Kuczaj II, S. A. (2012). The role of touch in the social interactions of Asian elephants (*Elephas maximus*). *International Journal of Comparative Psychology*, *25*(1), 60–82. https://doi.org/10.46867/IJCP.2012.25.01.01

Malhi, Y., Doughty, C. E., Galetti, M., Smith, F. A., Svenning, J.-C., & Terborgh, J. W. (2016). Megafauna and ecosystem function from the Pleistocene to the Anthropocene. *Proceedings of the National Academy of Sciences of the United States of America*, *113*(4), 838–846. https://doi.org/10.1073/pnas.1502540113

Manthi, F. K., Sanders, W. J., Plavcan, J. M., Cerling, T. E., & Brown, F. H. (2020). Late Middle Pleistocene elephants from Natodomeri, Kenya and the disappearance of *Elephas* (*Proboscidea, Mammalia*) in Africa. *Journal of Mammalian Evolution*, *27*(3), 483–495. https://doi.org/10.1007/s10914-019-09474-9

Mar, K. U., Lahdenperä, M., & Lummaa, V. (2012). Causes and correlates of calf mortality in captive Asian elephants (*Elephas maximus*). *PLOS One*, *7*(3), e32335. https://doi.org/10.1371/journal.pone.0032335

Marino, L. (2005). Big brains do matter in new environments. *Proceedings of the National Academy of Sciences of the United States of America*, *102*(15), 5306–5307. https://doi.org/10.1073/PNAS.0501695102

Marneweck, C., Jürgens, A., & Shrader, A. M. (2018). The role of middens in white rhino olfactory communication. *Animal Behaviour*, *140*, 7–18. https://doi.org/10.1016/j.anbehav.2018.04.001

Martínez-Abraín, A., Quevedo, M., & Serrano, D. (2022). Translocation in relict shy-selected animal populations: Program success versus prevention of wildlife-human conflict. *Biological Conservation*, *268*, 109519. https://doi.org/10.1016/J.BIOCON.2022.109519

Matsumura, S., Arlinghaus, R., & Dieckmann, U. (2010). Foraging on spatially distributed resources with sub-optimal movement, imperfect information, and travelling costs: Departures from the ideal free distribution. *Oikos*, *119*(9), 1469–1483. https://doi.org/10.1111/J.1600-0706.2010.18196.X

May, T. M., Page, M. J., & Fleming, P. A. (2016). Predicting survivors: Animal temperament and translocation. *Behavioral Ecology*, *27*(4), 969–977. https://doi.org/10.1093/BEHECO/ARV242

Mayr, E. (1942). *Systematics and the origin of species from the viewpoint of a zoologist*. Columbia University Press.

Mayr, E. (1961). Cause and effect in biology. *Science*, *134*, 1501–1506.

Mazur, R. L. (2010). Does aversive conditioning reduce human-black bear conflict? *Journal of Wildlife Management*, *74*(1), 48–54. https://doi.org/10.2193/2008-163

Mbamy, W., Beirne, C., Froese, G. Z., Ebanega, M. O., & Poulsen, J. R. (2024). Linking crop availability, forest elephant visitation and perceptions of human–elephant interactions in villages bordering Ivindo National Park, Gabon. *Oryx*, *58*(2), 261–268. https://doi.org/10.1017/S0030605323000704

McComb, K., Moss, C., Sayialel, S., & Baker, L. (2000). Unusually extensive networks of vocal recognition in African elephants. *Animal Behaviour*, *59*(6), 1103–1109. https://doi.org/10.1006/anbe.2000.1406

McComb, K., Moss, C., Durant, S. M., Baker, L., & Sayialel, S. (2001). Matriarchs as repositories of social knowledge in African elephants. *Science*, *292*(5516), 1–7.

McComb, K., Reby, D., Baker, L., Moss, C., & Sayialel, S. (2003). Long-distance communication of acoustic cues to social identity in African elephants. *Animal Behaviour*, *65*(2), 317–329. https://doi.org/10.1006/ANBE.2003.2047

McComb, K., Baker, L., & Moss, C. (2006). African elephants show high levels of interest in the skulls and ivory of their own species. *Biology Letters*, *2*(1), 26–28. https://doi.org/10.1098/RSBL.2005.0400

McComb, K., Shannon, G., Durant, S. M., Sayialel, K., Slotow, R., Poole, J., & Moss, C. (2011). Leadership in elephants: The adaptive value of age. *Proceedings of the Royal Society B: Biological Sciences*, *278*(1722), 3270–3276. https://doi.org/10.1098/rspb.2011.0168

McFarland, R., Murphy, D., Lusseau, D., Henzi, S. P., Parker, J. L., Pollet, T. V., & Barrett, L. (2017). The 'strength of weak ties' among female baboons: Fitness-related benefits of social bonds. *Animal Behaviour*, *126*, 101–106. https://doi.org/10.1016/j.anbehav.2017.02.002

McKay, G. M. (1973). Behavior and ecology of the Asiatic elephant in Southeastern Ceylon. *Smithsonian Contributions to Zoology*, *125*, 1–113.

Merte, C. E., Goodwin, T. E., & Schulte, B. A. (2009). Male and female developmental differences in chemosensory investigations by African elephants (*Loxodonta africana*)

approaching waterholes. *Behavioral Ecology and Sociobiology*, *64*(3), 401–408. https://doi.org/10.1007/S00265-009-0856-9

Meyer, M., Palkopoulou, E., Baleka, S., Stiller, M., Penkman, K. E. H., Alt, K. W., Ishida, Y., Mania, D., Mallick, S., Meijer, T., Meller, H., Nagel, S., Nickel, B., Ostritz, S., Rohland, N., Schauer, K., Schüler, T., Roca, A. L., Reich, D., … Hofreiter, M. (2017). Palaeogenomes of Eurasian straight-tusked elephants challenge the current view of elephant evolution. *ELife*, *6*, 1–14. https://doi.org/10.7554/eLife.25413

Midgley, J. J., Kruger, L. M., Viljoen, S., Bijl, A., & Steenhuisen, S.-L. (2015). Fruit and seed traits of the elephant-dispersed African savanna plant *Balanites maughamii*. *Journal of Tropical Ecology*, *31*(6), 557–561. https://doi.org/10.1017/s0266467415000437

Millot, S., Nilsson, J., Fosseidengen, J. E., Bégout, M. L., Fernö, A., Braithwaite, V. A., & Kristiansen, T. S. (2014). Innovative behaviour in fish: Atlantic cod can learn to use an external tag to manipulate a self-feeder. *Animal Cognition*, *17*(3), 779–785. https://doi.org/10.1007/S10071-013-0710-3/FIGURES/3

Milner-Gulland, E. J. (2011). *Animal migration: A synthesis*. Oxford University Press.

Miquelle, D. G. (1990). Why don't bull moose eat during the rut? *Behavioral Ecology and Sociobiology*, *27*(2), 145–151. https://doi.org/10.1007/BF00168458

Mithöfer, A., & Boland, W. (2012). Plant defense against herbivores: Chemical aspects. *Annual Review of Plant Biology*, *63*, 431–450. https://doi.org/10.1146/ANNUREV-ARPLANT-042110-103854

Moffett, M. (2013). Human identity and the evolution of societies. *Human Nature*, *24*(3), 219–267.

Moffett, M. (2020). Societies, identity and belonging. *Proceedings of the American Philosophical Society*, *164*(3), 1–9.

Momont, L. (2007). Sélection de l'habitat et organisation sociale de l'éléphant de forêt, *Loxodonta africana cyclotis* (Matschie 1900), au Gabon. Muséum National d'Histoire Naturelle.

Mortimer, B., Walker, J. A., Lolchuragi, D. S., Reinwald, M., & Daballen, D. (2021). Noise matters: Elephants show risk-avoidance behaviour in response to human-generated seismic cues. *Proceedings of the Royal Society B: Biological Sciences*, *288*(1953). https://doi.org/10.1098/RSPB.2021.0774

Moss, C. J., & Lee, P. C. (2011a). Female reproductive strategies: Individual life histories. In Moss, H. Croze, & P. C. Lee (Eds.), *The Amboseli Elephants: A long term perspective on a long-lived mammal* (pp. 187–204). University of Chicago Press.

Moss, C. J., & Lee, P. C. (2011b). Female social dynamics: Fidelity and flexibility. In Moss, H. Croze, & P. C. Lee (Eds.), *The Amboseli elephants: A long term perspective on a long-lived mammal* (pp. 205–223). University of Chicago Press.

Moss, C. J., & Poole, J. H. (1983). Relationships and social structure of African elephants. In R. A. Hinde (Ed.), *Primate social relationships: An integrated approach* (pp. 315–325). Blackwell Scientific Publications.

Moss, C. J., Croze, H., & Lee, P. (Eds.). (2011). *The Amboseli Elephants: A long term perspective on a long-lived mammal*. University of Chicago Press.

Moßbrucker, A. M., Apriyana, I., Fickel, J., Imron, M. A., & Pudyatmoko, S. (2015). Non-invasive genotyping of Sumatran elephants: Implications for conservation. *Tropical Conservation Science*, *8*(3), 745–759.

Moura, L. C., Scariot, A. O., Schmidt, I. B., Beatty, R., & Russell-Smith, J. (2019). The legacy of colonial fire management policies on traditional livelihoods and ecological sustainability in savannas: Impacts, consequences, new directions. *Journal of Environmental Management*, *232*, 600–606. https://doi.org/10.1016/J.JENVMAN.2018.11.057

Murphy, B. P., Andersen, A. N., & Parr, C. L. (2016). The underestimated biodiversity of tropical grassy biomes. *Philosophical Transactions of the Royal Society B: Biological Sciences, 371*(1703). https://doi.org/10.1098/RSTB.2015.0319

Mutinda, H., Poole, J. H., & Moss, C. J. (2011). Decision making and leadership in using the ecosytem. In H. Croze Moss, & P. C. Lee (Eds.), *The Amboseli elephants: A long term perspective on a long-lived mammal* (pp. 246–259). University of Chicago Press.

Mysterud, A., Bonenfant, C., Loe, L. E., Langvatn, R., Yoccoz, N. G., & Stenseth, N. C. (2008). Age-specific feeding cessation in male red deer during rut. *Journal of Zoology, 275*(4), 407–412. https://doi.org/10.1111/j.1469-7998.2008.00453.x

Nadal, A. (2008). *Rethinking macroeconomics for sustainability*. Bloomsbury Publishing.

Nagarajan B., Kanakasabai, R., Desai, A. (2018). Influence of ranging and hierarchy on the habitat use pattern by Asian Elephant (*Elephas maximus*) in the tropical forests of Southern India. In C. Sivaperuman, & K. Venkataraman (Eds.), *Indian hotspots: Vertebrate faunal diversity, conservation and management* (pp. 345–358). Springer. https://doi.org/10.1007/978-981-10-6605-4_17

Nagel, T. (1974). What is it like to be a bat? *Philosophical Review, 83*(4), 435–450. www.jstor.org/stable/2183914

Nair, S., Balakrishnan, R., Seelamantula, C. S., & Sukumar, R. (2009). Vocalizations of wild Asian elephants (*Elephas maximus*): Structural classification and social context. *The Journal of the Acoustical Society of America, 126*(5), 2768–2778. https://doi.org/10.1121/1.3224717

Nandini, S., Keerthipriya, P., & Vidya, T. N. C. (2017). Seasonal variation in female Asian elephant social structure in Nagarahole-Bandipur, southern India. *Animal Behaviour, 134*, 135–145. https://doi.org/10.1016/j.anbehav.2017.10.012

Nandini, S., Keerthipriya, P., & Vidya, T. N. C. (2018). Group size differences may mask underlying similarities in social structure: A comparison of female elephant societies. *Behavioral Ecology, 29*(1), 145–159. https://doi.org/10.1093/beheco/arx135

Nelson, X. J. (2014). Animal behavior can inform conservation policy, we just need to get on with the job – or can it? *Current Zoology, 60*(4), 479–485. https://doi.org/10.1093/czoolo/60.4.479

Nerlekar, A. N., Chorghe, A. R., Dalavi, J. V., Kusom, R. K., Karuppusamy, S., Kamath, V., Pokar, R., Rengaian, G., Sardesai, M. M., & Kambale, S. S. (2022). Exponential rise in the discovery of endemic plants underscores the need to conserve the Indian savannas. *Biotropica, 54*(2), 405–417. https://doi.org/10.1111/btp.13062

Neupane, D., Kwon, Y., Risch, T. S., & Johnson, R. L. (2020). Changes in habitat suitability over a two decade period before and after Asian elephant recolonization. *Global Ecology and Conservation, 22*, e01023. https://doi.org/10.1016/j.gecco.2020.e01023

Newsom, L. A., & Mihlbachler, M. C. (2006). Mastodons (*Mammut americanum*) diet foraging patterns based on analysis of dung deposits. In D. D. Webb (Ed.), *First Floridians and last mastodons: The page-ladson site in the Aucilla river* (pp. 263–331). Springer.

Nichols, E., & Gómez, A. (2011). Conservation education needs more parasites. *Biological Conservation, 144*(2), 937–941. https://doi.org/10.1016/j.biocon.2010.10.025

Nieman, W. A., Van Wilgen, B. W., & Lesliel, A. J. (2021). A review of fire management practices in African savanna-protected areas. *Koedoe, 63*(1), 1655. https://doi.org/10.4102/KOEDOE.V63I1.1655

Nikolskiy, P., & Pitulko, V. (2013). Evidence from the Yana Palaeolithic site, Arctic Siberia, yields clues to the riddle of mammoth hunting. *Journal of Archaeological Science, 40*(12), 4189–4197. https://doi.org/10.1016/j.jas.2013.05.020

Ning, H. (2016). *Asian elephants social structure and mineral lick usage in a Malaysian rainforest using camera traps*. University of Nottingham.

Nissani, M. (2006). Do Asian elephants (*Elephas maximus*) apply causal reasoning to tool-use tasks? *Journal of Animal Behavior Processes*, *32*(1), 91–96.

Norton-Griffiths, M. (1979). The influence of grazing, browsing and fire on the vegetation dynamics of the Serengeti. In A. R. E. Sinclair, & M. Norton-Griffiths (Eds.), *Serengeti: Dynamics of an ecosystem* (pp. 310–352). University of Chicago Press.

Ntumi, C., van Aarde, R., Fairall, N., & de Boer, W. (2005). Use of space and habitat by elephants in Maputo Elephant reserve, Mozambique. *African Journal of Wildlife Research*, *35*(2), 139–146.

Nyhus, P., & Tilson, R. (2004). Agroforestry, elephants, and tigers: Balancing conservation theory and practice in human-dominated landscapes of Southeast Asia. *Agriculture Ecosystems & Environment*, *104*(1), 87–97. https://doi.org/10.1016/j.agee.2004.01.009

O'Connell-Rodwell, C. E. (2007). Keeping an "ear" to the ground: Seismic communication in elephants. *Physiology*, *22*(4), 287–294. https://doi.org/10.1152/physiol.00008.2007

O'Connell-Rodwell, C. E., Arnason, B. T., & Hart, L. A. (2000). Seismic properties of Asian elephant (*Elephas maximus*) vocalizations and locomotion. *The Journal of the Acoustical Society of America*, *108*(6), 3066–3072. www.ncbi.nlm.nih.gov/pubmed/11144599

O'Connell-Rodwell, C. E., Wood, J. D., Gunther, R., Klemperer, S., Rodwell, T. C., Puria, S., Sapolsky, R., Kinzley, C., Arnason, B. T., & Hart, L. A. (2004). Elephant low-frequency vocalizations propagate in the ground and seismic playbacks of these vocalizations are detectable by wild African elephants (*Loxodonta africana*). *The Journal of the Acoustical Society of America*, *115*(5), 2554–2554. https://doi.org/10.1121/1.4783836

O'Connell-Rodwell, C. E., Wood, J. D., Rodwell, T. C., Puria, S., Partan, S. R., Keefe, R., Shriver, D., Arnason, B. T., & Hart, L. (2006). Wild elephant (*Loxodonta africana*) breeding herds respond to artificially transmitted seismic stimuli. *Behavioral Ecology and Sociobiology*, *59*(6), 842–850. https://doi.org/10.1007/s00265-005-0136-2

O'Connell-Rodwell, C. E., Wood, J. D., Kinzley, C., Rodwell, T. C., Poole, J. H., & Puria, S. (2007). Wild African elephants (*Loxodonta africana*) discriminate between familiar and unfamiliar conspecific seismic alarm calls. *Journal of the Acoustical Society of America*, *122*(2), 823–830. https://doi.org/10.1121/1.2747161

O'Connell-Rodwell, C. E., Wood, J. D., Kinzley, C., Rodwell, T. C., Alarcon, C., Wasser, S. K., & Sapolsky, R. (2011). Male African elephants (*Loxodonta africana*) queue when the stakes are high. *Ethology Ecology & Evolution*, *23*(4), 388–397. https://doi.org/10.1080/03949370.2011.598569

Oba, G., Stenseth, N. C., & Lusigi, W. J. (2000). New perspectives on sustainable grazing management in arid zones of sub-Saharan Africa. *BioScience*, *50*(1), 35–51. https://doi.org/10.1641/0006-3568(2000)050[0035:NPOSGM]2.3.CO;2

Obanda, V., Mutinda, N. M., Gakuya, F., & Iwaki, T. (2011). Gastrointestinal parasites and associated pathological lesions in starving free-ranging African elephants. *South African Journal of Wildlife Research*, *41*(2), 167–172. https://doi.org/10.10520/EJC117379

Obura, D. O., Declerck, F., Verburg, P. H., Gupta, J., Abrams, J. F., Bai, X., Bunn, S., Ebi, K. L., Gifford, L., Gordon, C., Jacobson, L., Lenton, T. M., & Liverman, D. (2022). Achieving a nature- and people-positive future. *One Earth*. https://doi.org/10.1016/j.oneear.2022.11.013

Odling-Smee, F. J., Laland, K. N., & Feldman, M. W. (2003). *Niche construction: The neglected process in evolution*. (No. 37). Princeton University Press.

Okita-Ouma, B., Koskei, M., Tiller, L., Lala, F., King, L., Moller, R., Amin, R., & Douglas-Hamilton, I. (2021). Effectiveness of wildlife underpasses and culverts in connecting elephant habitats: A case study of new railway through Kenya's Tsavo National Parks. *African Journal of Ecology*, *59*(3), 624–640. https://doi.org/10.1111/AJE.12873

Olivier, R. (1978). Distribution and Status of the Asian Elephant. *Oryx*, *14*(4), 379–424.

Olson J. et al., (2004). The spatial patterns and root causes of land use change in East Africa. LUCID Working Paper.

Orbach, D. N. (2019). Sexual strategies: Male and female mating tactics. In *Ethology and behavioral ecology of odontocetes* (pp. 75–93). Springer. https://doi.org/10.1007/978-3-030-16663-2_4

Othman, N. B. (2017). *Behaviour and Spatial Ecology of the Bornean elephant (Elephas maximus borneensis) in Lower Kinabatangan, Sabah, Malaysia Borneo*. Cardiff University.

Overington, S. E., Griffin, A. S., Sol, D., & Lefebvre, L. (2011). Are innovative species ecological generalists? A test in North American birds. *Behavioral Ecology*, *22*(6), 1286–1293. https://doi.org/10.1093/BEHECO/ARR130

Palao, E. (2021). Social animals and the potential for morality: On the cultural exaptation of behavioral capacities required for normativity. In *Synthese library* (Vol. 437, pp. 111–134). Springer Science and Business Media B.V. https://doi.org/10.1007/978-3-030-68802-8_6

Panichev, A. M., Golokhvast, K. S., Gulkov, A. N., & Chekryzhov, I. Y. (2013). Geophagy in animals and geology of kudurs (mineral licks): A review of Russian publications. *Environmental Geochemistry and Health*, *35*(1), 133–152. https://doi.org/10.1007/s10653-012-9464-0

Pardo, M. A., Poole, J. H., Stoeger, A. S., Wrege, P. H., O'Connell-Rodwell, C. E., Padmalal, U. K., & de Silva, S. (2019). Differences in combinatorial calls among the 3 elephant species cannot be explained by phylogeny. *Behavioral Ecology*, *30*(3), 809–820. https://doi.org/10.1093/beheco/arz018

Pardo, M. A., Fristrup, K., Lolchuragi, D. S., Poole, J. H., Granli, P., Moss, C., Douglas-Hamilton, I., & Wittemyer, G. (2024). African elephants address one another with individually specific name-like calls. *Nature Ecology & Evolution*, 1–12.

Pastorini, J., Nishantha, H. G., Janaka, H. K., Isler, K., & Fernando, P. (2010). Water-body use by Asian elephants in Southern Sri Lanka. *Tropical Conservation Science*, *3*(4), 412–422. https://doi.org/10.1177/194008291000300406

Pastorini, J., Janaka, H. K., Nishantha, H. G., Prasad, T., Leimgruber, P., & Fernando, P. (2013). A preliminary study on the impact of changing shifting cultivation practices on dry season forage for Asian elephants in Sri Lanka. *Tropical Conservation Science*, *6*(6), 770–780.

Payne, K. B., Langbauer, W. R., Thomas, E. M., & Jr, W. R. L. (1986). Infrasonic calls of the Asian elephant (*Elephas maximus*). *Behavioral Ecology and Sociobiology*, *18*(4), 297–301.

Pelletier, F., Mainguy, J., & Côté, S. D. (2009). Rut-Induced hypophagia in male bighorn sheep and mountain goats: Foraging under time budget constraints. *Ethology*, *115*(2), 141–151. https://doi.org/10.1111/J.1439-0310.2008.01589.X

Péron, F., Rat-Fischer, L., Lalot, M., Nagle, L., & Bovet, D. (2011). Cooperative problem solving in African grey parrots (*Psittacus erithacus*). *Animal Cognition*, *14*(4), 545–553. https://doi.org/10.1007/S10071-011-0389-2

Perrin, T. E., & Rasmussen, L. E. L. (1994). Chemosensory responses of female Asian elephants (*Elephas maximus*) to cyclohexanone. *Journal of Chemical Ecology*, *20*(11), 2857–2866. https://doi.org/10.1007/BF02098394

Peters, R. H. (1986). *The ecological implications of body size*. Cambridge University Press.

Pianka, E. R. (1970). On r-and K-selection. *The American Naturalist*, *104*(940), 592–597.

Pilot, M., Dahlheim, M. E., & Hoelzel, A. R. (2010). Social cohesion among kin, gene flow without dispersal and the evolution of population genetic structure in the killer whale (*Orcinus orca*). *Journal of Evolutionary Biology*, *23*(1), 20–31. https://doi.org/10.1111/j.1420-9101.2009.01887.x

Pinter-Wollman, N., Isbell, L. A., & Hart, L. A. (2009). The relationship between social behaviour and habitat familiarity in African elephants (*Loxodonta africana*). Proceedings. *Biological Sciences / The Royal Society*, *276*(1659), 1009–1014. https://doi.org/10.1098/rspb.2008.1538

Plaza, P. I., & Lambertucci, S. A. (2017). How are garbage dumps impacting vertebrate demography, health, and conservation? *Global Ecology and Conservation*, *12*, 9–20. https://doi.org/10.1016/j.gecco.2017.08.002

Plieninger, T., Ferranto, S., Huntsinger, L., Kelly, M., & Getz, C. (2012). Appreciation, use, and management of biodiversity and ecosystem services in California's working landscapes. *Environmental Management*, *50*(3), 427–440. https://doi.org/10.1007/s00267-012-9900-z

Plotnik, J. M., & de Waal, F. B. M. (2014). Asian elephants (*Elephas maximus*) reassure others in distress. *PeerJ*, *2*, e278. https://doi.org/10.7717/peerj.278

Plotnik, J. M., de Waal, F. B. M., & Reiss, D. (2006). Self-recognition in an Asian elephant. *Proceedings of the National Academy of Sciences of the United States of America*, *103*(45), 17053–17057. https://doi.org/10.1073/pnas.0608062103

Plotnik, J. M., Lair, R., Suphachoksahakun, W., & De Waal, F. B. M. (2011). Elephants know when they need a helping trunk in a cooperative task. *Proceedings of the National Academy of Sciences of the United States of America*, *108*(12), 5116–5121. https://doi.org/10.1073/PNAS.1101765108

Plotnik, J. M., Brubaker, D. L., Dale, R., Tiller, L. N., Mumby, H. S., & Clayton, N. S. (2019). Elephants have a nose for quantity. *Proceedings of the National Academy of Sciences of the United States of America*, *116*(25), 12566–12571. https://doi.org/10.1073/pnas.1818284116

Pokharel, S. S., Seshagiri, P. B., & Sukumar, R. (2017). Assessment of season-dependent body condition scores in relation to faecal glucocorticoid metabolites in free-ranging Asian elephants. *Conservation Physiology*, *5*(1). https://doi.org/10.1093/CONPHYS/COX039

Pokharel, S. S., Sharma, N., & Sukumar, R. (2022). Viewing the rare through public lenses: Insights into dead calf carrying and other thanatological responses in Asian elephants using YouTube videos. *Royal Society Open Science*, *9*(5), 211740. https://doi.org/10.1098/rsos.211740

Polansky, L., Kilian, W., & Wittemyer, G. (2015). Elucidating the significance of spatial memory on movement decisions by African savannah elephants using state–space models. *Proceedings of the Royal Society B: Biological Sciences*, *282*(1805). https://doi.org/10.1098/rspb.2014.3042

Polla, E. J., Grueter, C. C., & Smith, C. L. (2018). Asian Elephants (*Elephas maximus*) discriminate between familiar and unfamiliar human visual and olfactory cues. *Animal Behavior and Cognition*, *5*(3), 279–291.

Poole, J. H. (1987). Rutting behavior in elephants: The phenomenon of musth in African elephants. *Animal Behaviour*, *102*, 283–316.

Poole, J. H. (1989a). Announcing intent: The aggressive state of musth in African elephants. *Animal Behaviour*, *37*(PART 1), 153–155. https://doi.org/10.1016/0003-3472(89)90015-8

Poole, J. H. (1989b). Mate guarding, reproductive success and female choice in African elephants. *Animal Behaviour*, *37*, 842–849.

Poole, J. H. (1999). Signals and assessment in African elephants: Evidence from playback experiments. *Animal Behaviour*, *58*(1), 185–193. https://doi.org/10.1006/anbe.1999.1117

Poole, J. H., & Granli, P. (2021). The Elephant Ethogram. https://elephantvoices.org/elephant-ethogram.html

Poole, J. H., & Moss, C. J. (1981). Musth in the African elephant, *Loxodonta africana*. *Nature*, *292*(5826), 830–831. https://doi.org/10.1038/292830a0

Poole, J. H., Payne, K., Langbauer, W. R., & Moss, C. J. (1988). The social contexts of some very low-frequency calls of African elephants. *Behavioral Ecology and Sociobiology*, *22*, 385–392.

Poole, J. H., Tyack, P. L., Stoeger-Horwath, A. S., & Watwood, S. (2005). Elephants are capable of vocal learning. *Nature*, *434*(7032), 455–456. https://doi.org/10.1038/434455a

Poole, J. H., Lee, P. C., Njiraini, N., & Moss, C. J. (2011). Longevity, competition, and musth: A long-term perspective on male reproductive strategies. In Moss, H. Croze, & P. C. Lee (Eds.), *The Amboseli elephants: A long-term perspective on a long-lived mammal* (pp. 272–286). University of Chicago Press.

Pooley, S., Bhatia, S., & Vasava, A. (2021). Rethinking the study of human–wildlife coexistence. *Conservation Biology*, *35*(3), 784–793. https://doi.org/10.1111/cobi.13653

Potts, R. (1989). Olorgesailie: New excavations and findings in Early and Middle Pleistocene contexts, southern Kenya rift valley. *Journal of Human Evolution*, *18*(5), 477–484. https://doi.org/10.1016/0047-2484(89)90076-6

Povinelli, D. (1989). Failure to find self-recognition in Asian elephants (*Elephas maximus*) in contrast to their use of mirror cues to discover hidden food. *Journal of Comparative Psychology*, *103*(2), 122–131. https://psycnet.apa.org/record/1989-31932-001

Pozo, R. A., Coulson, T., McCulloch, G., Stronza, A., & Songhurst, A. (2019). Chilli-briquettes modify the temporal behaviour of elephants, but not their numbers. *Oryx*, *53*(1), 100–108. https://doi.org/10.1017/S0030605317001235

Pringle, R. M., Goheen, J. R., Palmer, T. M., Charles, G. K., DeFranco, E., Hohbein, R., Ford, A. T., & Tarnita, C. E. (2014). Low functional redundancy among mammalian browsers in regulating an encroaching shrub (*Solanum campylacanthum*) in African savannah. *Proceedings of the Royal Society B: Biological Sciences*, *281*(1785), 20140390.

Purdon, A., & van Aarde, R. J. (2017). Water provisioning in Kruger National Park alters elephant spatial utilisation patterns. *Journal of Arid Environments*, *141*, 45–51. https://doi.org/10.1016/J.JARIDENV.2017.01.014

Pusey, A. (1987). Sex-biased dispersal and inbreeding avoidance in birds and mammals. *Trends in Ecology and Evolution*, *2*(10), 295–299. https://doi.org/10.1016/0169-5347(87)90081-4

Pusey, A., & Wolf, M. (1996). Inbreeding avoidance in animals. *Trends in Ecology and Evolution*, *11*(5), 201–206. https://doi.org/10.1016/0169-5347(96)10028-8

Qiu, J. (2016). How China is rewriting the book on human origins. *Nature*, *535*(7611), 218–220. https://doi.org/10.1038/535218a

Ram, A. K., Yadav, N. K., Kandel, P. N., Mondol, S., Pandav, B., Natarajan, L., Subedi, N., Naha, D., Reddy, C. S., & Lamichhane, B. R. (2021). Tracking forest loss and fragmentation between 1930 and 2020 in Asian elephant (*Elephas maximus*) range in Nepal. *Scientific Reports*, *11*(1), 19514. https://doi.org/10.1038/s41598-021-98327-8

Ramsey, F. P. (1928). A mathematical theory of saving. *The Economic Journal*, *38*(152), 543–559.

Ranganathan, J., Daniels, R. J. R., Chandran, M. D. S., Ehrlich, P. R., & Daily, G. C. (2008). Sustaining biodiversity in ancient tropical countryside. *Proceedings of the National Academy of Sciences of the United States of America*, *105*(46), 17852–17854. https://doi.org/10.1073/pnas.0808874105

Rasmussen, H. B. (2005). *Reproductive tactics of male African savannah elephants (Loxodonta africana)*. University of Oxford.

Rasmussen, H. B., Okello, J. B. A., Wittemyer, G., Siegismund, H. R., Arctander, P., Vollrath, F., & Douglas-Hamilton, I. (2007). Age- and tactic-related paternity success in male African elephants. *Behavioral Ecology*, *19*(1), 9–15. https://doi.org/10.1093/beheco/arm093

Rasmussen, L. E. L. (1988). Chemosensory responses in two species of elephants to constituents of temporal gland secretion and musth urine. *Journal of Chemical Ecology*, *14*(8), 1687–1711. https://doi.org/10.1007/BF01014552

Rasmussen, L. E. L. (2001). Source and cyclic release pattern of (Z)-7-dodecenyl acetate, the pre-ovulatory pheromone of the female Asian elephant. *Chemical Senses*, *26*(6), 611–623. https://doi.org/10.1093/CHEMSE/26.6.611

Rasmussen, L. E. L., & Munger, B. L. (1996). The sensorineural specializations of the trunk tip (finger) of the Asian elephant, *Elephas maximus*. *The Anatomical Record*, *246*, 127–134. https://doi.org/10.1002/(SICI)1097-0185(199609)246:1

Rasmussen, L. E. L., & Greenwood, D. R. (2003). Frontalin: A chemical message of musth in Asian elephants (*Elephas maximus*). *Chemical Senses*, *28*(5), 433–446. https://doi.org/10.1093/CHEMSE/28.5.433

Rasmussen, L. E. L., Schmidt, M. J., Henneous, R., Groves, D., & Daves, G. D. (1982). Asian Bull elephants: Flehmen-like responses to extractable components in female elephant estrous urine. *Science*, *217*(4555), 159–162. https://doi.org/10.1126/SCIENCE.7089549

Rathnayake, C. W. M, Jones, S., Soto-Berelov, M., & Wallace, L. (2022a). Assessing protected area networks in the conservation of elephants (*Elephas maximus*) in Sri Lanka. *Environmental Challenges*, *9*, 100625. https://doi.org/10.1016/j.envc.2022.100625

Rathnayake, C. W. M., Jones, S., Soto-Berelov, M., & Wallace, L. (2022b). Human–elephant conflict and land cover change in Sri Lanka. *Applied Geography*, *143*, 102685. https://doi.org/10.1016/j.apgeog.2022.102685

Ratnam, J., Bond, W. J., Fensham, R. J., Hoffmann, W. A., Archibald, S., Lehmann, C. E. R., Anderson, M. T., Higgins, S. I., & Sankaran, M. (2011). When is a “forest” a savanna, and why does it matter? *Global Ecology and Biogeography*, *20*(5), 653–660. https://doi.org/10.1111/j.1466-8238.2010.00634.x

Ratnam, J., Tomlinson, K. W., Rasquinha, D. N., & Sankaran, M. (2016). Savannahs of Asia: Antiquity, biogeography, and an uncertain future. *Philosophical Transactions of the Royal Society B: Biological Sciences*, *371*(1703). The Royal Society. https://doi.org/10.1098/rstb.2015.0305

Raubenheimer, E. J., & Miniggio, H. D. (2016). Ivory harvesting pressure on the genome of the African elephants: A phenotypic shift to tusklessness. *Head and Neck Pathology*, *10*(3), 332–335. https://doi.org/10.1007/S12105-016-0704-Y

Reich, D., Green, R. E., Kircher, M., Krause, J., Patterson, N., Durand, E. Y., Viola, B., Briggs, A. W., Stenzel, U., Johnson, P. L. F., Maricic, T., Good, J. M., Marques-Bonet, T., Alkan, C., Fu, Q., Mallick, S., Li, H., Meyer, M., Eichler, E. E., … Pääbo, S. (2010). Genetic history of an archaic hominin group from Denisova Cave in Siberia. *Nature*, *468*(7327), 1053–1060. https://doi.org/10.1038/nature09710

Reid, A. (1995). Humans and forests in pre-colonial Southeast Asia. *Environment & History*, *1*(1), 93–110. https://doi.org/10.3197/096734095779522717

Rendell, L., Mesnick, S. L., Dalebout, M. L., Burtenshaw, J., & Whitehead, H. (2012). Can genetic differences explain vocal dialect variation in sperm whales, *Physeter macrocephalus*? *Behavior Genetics*, *42*(2), 332–343. https://doi.org/10.1007/S10519-011-9513-Y/FIGURES/3

Ripple, W. J., & Beschta, R. L. (2012). Trophic cascades in Yellowstone: The first 15 years after wolf reintroduction. *Biological Conservation*, *145*(1), 205–213. https://doi.org/10.1016/j.biocon.2011.11.005

Ripple, W. J., Newsome, T. M., Wolf, C., Dirzo, R., Everatt, K. T., Hayward, M. W., Kerley, G. I. H., Levi, T., Lindsey, P. A., Macdonald, D. W., Malhi, Y., Painter, L. E., Sandom, C. J.,

Terborgh, J., & Valkenburgh, B. Van. (2015). Collapse of the world's largest herbivores. *Science Advances*, *1*(4), e1400103. https://doi.org/10.1126/sciadv.1400103

Robinson, M. R., Mar, K. U., & Lummaa, V. (2012). Senescence and age-specific trade-offs between reproduction and survival in female Asian elephants. *Ecology Letters*, *15*, 260–266. https://doi.org/10.1111/j.1461-0248.2011.01735.x

Roca, A. L. (2001). Genetic evidence for two species of elephant in Africa. *Science*, *293*(5534), 1473–1477. https://doi.org/10.1126/science.1059936

Roca, A. L, Georgiadis, N., & O'Brien, S. J. (2005). Cytonuclear genomic dissociation in African elephant species. *Nature Genetics*, *37*(1), 96–100. https://doi.org/10.1038/ng1485

Rohland, N., Malaspinas, A.-S., Pollack, J. L., Slatkin, M., Matheus, P., & Hofreiter, M. (2007). Proboscidean mitogenomics: Chronology and mode of elephant evolution using mastodon as outgroup. *PLOS Biology*, *5*(8), e207. https://doi.org/10.1371/journal.pbio.0050207

Rohland, N., Reich, D., Mallick, S., Meyer, M., Green, R. E., Georgiadis, N. J., Roca, A. L., & Hofreiter, M. (2010). Genomic DNA sequences from mastodon and woolly mammoth reveal deep speciation of forest and savanna elephants. *PLOS Biology*, *8*(12), e1000564. https://doi.org/10.1371/journal.pbio.1000564

Romero, L., & Beattie, U. (2022). Common myths of glucocorticoid function in ecology and conservation. *Journal of Experimental Zoology Part A: Ecological and Integrative Physiology*, *337*(1), 7–14. https://doi.org/10.1002/jez.2459

Rood, E., Ganie, A. A., & Nijman, V. (2010). Using presence-only modelling to predict Asian elephant habitat use in a tropical forest landscape: Implications for conservation. *Diversity and Distributions*, *16*(6), 975–984. https://doi.org/10.1111/j.1472-4642.2010.00704.x

Rudel, T. K. (2007). Changing agents of deforestation: From state-initiated to enterprise driven processes, 1970–2000. *Land Use Policy*, *24*(1), 35–41. https://doi.org/10.1016/J.LANDUSEPOL.2005.11.004

Saif, O., Kansky, R., Palash, A., Kidd, M., & Knight, A. (2019). Costs of coexistence: Understanding the drivers of tolerance towards Asian elephants *Elephas maximus* in rural Bangladesh. *Oryx*, *54*(5), 1–9. https://doi.org/10.1017/S0030605318001072

Sanderson, G. P. (1882). Thirteen years among the wild beasts of India: Their haunts and habits from personal observation; With an account of the modes of capturing and taming elephants. First published: London, 1882. Reprint: Asian Educational Services, New Delhi, 2000.

Sandom, C., Faurby, S., Sandel, B., & Svenning, J.-C. (2014). Global late Quaternary megafauna extinctions linked to humans, not climate change. *Proceedings of the Royal Society B: Biological Sciences*, *281*(1787), 20133254–20133254. https://doi.org/10.1098/rspb.2013.3254

Sankaran, M., Hanan, N. P., Scholes, R. J., Ratnam, J., Augustine, D. J., Cade, B. S., Gignoux, J., Higgins, S. I., Le Roux, X., Ludwig, F., Ardo, J., Banyikwa, F., Bronn, A., Bucini, G., Caylor, K. K., Coughenour, M. B., Diouf, A., Ekaya, W., Feral, C. J., … Zambatis, N. (2005). Determinants of woody cover in African savannas. *Nature*, *438*, 846–849.

Santiapillai, C., & Read, B. (2010). Would masking the smell of ripening paddy-fields help mitigate human-elephant conflict in Sri Lanka? *Oryx*, *44*(4), 509–511. https://doi.org/10.1017/s0030605310000906

Sapolsky, R. M. (2005). The influence of social hierarchy on primate health. *Science*, *308*(2005), 648–652. https://doi.org/10.1126/science.1106477

Sapolsky, R. M. (2007). *A primate's memoir*. Simon and Schuster.

Sapolsky, R. M., & Else, J. G. (1987). Bovine tuberculosis in a wild baboon population: Epidemological aspects. *Journal of Medical Primatology*, *16*(4), 229–235.

Schmelz, M., Duguid, S., Bohn, M., & Völter, C. J. (2017). Cooperative problem solving in giant otters (*Pteronura brasiliensis*) and Asian small-clawed otters (*Aonyx cinerea*). *Animal Cognition*, *20*(6), 1107–1114. https://doi.org/10.1007/S10071-017-1126-2

Schmitt, M. H., Shuttleworth, A., Shrader, A. M., & Ward, D. (2020). The role of volatile plant secondary metabolites as pre-ingestive cues and potential toxins dictating diet selection by African elephants. *Oikos*, *129*(1), 24–34. https://doi.org/10.1111/OIK.06665

Schmitz, O. J., Beckerman, A. P., & O'Brien, K. M. (1999). Behaviorally mediated trophic cascades. *Ecology*, *78*(5), 1388–1399.

Schröder, A., Persson, L., & De Roos, A. M. (2005). Direct experimental evidence for alternative stable states: A review. *Oikos*, *110*(1), 3–19. https://doi.org/10.1111/j.0030-1299.2005.13962.x

Schulte, B. A., & LaDue, C. A. (2021). The chemical ecology of elephants: 21st century additions to our understanding and future outlooks. *Animals 2021*, *11*(10), 2860. https://doi.org/10.3390/ANI11102860

Schulte, B. A., & Rasmussen, L. E. L. (1999). Signal-receiver interplay in the communication of male condition by Asian elephants. *Animal Behaviour*, *57*(6), 1265–1274. https://doi.org/10.1006/anbe.1999.1092

Schulte, B. A., Freeman, E. W., Goodwin, T. E., Hollister-Smith, J., & Rasmussen, L. E. L. (2007). Honest signalling through chemicals by elephants with applications for care and conservation. *Applied Animal Behaviour Science*, *102*(3–4), 344–363. https://doi.org/10.1016/J.APPLANIM.2006.05.035

Schuttler, S. G., Philbrick, J. A., Jeffery, K. J., & Eggert, L. S. (2014a). Fine-scale genetic structure and cryptic associations reveal evidence of kin-based sociality in the African forest elephant. *PLOS One*, *9*(2). https://doi.org/10.1371/journal.pone.0088074

Schuttler, S. G., Whittaker, A., Jeffery, K. J., & Eggert, L. S. (2014b). African forest elephant social networks: Fission-fusion dynamics, but fewer associations. *Endangered Species Research*, *25*(2), 165–173. https://doi.org/10.3354/esr00618

Seaworld. (2019). Diet and eating habits. https://seaworld.org/animals/all-about/elephants/diet/

Seed, A. M., Clayton, N. S., & Emery, N. J. (2008). Cooperative problem solving in rooks (*Corvus frugilegus*). *Proceedings. Biological Sciences / The Royal Society*, *275*(1641), 1421–1429. https://doi.org/10.1098/rspb.2008.0111

Sekar, N., & Sukumar, R. (2013). Waiting for Gajah: An elephant mutualist's contingency plan for an endangered megafaunal disperser. *Journal of Ecology*, *101*(6), 1379–1388. https://doi.org/10.1111/1365-2745.12157

Seltmann, M. W., Helle, S., Htut, W., & Lahdenperä, M. (2019). Males have more aggressive and less sociable personalities than females in semi-captive Asian elephants. *Scientific Reports*, *9*(1), 1–7. https://doi.org/10.1038/s41598-019-39915-7

Shaffer, L. J. (2010). Indigenous fire use to manage savanna landscapes in southern mozambique. *Fire Ecology*, *6*(2), 43–59. https://doi.org/10.4996/fireecology.0602043

Shaffer, L. J., Khadka, K. K., Van Den Hoek, J., & Naithani, K. J. (2019). Human-elephant conflict: A review of current management strategies and future directions. *Frontiers in Ecology and Evolution*, *6*(JAN), 1–12. https://doi.org/10.3389/fevo.2018.00235

Shannon, G., Slotow, R., Durant, S. M., Sayialel, K. N., Poole, J., Moss, C., & McComb, K. (2013). Effects of social disruption in elephants persist decades after culling. *Frontiers in Zoology*, *10*(1). https://doi.org/10.1186/1742-9994-10-62

Shannon, G., Cordes, L. S., Hardy, A. R., Angeloni, L. M., & Crooks, K. R. (2014). Behavioral responses associated with a human-mediated predator shelter. *PLOS One*, *9*(4), e94630. https://doi.org/10.1371/JOURNAL.PONE.0094630

Shannon, G., Cordes, L. S., Slotow, R., Moss, C., & McComb, K. (2022). Social disruption impairs predatory threat assessment in African elephants. *Animals*, *12*(4), 495. https://doi.org/10.3390/ANI12040495/S1

Sharma, R., Goossens, B., Heller, R., Rasteiro, R., Othman, N., W., M. W. B., & Chikhi, L. (2018). Genetic analyses favour an ancient and natural origin of elephants on Borneo. *Scientific Reports*, *8*(1), 880.

Sharma, N., Pokharel, S. S., Shiro, K., & Sukumar, R. (2019). Behavioural responses of free-ranging Asian elephants (*Elephas maximus*) towards dying and dead conspecifics. *Primates*, *61*, 129–138. https://doi.org/10.1007/s10329-019-00739-8

Sharma, N., Prakash S, V., Kohshima, S., & Sukumar, R. (2020). Asian elephants modulate their vocalizations when disturbed. *Animal Behaviour*, *160*, 99–111. https://doi.org/10.1016/J.ANBEHAV.2019.12.004

Shipman, P. (2015). *The Invaders: How humans and their dogs drove neanderthals to extinction*. The Belknap Press of Harvard University Press Cambridge, England.

Shoshani, J., & Tassy, P. (1996). *The proboscidea: Evolution and palaeoecology of elephants and their relatives*. Oxford University Press.

Shoshani, J., & Tassy, P. (2005). Advances in proboscidean taxonomy & classification, anatomy & physiology, and ecology & behavior. *Quaternary International*, *126–128*, 5–20. https://doi.org/10.1016/j.quaint.2004.04.011

Shoshani, J., Ferretti, M. P., Lister, A. M., Agenbroad, L. D., Saegusa, H., Mol, D., & Takahashi, K. (2007). Relationships within the *Elephantidae* using hyoid characters. *Quaternary International*, *169–170*(SPEC. ISS.), 174–185. https://doi.org/10.1016/j.quaint.2007.02.003

Shyan-Norwalt, M. R., Peterson, J., King, B. M., Staggs, T. E., & Dale, R. H. I. (2010). Initial findings on visual acuity thresholds in an African elephant (*Loxodonta africana*). *Zoo Biology*, *29*(1), 30–35. https://doi.org/10.1002/ZOO.20259

Sih, A., Mathot, K. J., Moirón, M., Montiglio, P. O., Wolf, M., & Dingemanse, N. J. (2015). Animal personality and state–behaviour feedbacks: A review and guide for empiricists. *Trends in Ecology & Evolution*, *30*(1), 50–60. https://doi.org/10.1016/J.TREE.2014.11.004

Silk, J. B. (2007). The adaptive value of sociality in mammalian groups. *Philosophical Transactions of the Royal Society of London, Series B, Biological Sciences*, *362*(1480), 539–559. https://doi.org/10.1098/rstb.2006.1994

Silva, K. A. I. D. (2011). *Action plan for implementing the Convention on Biological Diversity's programme of work on protected areas, Sri Lanka*. Sri Lanka Ministry of Environment.

Sirua, H. (2006). Nature above people: Rolston and "fortress" conservation in the South. *Ethics and the Environment*, *11*(1), 71–96. www.jstor.org/stable/40339115

Sitters, J., Kimuyu, D., Young, T., Claeys, P., & Olde Venterink, H. (2020). Negative effects of cattle on soil carbon and nutrient pools reversed by megaherbivores. *Nature*, *3*(5), 360–366. https://doi.org/10.1038/s41893-020-0490-0

Skinner, J. D., Carruth, B. R., Bounds, W., & Ziegler, P. J. (2002). Children's food preferences: A longitudinal analysis. *Journal of the American Dietetic Association*, *102*(11), 1638–1647. https://doi.org/10.1016/S0002-8223(02)90349-4

Slotow, R., & van Dyk, G. (2001). Role of delinquent young "orphan" male elephants in high mortality of white rhinoceros in Pilanesberg National Park, South Africa. *Koedoe*, *44*(1), 85–94.

Slotow, R., Dyk, G. Van, Poole, J., Page, B., Klocke, A., Wang, C., & Kai, T. (2000). Older bull elephants control young males. *Nature*, *408*(November), 425–426.

Small, E. (2019). In defence of the world's most reviled invertebrate "bugs." *Biodiversity*, *20*(4), 168–221. https://doi.org/10.1080/14888386.2019.1663636

Smith, J. M. B. (2016). *Savannahs*. Encyclopædia Britannica. www.britannica.com/science/savanna

Smithsonian Museum of Natural History. *Homo heidelbergensis*. What does it mean to be human? http://humanorigins.si.edu/evidence/human-fossils/species/homo-heidelbergensis

Smuts, B. B., Cheney, D. L., Seyfarth, R. M., Wrangham, R. W., & Struhsaker, T. T. (1987). *Primate societies*. University of Chicacgo Press.

Sol, D., Duncan, R. P., Blackburn, T. M., Cassey, P., & Lefebvre, L. (2005). Big brains, enhanced cognition, and response of birds to novel environments. *Proceedings of the National Academy of Sciences of the United States of America*, *102*(15), 5460–5465. https://doi.org/10.1073/PNAS.0408145102

Soltis, J. (2010). Vocal communication in African Elephants (*Loxodonta africana*). *Zoo Biology*, *29*(2), 192–209. https://doi.org/10.1002/zoo.20251

Soltis, J., Leong, K., & Savage, A. (2005). African elephant vocal communication II: Rumble variation reflects the individual identity and emotional state of callers. *Animal Behaviour*, *70*(3), 589–599. https://doi.org/10.1016/j.anbehav.2004.11.016

Soltis, J., Leighty, K. A., Wesolek, C. M., & Savage, A. (2009). The expression of affect in African Elephant (*Loxodonta africana*) rumble vocalizations. *Journal of Comparative Psychology*, *123*(2), 222–225. https://doi.org/10.1037/A0015223

Soltis, J., Blowers, T. E., & Savage, A. (2011). Measuring positive and negative affect in the voiced sounds of African elephants (*Loxodonta africana*). *The Journal of the Acoustical Society of America*, *129*(2), 1059–1066. https://doi.org/10.1121/1.3531798

Sordello, R., Ratel, O., De Lachapelle, F. F., Leger, C., Dambry, A., & Vanpeene, S. (2020). Evidence of the impact of noise pollution on biodiversity: A systematic map. *Environmental Evidence*, *9*(1), 1–27. https://doi.org/10.1186/S13750-020-00202-Y

Soulsbury, C. D. (2019). Income and capital breeding in males: Energetic and physiological limitations on m ale mating strategies. *Journal of Experimental Biology*, *222*(1). https://doi.org/10.1242/jeb.184895

Spiegel, O., Leu, S., Bull, C., & Sih, A. (2017). What's your move? Movement as a link between personality and spatial dynamics in animal populations. *Ecology Letters*, *20*(1), 3–18. https://doi.org/10.1111/ele.12708

Srinivasaiah, N., Kumar, V., Vaidyanathan, S., Sukumar, R., & Sinha, A. (2019). All-male groups in Asian elephants: A novel, adaptive social strategy in increasingly anthropogenic landscapes of southern India. *Scientific Reports*, *9*(1), 8678. https://doi.org/10.1038/s41598-019-45130-1

Stamps, J., & Groothuis, T. G. G. (2010). The development of animal personality: Relevance, concepts and perspectives. *Biological Reviews*, *85*(2), 301–325. https://doi.org/10.1111/J.1469-185X.2009.00103.X

Stankowich, T., & Caro, T. (2009). Evolution of weaponry in female bovids. *Proceedings of the Royal Society B: Biological Sciences*, *276*(1677), 4329–4334. https://doi.org/10.1098/rspb.2009.1256

Staver, A. C., Bond, W. J., Stock, W. D., Van Rensburg, S. J., & Waldram, M. S. (2009). Browsing and fire interact to suppress tree density in an African savanna. *Ecological Applications*, *19*(7), 1909–1919. https://doi.org/10.1890/08-1907.1

Staver, A. C., Archibald, S., & Levin, S. (2011a). Tree cover in sub-Saharan Africa: Rainfall and fire constrain forest and savanna as alternative stable states. *Ecology*, *92*(5), 1063–1072. https://doi.org/10.1890/10-1684.1

Staver, A. C., Archibald, S., & Levin, S. A. (2011b). The global extent and determinants of savanna and forest as alternative biome states. *Science*, *334*(6053), 230–232. https://doi.org/10.1126/science.1210465

Steinheim, G., Wegge, P., Fjellstad, J. I., Jnawali, S. R., & Weladji, R. B. (2005). Dry season diets and habitat use of sympatric Asian elephants (*Elephas maximus*) and greater one-horned rhinoceros (*Rhinocerus unicornis*) in Nepal. *Journal of Zoology*, *265*(4), 377–385.

Stephens, P. A., Boyd, I. L., McNamara, J. M., & Houston, A. I. (2009). Capital breeding and income breeding: Their meaning, measurement and worth. *Ecology*, *90*(8), 2057–2067.

Sterck, E. H. M., & Watts, D. P. (1997). The evolution of female social relationships in nonhuman primates. *Behavioral Ecology and Sociobiology*, *41*, 291–309.

Štibrániová, I., Bartíková, P., Holíková, V., & Kazimírová, M. (2019). Deciphering biological processes at the tick-host interface opens new strategies for treatment of human diseases. *Frontiers in Physiology*, *10*, 1–21. https://doi.org/10.3389/fphys.2019.00830

Stimpson, C. M., Lister, A., Parton, A., Clark-Balzan, L., Breeze, P. S., Drake, N. A., Groucutt, H. S., Jennings, R., Scerri, E. M., White, T. S., & Zahir, M. (2016). Middle Pleistocene vertebrate fossils from the Nefud Desert, Saudi Arabia: Implications for biogeography and palaeoecology. *Quaternary Science Reviews*, *143*, 13–36.

Stoeger, A. S. (2021). Chapter 12 – Elephant sonic and infrasonic sound production, perception, and processing. In Cheryl S. Rosenfeld, & Frauke Hoffmann (Eds.), *Neuroendocrine regulation of animal vocalization: Mechanisms and anthropogenic factors in animal communication* (pp. 189–199). Academic Press. https://doi.org/10.1016/B978-0-12-815160-0.00023-2

Stoeger, A. S., & de Silva, S. (2013). African and Asian elephant vocal communication: A cross-species comparison. In G. Witzany (Ed.), *Biocommunication in animals* (pp. 21–39). Springer.

Stoeger, A. S., & Baotic, A. (2016). Information content and acoustic structure of male African elephant social rumbles. *Scientific Reports*, *6*(1), 27585. https://doi.org/10.1038/srep27585

Stoeger, A. S., & Baotic, A. (2021). Operant control and call usage learning in African elephants. *Philosophical Transactions of the Royal Society B, Biological Sciences*, *376*(1836). https://doi.org/10.1098/RSTB.2020.0254

Stoeger, A. S., Heilmann, G., Zeppelzauer, M., Ganswindt, A., Hensman, S., & Charlton, B. D. (2012a). Visualizing sound emission of elephant vocalizations: Evidence for two rumble production types. *PLOS One*, *7*(11), e48907. https://doi.org/10.1371/journal.pone.0048907

Stoeger, A. S., Mietchen, D., Oh, S., de Silva, S., Herbst, C. T., Kwon, S., & Fitch, W. T. (2012b). An Asian elephant imitates human speech. *Current Biology*, *22*(22), 2144–2148. https://doi.org/10.1016/j.cub.2012.09.022

Stoeger, A. S., Zeppelzauer, M., & Baotic, A. (2014). Age-group estimation in free-ranging African elephants based on acoustic cues of low-frequency rumbles. *Bioacoustics*, *23*(3), 231–246. https://doi.org/10.1080/09524622.2014.888375

Stoeger-Horwath A. S., Stoeger, S., Schwammer, H. M., & Kratochvil, H. (2007). Call repertoire of infant African elephants: First insights into the early vocal ontogeny. *Journal of the Acoustical Society of America*, *121*(6), 3922–3931.

Stone, J., & Halasz, P. (1989). Topography of the retina in the elephant *Loxodonta africana*. *Brain, Behavior and Evolution*, *34*(2), 84–95. www.karger.com/Article/Abstract/116494

Stuart, A. J. (2005). The extinction of woolly mammoth (*Mammuthus primigenius*) and straight-tusked elephant (*Palaeoloxodon antiquus*) in Europe. *Quaternary International*, *126*, 171–177. https://doi.org/10.1016/j.quaint.2004.04.021

Sukumar, R. (1989). *The Asian elephants: Ecology and management*. Cambridge University Press.

Sukumar, R. (1995). Elephant raiders and rogues. *Natural History*, *104*, 52–61.

Sukumar, R. (2003). *The Living Elephants*. Oxford University Press.

Sundaresan, S. R., Fischhoff, I. R., & Rubenstein, D. I. (2007). Male harassment influences female movements and associations in Grevy's zebra (*Equus grevyi*). *Behavioral Ecology*, *18*(5), 860–865. https://doi.org/10.1093/beheco/arm055

Surovell, T. A., & Waguespack, N. M. (2008). How many elephant kills are 14: Clovis mammoth and mastodon kills in context. *Quaternary International*, *191*(1), 82–97. https://doi.org/10.1016/j.quaint.2007.12.001

Surovell, T. A., Waguespack, N., & Brantingham, P. J. (2005). Global archaeological evidence for proboscidean overkill. *Proceedings of the National Academy of Sciences of the United States of America*, *102*(17), 6231–6236. https://doi.org/10.1073/pnas.0501947102

Sutherland, W. J. (1996). *From individual behaviour to population ecology*. Oxford University Press.

Tan, J. Z., & Hare, B. (2013). Bonobos share with strangers. *PLOS One*, *8*(1). https://doi.org/10.1371/journal.pone.0051922

Tassy, P. (1995). Les Proboscidiens (Mammalia) fossiles du Rift occidental, Ouganda. In B. Senut, & M. Pickford (Eds.), *Geology and palaeobiology of the Albertine Rift Valley, Uganda- Zaire*, Vol. 2, Palaeobiology/Pale´obiologie (Occasional, pp. 215–255). Centre International pour la Formation et les Echanges Ge´ologiques.

Teale, C. L., & Miller, N. G. (2012). Mastodon herbivory in mid-latitude late-Pleistocene boreal forests of eastern North America. *Quaternary Research*, *78*(1), 72–81. https://doi.org/10.1016/j.yqres.2012.04.002

Thuppil, V., & Coss, R. G. (2013). Wild Asian elephants distinguish aggressive tiger and leopard growls according to perceived danger. *Biology Letters*, *9*(5). https://doi.org/10.1098/RSBL.2013.0518

Tiller, L. N., Oniba, E., Opira, G., Brennan, E. J., King, L. E., Ndombi, V., Wanjala, D., & Robertson, M. R. (2022). "Smelly" elephant repellent: Assessing the efficacy of a novel olfactory approach to mitigating elephant crop raiding in Uganda and Kenya. *Diversity*, *14*(7), 509. https://doi.org/10.3390/D14070509

Tinbergen, N. (1963). On aims and methods of ethology. *Zeitschrift Für Tierpsychologie*, *20*(4), 410–433.

Tobias, J. A., Montgomerie, R., & Lyon, B. E. (2012). The evolution of female ornaments and weaponry: Social selection, sexual selection and ecological competition. *Philosophical Transactions of the Royal Society B: Biological Sciences*, *367*(1600), 2274–2293. https://doi.org/10.1098/rstb.2011.0280

Trivers, R. (1972). Parental investment and sexual selection. In B. Campbell (Ed.), *Sexual selection and the descent of man* (pp. 136–179). Aldine-Atherton.

Tsalyuk, M., Kilian, W., Reineking, B., & Getz, W. M. (2019). Temporal variation in resource selection of African elephants follows long-term variability in resource availability. *Ecological Monographs*, *89*(2), e01348. https://doi.org/10.1002/ECM.1348

Tshipa, A., Valls-Fox, H., Fritz, H., Collins, K., Sebele, L., Mundy, P., & Chamaillé-Jammes, S. (2017). Partial migration links local surface-water management to large-scale elephant conservation in the world's largest transfrontier conservation area. *Biological Conservation*, *215*(May), 46–50. https://doi.org/10.1016/j.biocon.2017.09.003

Tucker, M. A., Böhning-Gaese, K., Fagan, W. F., Fryxell, J. M., Moorter, B. Van, Alberts, S. C., Ali, A. H., Allen, A. M., Attias, N., Avgar, T., Bartlam-Brooks, H., Bayarbaatar, B., Belant, J. L., Bertassoni, A., Beyer, D., Bidner, L., Beest, F. M. van, Blake, S., Blaum, N., … Kranstauber, T. M. (2018). Moving in the Anthropocene: Global reductions in terrestrial mammalian movements. *Science*, *359*, 466–469.

Tudge, C. (2009). *The secret life of birds: Who they are and what they do*. Penguin.

Turkalo, A. K. (2001). Forest elephant behavior and ecology: Observations from the Dzanga saline. In W. Weber, L. J. T. White, A. Vedder, & L. Naughton-Treves (Eds.), *African Rainforest Ecology and Conservation* (pp. 207–213). Yale University Press.

Turkalo, A. K., Wrege, P. H., & Wittemyer, G. (2013). Long-term monitoring of Dzanga Bai forest elephants: Forest clearing use patterns. *PLOS One*, *8*(12), e85154. https://doi.org/10.1371/journal.pone.0085154

Turkalo, A. K., Wrege, P. H., & Wittemyer, G. (2017). Slow intrinsic growth rate in forest elephants indicates recovery from poaching will require decades. *Journal of Applied Ecology*, *54*(1), 153–159. https://doi.org/10.1111/1365-2664.12764

van Aarde, R. J., Pimm, S. L., Guldemond, R., Huang, R., & Maré, C. (2021). The 2020 elephant die-off in Botswana. *PeerJ*, *9*, e10686. https://doi.org/10.7717/PEERJ.10686/SUPP-2

van de Waal, E., Borgeaud, C., & Whiten, A. (2013). Potent social learning and conformity shape a wild primate's foraging decisions. *Science*, *340*(6131), 483–485. https://doi.org/10.1126/science.1232769

van Schaik, C. P., & van Hooff, J. (1983). On the ultimate causes of primate social systems. *Behaviour*, *85*(1–2), 91–117. https://doi.org/10.1163/156853983X00057

van Schaik, C. P., Pradhan, G. R., & van Noordwijk, M. A. (2004). Mating conflict in primates: Infanticide, sexual harassment and female sexuality. In P. M. Kappeler, & C. P. van Schaik (Eds.), *Sexual selection in primates: New and comparative perspectives* (Issue November, pp. 141–163). https://doi.org/10.1017/CBO9780511542459.010

van Woerden, J. T., Willems, E. P., van Schaik, C. P., & Isler, K. (2012). Large brains buffer energetic effects of seasonal habitats in catarrhine primates. *Evolution*, *66*(1), 191–199. https://doi.org/10.1111/J.1558-5646.2011.01434.X

Vanhooland, L. C., Szabó, A., Bugnyar, T., & Massen, J. J. M. (2022). A comparative study of mirror self-recognition in three corvid species. *Animal Cognition*, *26*(1), 229–248. https://doi.org/10.1007/S10071-022-01696-4

Veerman, J., Kumar, A., & Mishra, D. R. (2022). Exceptional landscape-wide cyanobacteria bloom in Okavango Delta, Botswana in 2020 coincided with a mass elephant die-off event. *Harmful Algae*, *111*, 102145. https://doi.org/10.1016/J.HAL.2021.102145

Vehrencamp, S. L. (1983). Optimal degree of skew in cooperative societies. *American Zoologist*, *23*(2), 327–335. https://doi.org/10.1093/icb/23.2.327

Veldman, J. W., Buisson, E., Durigan, G., Fernandes, G. W., Le Stradic, S., Mahy, G., Negreiros, D., Overbeck, G. E., Veldman, R. G., Zaloumis, N. P., Putz, F. E., & Bond, W. J. (2015). Toward an old-growth concept for grasslands, savannas, and woodlands. *Frontiers in Ecology and the Environment*, *13*(3), 154–162. https://doi.org/10.1890/140270

Ventura, A. K., & Worobey, J. (2013). Early influences on the development of food preferences. *Current Biology*, *23*(9), 401–408. https://doi.org/10.1016/j.cub.2013.02.037

Ventura, A. K., Phelan, S., & Silva Garcia, K. (2021). Maternal diet during pregnancy and lactation and child food preferences, dietary patterns, and weight outcomes: A review of recent research. *Current Nutrition Reports*, *10*, 413–426. https://doi.org/10.1007/S13668-021-00366-0

Verlinden, A., & Gavor, I. K. N. (1998). Satellite tracking of elephants in northern Botswana. *African Journal of Ecology*, *36*, 105–116.

Vetter, S. (2005). Rangelands at equilibrium and non-equilibrium: Recent developments in the debate. *Journal of Arid Environments*, *62*(2), 321–341. https://doi.org/10.1016/j.jaridenv.2004.11.015

Vidya, T. N. C., & Sukumar, R. (2002). The effect of some ecological factors on the intestinal parasite loads of the Asian elephant (*Elephas maximus*) in southern India. *Journal of Biosciences*, *27*(5), 521–528. https://doi.org/10.1007/BF02705050/METRICS

Vidya, T. N. C., & Sukumar, R. (2005). Social organization of the Asian elephant (*Elephas maximus*) in southern India inferred from microsatellite DNA. *Journal of Ethology*, *23*, 205–210.

Vidya, T. N. C., Fernando, P., Melnick, D. J., & Sukumar, R. (2005). Population differentiation within and among Asian elephant (*Elephas maximus*) populations in southern India. *Heredity*, *94*(1), 71–80. https://doi.org/10.1038/sj.hdy.6800568

Vidya, T. N. C., Sukumar, R., & Melnick, D. J. (2009). Range-wide mtDNA phylogeography yields insights into the origins of Asian elephants. *Proceedings of the Royal Society B: Biological Sciences*, *278*(1706), 798–798. https://doi.org/10.1098/rspb.2010.2088

Viljoen, P. J. (1988). *The ecology of the desert-dwelling elephants, Loxodonta africana (Blumenbach, 1797) of western Damaraland and Kaokoland*. University of Pretoria.

Viljoen, P. J. (1989). Habitat selection and preferred food plants of a desert-dwelling elephant population in the northern Namib Desert, South West Africa/Namibia. *African Journal of Ecology*, *27*(3), 227–240. https://doi.org/10.1111/j.1365-2028.1989.tb01016.x

Vincent, S., Ring, R., & Andrews, K. (2018). Normative practices of other animals. In A. Zimmerman, K. Jones, & M. Timmons (Eds.), *The Routledge handbook of moral epistemology* (pp. 57–83). Oxford. https://doi.org/10.4324/9781315719696-4

Von Gerhardt, K., Van Niekerk, A., Kidd, M., Samways, M., & Hanks, J. (2014). The role of elephant *Loxodonta africana* pathways as a spatial variable in crop-raiding location. *Oryx*, *48*(3), 436–444. https://doi.org/10.1017/S003060531200138X

Wall, J., Wittemyer, G., Klinkenberg, B., Lemay, V., & Douglas-hamilton, I. (2013). Characterizing properties and drivers of long distance movements by elephants (*Loxodonta africana*) in the Gourma, Mali. *Biological Conservation*, *157*, 60–68. https://doi.org/10.1016/j.biocon.2012.07.019

Wan, H. Y., Olson, A. C., Muncey, K. D., & St. Clair, S. B. (2014). Legacy effects of fire size and severity on forest regeneration, recruitment, and wildlife activity in aspen forests. *Forest Ecology and Management*, *329*, 59–68. https://doi.org/10.1016/J.FORECO.2014.06.006

Wang, H., Wang, P., Zhao, X., Zhang, W., Li, J., Xu, C., & Xie, P. (2021). What triggered the Asian elephant's northward migration across southwestern Yunnan? *The Innovation*, *2*(3), 1–2. https://doi.org/10.1016/j.xinn.2021.100142

Wasser, S. K., Hunt, K. E., Brown, J. L., Cooper, K., Crockett, C. M., Bechert, U., Millspaugh, J. J., Larson, S., & Monfort, S. L. (2000). A generalized fecal glucocorticoid assay for use in a diverse array of nondomestic mammalian and avian species. *General and Comparative Endocrinology*, *120*(3), 260–275. https://doi.org/10.1006/gcen.2000.7557

Watson, J. E. M., Shanahan, D. F., Di Marco, M., Allan, J., Laurance, W. F., Sanderson, E. W., Mackey, B., & Venter, O. (2016). Catastrophic declines in wilderness areas undermine global environment targets. *Current Biology*, *26*(21), 2929–2934. https://doi.org/10.1016/j.cub.2016.08.049

Watson, N. L., & Simmons, L. W. (2010). Reproductive competition promotes the evolution of female weaponry. *Proceedings of the Royal Society of London B: Biological Sciences*, *277*(1690), 2035–2040.

Watve, M. G., & Sukumar, R. (1997). Asian elephants with longer tusks have lower parasite loads. *Current Science*, *7*(11), 885–889. www.jstor.org/stable/24100035

Weaver, J. L., Paquet, P. C., & Ruggiero, L. F. (1996). Resilience and conservation of large carnivores in the Rocky Mountains. *Conservation Biology*, *10*, 964–976.

Webber, C. E., Sereivathana, T., Maltby, M. P., & Lee, P. C. (2011). Elephant crop-raiding and human–elephant conflict in Cambodia: Crop selection and seasonal timings of raids. *Oryx*, *45*(2), 243–251. https://doi.org/10.1017/S0030605310000335

Weinbaum, K., Nzooh, Z., Usongo, L., & Laituri, M. (2007). Preliminary survey of forest elephant crossings in Sangha Trinational Park, central Africa. *Pachyderm*, *0*(43), 52–62.

Weiss, M. N., Ellis, S., Franks, D. W., Ellifrit, D. K., Balcomb, K. C., & Croft Correspondence, D. P. (2023). Costly lifetime maternal investment in killer whales. *Current Biology*, *33*, 744–748.e3. https://doi.org/10.1016/j.cub.2022.12.057

Weissengruber, G. E., Egger, G. F., Hutchinson, J. R., Groenewald, H. B., Elsässer, L., Famini, D., & Forstenpointner, G. (2006). The structure of the cushions in the feet of African elephants (*Loxodonta africana*). *Journal of Anatomy*, *209*(6), 781–792. https://doi.org/10.1111/J.1469-7580.2006.00648.X

Wemmer, C. E., & Christen, C. A. (2008). *Elephants and ethics*. Johns Hopkins University Press.

West-Eberhard, M. J. (1979). Sexual selection, social competition, and evolution. *Proceedings of the American Philosophical Society*, *123*(4), 222–234.

West-Eberhard, M. J. (1983). Sexual selection, social competition, and speciation. *Quarterly Review of Biology*, *58*(2), 155–183. https://doi.org/10.1086/413215

Western, D., Tyrrell, P., Brehony, P., Russell, S., Western, G., & Kamanga, J. (2020). Conservation from the inside-out: Winning space and a place for wildlife in working landscapes. *People and Nature*, *2*(2), 279–291. https://doi.org/10.1002/pan3.10077

White, L. J. T. (1994). Sacoglottis gabonensis fruiting and the seasonal movements of elephants in the Lope Reserve, Gabon. *Journal of Tropical Ecology*, *10*(1), 121–125. https://doi.org/10.1017/S0266467400007768

Whiten, A. (2021). The burgeoning reach of animal culture. *Science*, *372*(6537), eabe6514. https://doi.org/10.1126/science.abe6514

Widén, A., Clinchy, M., Felton, A. M., Hofmeester, T. R., Kuijper, D. P. J., Singh, N. J., Widemo, F., Zanette, L. Y., & Cromsigt, J. P. G. M. (2022). Playbacks of predator vocalizations reduce crop damage by ungulates. *Agriculture, Ecosystems & Environment*, *328*, 107853. https://doi.org/10.1016/J.AGEE.2022.107853

Wijayagunawardane, M. P. B., Short, R. V., Samarakone, T. S., Nishany, K. B. M., Harrington, H., Perera, B. V. P., Rassool, R., & Bittner, E. P. (2016). The use of audio playback to deter crop-raiding Asian elephants. *Wildlife Society Bulletin*, *40*(2), 375–379. https://doi.org/10.1002/WSB.652

Wijesinha, R., Hapuarachchi, N., Abbott, B., Pastorini, J., & Fernando, P. (2013). Disproportionate dwarfism in a wild Asian elephant. *Gajah*, *38*, 30–32.

Wilkie, D. S., Bennett, E. L., Peres, C. A., & Cunningham, A. A. (2011). The empty forest revisited. *Annals of the New York Academy of Sciences*, *1223*(1), 120–128. https://doi.org/10.1111/J.1749-6632.2010.05908.X

Williams, C., Tiwari, S. K., Goswami, V. R., de Silva, S., Kumar, A., Baskaran, N., Yoganand, K., & Menon, V. (2020). *Elephas maximus*. The IUCN Red List of Threatened Species. e.T7140A45818198. Accessed on 15 February 2022. https://doi.org/https://dx.doi.org/10.2305/IUCN.UK.2020-3.RLTS.T7140A45818198.en

Williams, E., Carter, A., Hall, C., & Bremner-Harrison, S. (2019). Exploring the relationship between personality and social interactions in zoo-housed elephants: Incorporation of keeper expertise. *Applied Animal Behaviour Science*, *221*, 104876. https://doi.org/10.1016/J.APPLANIM.2019.104876

Williams, G. C. (1957). Pleiotropy, natural selection, and the evolution of senescence. *Evolution*, *11*(4), 398–411. https://doi.org/10.2307/2406060

Wilson, G., Gray, R. J., Radinal, R., Hasanuddin, H., Azmi, W., Sayuti, A., Muhammad, H., Abdullah, A., Nazamuddin, B. S., Sofyan, H., Riddle, H. S., Stremme, C., & Desai, A. A. (2021). Between a rock and a hard place: Rugged terrain features and human disturbance affect behaviour and habitat use of Sumatran elephants in Aceh, Sumatra, Indonesia. *Biodiversity and Conservation*, *30*(3), 597–618. https://doi.org/10.1007/S10531-020-02105-3

Wing, L. D., & Buss, I. O. (1970). Elephants and forests. *Wildlife Monographs*, *19*, 1–92.

Winkler, K., Fuchs, R., Rounsevell, M., & Herold, M. (2021). Global land use changes are four times greater than previously estimated. *Nature Communications*, *12*(1), 2501. https://doi.org/10.1038/S41467-021-22702-2

Wittemyer, G., & Getz, W. M. (2007). Hierarchical dominance structure and social organization in African elephants, *Loxodonta africana*. *Animal Behaviour*, *73*(4), 671–681. https://doi.org/10.1016/j.anbehav.2006.10.008

Wittemyer, G., Daballen, D., Rasmussen, H., Kahindi, O., & Douglas-Hamilton, I. (2005a). Demographic status of elephants in the Samburu and Buffalo Springs National Reserves, Kenya. *African Journal of Ecology*, *43*(1), 44–47. https://doi.org/10.1111/j.1365-2028.2004.00543.x

Wittemyer, G, Douglas-Hamilton, I., & Getz, W. M. (2005b). The socioecology of elephants: Analysis of the processes creating multitiered social structures. *Animal Behaviour*, *69*, 1357–1371. https://doi.org/10.1016/j.anbehav.2004.08.018

Wittemyer, G., Getz, W. M., Vollrath, F., & Douglas-Hamilton, I. (2007). Social dominance, seasonal movements, and spatial segregation in African elephants: A contribution to conservation behavior. *Behavioral Ecology and Sociobiology*, *61*(12), 1919–1931. https://doi.org/10.1007/s00265-007-0432-0

Wittemyer, G., Polansky, L., Douglas-hamilton, I., & Getz, W. M. (2008). Disentangling the effects of forage, social rank, and risk on movement autocorrelation of elephants using Fourier and wavelet analyses. *Proceedings of the National Academy of Sciences of the United States of America*, *105*(49), 1–6. https://doi.org/10.1073/pnas.0801744105

Wittemyer, G., Okello, J. B. a, Rasmussen, H. B., Arctander, P., Nyakaana, S., Douglas-Hamilton, I., & Siegismund, H. R. (2009). Where sociality and relatedness diverge: The genetic basis for hierarchical social organization in African elephants. *Proceedings. Biological Sciences / The Royal Society*, *276*(1672), 3513–3521. https://doi.org/10.1098/rspb.2009.0941

Wolf, M., & Weissing, F. J. (2012). Animal personalities: Consequences for ecology and evolution. *Trends in Ecology & Evolution*, *27*(8), 452–461. https://doi.org/10.1016/J.TREE.2012.05.001

Wolf, M., Van Doorn, G. S., Leimar, O., & Weissing, F. J. (2007). Life-history trade-offs favour the evolution of animal personalities. *Nature*, *447*(7144), 581–584. https://doi.org/10.1038/nature05835

Wong, E. P., Yon, L., Purcell, R., Walker, S. L., Othman, N., Saaban, S., & Campos-Arceiz, A. (2016). Concentrations of faecal glucocorticoid metabolites in Asian elephant's dung are stable for up to 8 h in a tropical environment. *Conservation Physiology*, *4*(1), cow070. https://doi.org/10.1093/CONPHYS/COW070

Woodroffe, R. (2011). Demography of a recovering African wild dog (*Lycaon pictus*) population. *Journal of Mammalogy*, *92*(2), 305–315. https://doi.org/10.1644/10-mamm-a-157.1

Wrege P. H., (2015). Why elephants come to bais. In B. Fishlock (Ed.), *Studying Forest Elephants* (pp. 88–89). Neuer Sportverlag.

Wrege, P. H., Rowland, E. D., Bout, N., & Doukaga, M. (2012). Opening a larger window onto forest elephant ecology. *African Journal of Ecology*, *50*(2), 176–183. https://doi.org/10.1111/j.1365-2028.2011.01310.x

Wrege, P. H., Rowland, E. D., Keen, S., & Shiu, Y. (2017). Acoustic monitoring for conservation in tropical forests: examples from forest elephants. *Methods in Ecology and Evolution*, *8*(10), 1292–1301.

Wrege, P. H., Rowland, E. D., Thompson, B. G., & Batruch, N. (2010). Use of acoustic tools to reveal otherwise cryptic responses of forest elephants to oil exploration. *Conservation Biology*, *24*(6), 1578–1585.

Wright, T. F., Eberhard, J. R., Hobson, E. A., Avery, M. L., & Russello, M. A. (2010). Behavioral flexibility and species invasions: The adaptive flexibility hypothesis. *Ethology Ecology & Evolution*, *22*(4), 393–404. https://doi.org/10.1080/03949370.2010.505580

Wyse, J. M., Hardy, I. C. W., Yon, L., & Mesterton-Gibbons, M. (2017). The impact of competition on elephant musth strategies: A game–theoretic model. *Journal of Theoretical Biology*, *417*, 109–130. https://doi.org/10.1016/j.jtbi.2017.01.025

Yoccoz, N. G., Mysterud, A., Langvatn, R., & Stenseth, N. C. (2002). Age- and density-dependent reproductive effort in male red deer. *Proceedings of the Royal Society B: Biological Sciences*, *269*(1500), 1523–1528. https://doi.org/10.1098/rspb.2002.2047

Yokoyama, S., Takenaka, N., Agnew, D. W., & Shoshani, J. (2005). Elephants and human color-blind deuteranopes have identical sets of visual pigments. *Genetics*, *170*(1), 335–344. https://doi.org/10.1534/GENETICS.104.039511

Yravedra, J., Rubio-Jara, S., Panera, J., Uribelarrea, D., & Pérez-González, A. (2012). Elephants and subsistence. Evidence of the human exploitation of extremely large mammal bones from the Middle Palaeolithic site of PRERESA (Madrid, Spain). *Journal of Archaeological Science*, *39*(4), 1063–1071. https://doi.org/10.1016/j.jas.2011.12.004

Zahavi, A. (1975). Mate selection – A selection for a handicap. *Journal of Theoretical Biology*, *53*(1), 205–214.

Zhang, H., Wang, Y., Janis, C. M., Goodall, R. H., & Purnell, M. A. (2017). An examination of feeding ecology in Pleistocene proboscideans from southern China (*Sinomastodon, Stegodon, Elephas*), by means of dental microwear texture analysis. *Quaternary International*, *445*, 60–70. https://doi.org/10.1016/j.quaint.2016.07.011

Zuberbühler, K., Noë, R., & Seyfarth, R. M. (1997). Diana monkey long-distance calls: Messages for conspecifics and predators. *Animal Behaviour*, *53*(3), 589–604. https://doi.org/10.1006/ANBE.1996.0334

Zwolak, R., & Sih, A. (2020). Animal personalities and seed dispersal: A conceptual review. *Functional Ecology*, *34*(7), 1294–1310. https://doi.org/10.1111/1365-2435.13583

Index

Printed in the United States
by Baker & Taylor Publisher Services